NATHANAEL-ISRAEL ISRAEL, PhD

Turbulent Origin of Life

OTHER BOOKS BY NATHANAEL-ISRAEL ISRAEL

Get them at your local bookstore, or online (e.g. on Amazon, Science180.com/books)

Turbulent Origin of the Universe
There is Only One Scientific, Simple, Safe, Trustworthy, Unexpensive, Brave, Practical, Nonconformist, Universal, Verifiable Formula that Accurately Decodes the Universe Formation ... But You Are Not Using It

Reconciling Science and Creation Accurately
What Science Accurately Teaches about Creation and God's Existence that Atheists, Freethinkers, and even Most Christians Ignore ... And How to Demonstrate it Without Taking Sides Between Rationality and Faith

Turbulent Origin of Chemical Particles
Why You Don't Have to Embrace Evolution, Big Bang, or Deny God to Scientifically Prove the Formation of All Chemical Particles

Origin of the Spiritual World
Top Secrets about the Origin of Everything in the Universe that Some Elites Have Hidden from You for Thousands of Years

From Science to Bible's Conclusions
How Decoding the Universe-Origin by Properly Revisiting Scientific Data—That Top Scientists Collected but Wrongly Analyzed—Bizarrely led to the 3500 Years Old Biblical Account of Creation

How God Created Baby Universe
What Children Must Scientifically Learn Early about the Universe Formation to Avoid Dangerously Abandoning God Later in Life Just Like Most College Students Who Embrace Evolution and Big Bang That Deny Biblical Creation

How Baby Universe Was Born
How to Scientifically Talk to Children about the Universe Formation and They will Know Forever How to Correctly Test the Intersection of Science and Faith

Science180 Accurate Scientific Proof of God
Can We Scientifically Explain the Formation of the Universe Through Natural Processes Without Evoking Evolution and Big Bang?

More books written by Nathanael-Israel Israel can be found at Israel120.com/books

NATHANAEL-ISRAEL ISRAEL, PhD
Founder of Science180: www.Science180.com
Father of Science180 Cosmology, of Science180 Model of the Origin of Life, and of Science180 Model of the Origin of Chemical Particles. Creator of the Universe Turbulent Origin Formula

Turbulent Origin of Life

Why You Don't Have to Embrace Evolutionism or Check Your Brain at the Door in the Name of Faith or Science to Accurately Decrypt the Origin of Life Using the Historic Formula of the Universe Formation

Science180
Augusta, Georgia
United States of America
www.Science180Publishing.com

Copyright © 2025 by Nathanael-Israel Israel
Visit the author's website at Israel120.com

Turbulent Origin of Life
Why You Don't Have to Embrace Evolutionism or Check Your Brain at the Door in the Name of Faith or Science to Accurately Decrypt the Origin of Life Using the Historic Formula of the Universe Formation

First edition: October 2025

Published by Science180
Augusta, Georgia (USA)
www.Science180Publishing.com

Book Cover and Illustrations by Nathanael-Israel Israel

ISBN: 979-8-9932150-6-8

Library of Congress Control Number: 2025920906

All rights reserved. For permission requests, please visit Science180.com/permissions. Thank you for your support of the author's rights.

Neither the publisher nor the author shall be liable for any loss of profit or any other damages, including but not limited to special, incidental, consequential, personal, commercial, or other damages. More about copyright and disclaimers at Science180.com/copyright

More books by the same author can be found at Israel120.com and Science180.com

For information about special discounts available for bulk purchases, please visit Science180.com/discount for more details.

Science180 can bring authors including Dr. Nathanael-Israel Israel to your live or recorded events. For more information or to book an event, please visit Science180.com/speaking

For any questions, please visit Science180.com/contact

To publish your book(s) with Science180 Publishing, go to Science180Publishing.com

To interview the author of this book, visit Israel120.com/interview
To donate, please visit Israel120.com/donate or Science180.com/donate.

Published in Augusta, United States of America.

CONTENT

1. What is preventing scientists from properly defining life?........................... 1
2. Can you ignore the types of organisms on earth and still be able to decode their origin accurately?........................... 15
3. From cracking the dna of the universe-origin to unlocking the code of the formation of life and of chemical particles, this scientist and mathematician has done it all........................... 25
4. To never get the origin of life wrong, pay attention to this scientific breakthrough on the study of turbulence in the early universe........................... 28
5. Why don't people take seriously this summary of the turbulent origin of the universe that shook the foundation of all existing cosmological theories … so they can live happily and think more clearly?........................... 33
6. How nathanael-israel israel discovered turbulent trees, turbulent branches, and turbulent leaves in the precursors of celestial bodies … (and how they will enlighten your path to finally decode the origin of life)........................... 41
7. Uncover how the features and patterns on living things point at the biological turbulence that shaped them … but top scientists have struggled to discover them........................... 49
8. Can the length of the neck hold a key to decoding how an animal was formed? This expert says yes … and here's how to properly understand what he termed "turbulent neck"........................... 53
9. To know how great scientists challenge conventional science while investigating the anatomy of living things, first decrypt the secrets of the turbulent head, belly, tail, spine, arms, and legs........................... 61
10. Extraordinary scientific advancement that changed the secular explanation of life forever using the turbulent geometry of leaves and leaf-like shapes of organisms........................... 67
11. How to quickly defeat the biggest lies people spewed about the formation of life by knowing that developmental biology points to a biological turbulence—did darwin miss this or what happened?........................... 75
12. Can you scientifically connect the spatial distribution of all forms of life to a biological intermittence of size during the formation of life? Checkout why this matters a lot!........................... 83
13. If you think that size doesn't matter or that its impartation on living things was by chance, then pay attention to this interesting detail about the largest and smallest organisms........................... 89
14. Can you judge the origin of an organism by its speed?........................... 91
15. Can you really decode the origin of life while neglecting its spiritual component or what must change in the believers-nonbelievers relationship before rationalists and freethinkers take the spiritual differently?........................... 93

16. How to scientifically talk about the formation of life and have everybody bow to the universal pool of qualities and the universal holistic being (that has nothing to do with religion or faith) 99
17. Can you accurately define and decode life by ignoring the turbulent program of life that is more advanced than any computer program? 107
18. Why aren't scientists across the globe thinking about the split-gathering of the universal pool of abilities in living and nonliving things before wasting billions of dollars on life-origin researches we don't need? 110
19. What rapid paradigm shift can we expect from the demonstration of the turbulent law of mission and calibration of life? Don't say i did not warn you! 116
20. Gigantic errors made about life-origin and how this scientific formula accurately proves the generic turbulent process of the formation of all forms of life ... (is there any nobelist not deeply shocked yet?) 126
21. 'i will not answer that reckless mistake people made about the formation of macromolecules': the 180scientist has no time for this irrational and nonsense argument 147
22. What is the accurate, simple, straightforward way to use modern science to quickly improve your understanding of the formation of organelles, cells, tissues, organs, and apparatuses ... so you can save time and money? 156
23. Can a simple formula accurately crack the code of the formation of all plants and reveal the one thing that scientists have missed and that has been causing them headaches, overwhelm, and burnout? 166
24. The only scientific step-by-step pathway you need to accurately decode the formation of all animals and get the power, freedom, and boldness to take advantage of new opportunities 218
25. The easy yet accurate theory that scientists trust to quickly understand the formation of fungi, protists, archaea, bacteria, and viruses so they can become fulfilled thought leaders in their field of expertise 257
26. How to talk to smart people about the characteristics of celestial bodies and they will implore to review whether the similarities between living things is sufficient to imply similar descendances 268
27. Do we have to deny god to scientifically test whether he created life—or is there any rational explanation of the biblical creation of life that can scientifically bail anti-creationists out of any doubt? 275
28. Is the bible irrational or at war with science when it revealed that some animals are clean and others unclean or is there any encrypted scientific code behind the leviticus 11 laws? 300
29. I asked smart scientists the best way to scientifically test evolution—they all said almost the same thing ... except this scientist who came out of nowhere and shockingly proved something you can't learn at any church or public school 312

30. To be scientifically 100% sure about whether god created life or evolution produced it, pay attention to "science180 model of the origin of life" .. 331
Next steps of the journey 343
References 351
Index 360
About the author 369

CHAPTER 1: GENERALITIES AND SECULAR DEFINITION OF LIFE

CHAPTER 1

WHAT IS PREVENTING SCIENTISTS FROM PROPERLY DEFINING LIFE?

1.1. Introductory remark on the real definition of life

Can we accurately explain the formation of life through natural processes that completely excluded Evolution and the Big Bang theory?

- Where do we come from?
- Why are we here?
- Where did all the forms of life come from?
- Do living things descend from one other or were they created by God?
- If God created life, how did He do it?
- If God did not create life, what process did it then?
- Did evolution really create life or did everything come from chemical reactions?
- No matter how life arose, what was the mechanism of the formation of all forms of life?
- Are living things or beings the products of a long evolutionist process just as the evolutionism theory claims or did God spontaneously create life like the Bible said?
- Using pure science, can the scientific data be reconciled with the creation story?
- How can we know for sure which story of the formation or creation of life is correct?

People of all nations have pondered on the answers of these questions, and books were written about life. Yet, many questions remain and the scientific process of the formation of life is not settled yet. Likewise, many people also wondered about the process of the origin of the universe. For those of you who don't know,

TURBULENT ORIGIN OF LIFE

after my doctorate degree in the US, I spent 12 years decoding the formation of the universe and writing books about it. During those years, I also investigated the process by which living things were formed. My discovery of the real process of the formation of the universe (through the perspective of turbulence) gave me a unique insight into the formation of life. Using the code that helped me to decipher the formation of the universe, I unlocked the process by which all forms of life were formed. I originally presented my findings as a chapter in the draft of *"Turbulent Origin of the Universe"*, but the details were too much to be contained in that book. Therefore, I decided to devote a completely different book to the origin of life and to present my findings to the world, for the benefit of all, hence, the birth of this book. By the time you will finish reading this book, you will fully comprehend the process by which all forms of life came into existence.

As the Father of Science180 Cosmology and the Founder of Science180 Academy, I am fortunate to be known as the source of unconventional wisdom and knowledge that help people accurately crack the code of the formation of the universe, of life, and of chemicals. I know that every human being will benefit from understanding the real origin of life. But the problem is that most efforts to explain the origin of life are complex, inaccurate, confusing, partisan, complicated, therefore, creating serious challenges to those who are eager to scientifically decrypt where all forms of life came from. Most people want an accurate, simple, straightforward, nonpartisan life-origin book that is free from jargons and difficult concepts only known by the experts. This elegant scientific book breaks down the technicality of the origin of life in a language that even the nonscientists can easily comprehend. It is a trustworthy book that will help you to quickly, cheaply, easily, and efficiently navigate everything you need to know to finally decode and solve the puzzling problems about the origin of life, while also giving you a crash course on the universe-origin. Unlike any book you have ever read on the origin of life, this historic masterpiece (that distills complex scientific data down to simple explanations that make sense) is the starting point of any smart person wanting to rationally understand the formation of all living things.

By the time you finish reading *"Turbulent Origin of Life"*, you will discover:
- Why in spite of the massive amount of scientific data collected on living things, scientists have misunderstood the formation of life until now, and then uncover in a simple language the one thing that was needed to accurately crack the code of life but that scientists have missed and that has been causing them headaches, overwhelm, and burnout
- Step-by-step pathway to decode the origin of life and get the power, freedom, and boldness to take advantage of the opportunities that accurate understanding of the origin of life creates (see more at *Science180.com*/life)
- The high connection between the code of the universe formation and the process by which life on Earth was formed, so you can become a fulfilled thought leader in your field of expertise

CHAPTER 1: GENERALITIES AND SECULAR DEFINITION OF LIFE

- Tools to stand as a lighting bolt that electrifies those who are still struggling to understand the formation of all forms of life in the universe
- Strategies to push the boundaries of human abilities to properly understand what is perceived as un-understandable, mysterious, supernatural, unimaginable, impossible, and unthinkable that hold people back
- Scientific approach to holistically detect, correct, and remove all misinformation, ambiguity, and misleading claims and theories surrounding the origin of life

With his book, you will also:

1. Become the leader that captures the heart of your followers, prospects, and customers craving for an unconventional explanation of the origin of life
2. Benefit from continual updates and assistance during your journey to decode life-origin … all to clear the way for the freedom, power, technology, innovation, and breakthroughs of the future
3. Bypass technical knowledge that restricts non-experts from accessing the life-origin truth contained in the massive scientific data, and get to the bottom of scientifically-locked origin-related secrets regardless of your background
4. Discover and understand the complex formation of life without leaving out the challenging questions that people of all ages have been struggling to answer for thousands of years!
5. Empower and align yourself with the historic breakthrough that has done what no other discovery has ever done: accurately decode the origin of the cosmos and life
6. Help you in your personal and professional life by teaching you, and answering your questions about the origin of life, and how to transform that knowledge into insight to significantly add value to your life and to that of others in less time
7. Learn great lessons from some of top scientists, philosophers, thinkers, and public figures who have realized historic mistakes they made in life (concerning the origin of the universe, life, and chemicals), and that they corrected thanks to the discoveries of Nathanael-Israel Israel
8. Protect yourself and loved ones by keeping all of you secured and empowered with the true knowledge of the origin of life
9. Reliably access to the world's authority on origin-related matters and get your origin questions professionally answered with the truth
10. Revolutionize every origin-related domain in the world with the accurate understanding of the origin of life

Whether you are a scientist or a layperson, a believer, or a skeptic, you cannot afford to ignore the greater, better, faster, simpler, cheaper, easier, and accurate formula unlocked in this important book that successfully decoded the origin of life. Keep reading *"Turbulence Origin of Life"* today and change lives!

TURBULENT ORIGIN OF LIFE

Before embarking on the journey of explaining the origin of life, I felt impelled to first present what the world understands by life. Because humankind is not made up of scientists only, but also of laypeople (meaning people who never went to school or who did not go far in the secular education to the point that they could formulate a scientific definition of life besides what they have learned according to their culture), I will present not only a secular definition of life, but also a non-secular one. However, due to the complexity of the philosophies related to life and the strategic thinking I used to put this book together, in this chapter, I will focus on the scientific definition, and later in this book, I will come back to other definitions coined across the globe according to some key cultures. Due to the complexity of the task I embraced concerning the comprehensive explanation of the origin of life, at one point during my journey of reporting my findings, I noticed weaknesses in the academic definitions of life, and therefore, I felt obliged to coin my own definition of life. However, for the sake of clarity, methodology, and to allow the readers to understand how I came up with my definition of life, I will not present such a definition in this chapter, but I will wait for a few more chapters (in which I critically present the implications of what is already known about life) before I elaborate on my definition.

The definition of life is a challenging task I never knew could be so hard to apprehend. Living things or beings have different forms, shapes, and composition. They live in various environments ranging from one negative extreme to one positive extreme, which sometimes contradict one another. With so many characteristics, living things behave so differently that it is impossible for scientists to put a definition for all of them under one umbrella. In other words, it has been difficult for scientists to define all forms of life using one simple comprehensive sentence. In general, scientists define life as: *"a characteristic that distinguishes physical entities that have biological processes (e.g. signaling) and self-sustaining processes, from those that do not, either because such functions have ceased (they have died), or because they never had such functions and are classified as inanimate"* (Wikipedia, 2021a). As of 2025, the year I published this book after pondering on it for 12 years, the scientific community has summarized all forms of life that the scientists think exist into six forms of life (that I later detailed in this book):

1. plants,
2. animals,
3. fungi (e.g. mushroom),
4. protists (e.g. algae),
5. archaea, and
6. bacteria.

Some people even consider non-cellular life forms such as viruses and viroids as living organisms, yet they lack many characteristics found with the aforementioned 6 forms of life. Nevertheless, due to what they are accustomed to in their daily life, most people (even the most secularly educated ones) would define life by mainly referring to living things such as:

CHAPTER 1: GENERALITIES AND SECULAR DEFINITION OF LIFE

1. plants (like the vegetables that they eat),
2. animals (e.g. their dogs or cats, or animals they eat),
3. fungi (some of which they buy like mushrooms in the store),
4. bacteria (that most people assume as microbes causing diseases).

In other words, if you ask even most of those who went to school and who even got a doctorate degree (but not in biological fields), they will likely not know what protists and archaea mean. Truth be told, if a survey could be done among the general public, just as the Greek philosopher Aristotle (384–322 BC) did it more than 2000 years ago, most people living today would classify living organisms in 2 groups: plants or animals, animals being generally defined as living things that move, while plants do not move by themselves (although some can move a few of their organs to some extent).

Despite the lack of consensus on the definition of life, and although all living things cannot fulfill all of the following requirements, living organisms are defined as "open systems capable of maintaining homeostasis (internal regulation to keep a constant state), composed of cells, have a life cycle, undergo metabolism, can grow (increase in size), adapt to their environment (ability to change in response to environment), respond to stimuli, reproduce (asexually or sexually) and evolve" (Wikipedia, 2021a). Here, metabolism is a 2-way process: 1) anabolism through which organisms convert chemicals and energy into cellular components and 2) catabolism which is about the breaking down or decomposition of organic matters. Some theorists have been trying to understand which life function is more important than the others as if some functions are useless and others more indispensable. But in reality, as you will read in this book later, what gives life is not the functions or processes associated to it.

Although they have their own genes, and are able to make multiple copies of themselves, viruses are unable to metabolize and they need a host cell before they can make new products. In other words, because viruses cannot perform most of the metabolic tasks like cells do, they have to rely on the cellular machinery of their hosts to satisfy some of their "needs" such as the transcription of DNA and the translation of RNA into protein. They achieve that goal by releasing their genetic material into cells that they manage to enter, and get it into the machinery of the host cells, which can transcribe it and translate it accordingly, therefore being able to damage the host cells. Because they "fail" to meet some of the key characteristics of the commonly accepted living organisms, viruses are not usually considered as living things. Yet, seeing the activity of some of them (e.g. catalysis), some people think that viruses can act as living things. My goal in this book is not to prove whether viruses are organisms or not, but to explain how they and the commonly accepted aforementioned 6 forms of life originated.

Also termed biological form, form of life, living creatures, and sometimes as fauna or flora according to their types, an organism is defined as "any organic, living system that functions as an individual entity" (Mosby's Dictionary of Medicine, 2017). Some people perceived them as an assembly of molecules operating altogether and displaying properties of life. As of 2018, more than 1.7 million species of organisms have been documented (Anderson, 2018). As of 2021, Encyclopedia Britannica reported that about 1.9 million species were described (Sagan, 2021).

Based on the functions that organisms can perform, their physical appearances and all of the "benefits" that human beings can withdraw from them, some human beings have classified some organisms more valuable or more interesting than others. Some which were thought to be less valuable in the past were later proven to be very appreciated based on new utilities and knowledge human beings gained about them. Likewise, some sequences of DNA were deemed useless by human beings for years to the point that some people even considered them as junk DNA until the days they were established very useful. In other words, what human beings do not understand, they think it is useless and what they think they understand, they try to denature or re-engineer as if they know everything and yet they lock themselves from being able to open their mind and learn. In the same manner, because some forms of existence seem not enough for some human beings, they are not classified as living. When it comes to spiritual things, the confusion is worse as some beings that exist but which cannot be physically felt using conventional means, are not believed to exist by some people. Furthermore, although the definition of life can be summarized to respiration and gas exchange, but the difficult part is that scientists are not able yet to create the breath of life. I will revisit this topic later in this book.

Before I close this segment, it is important to mention that the secular definitions of life are dominated by Darwinism, a theory of evolution that I will present later in this book. In other words, it has become increasingly difficult for most people in the academia to be "accepted" by most of their peers if they define life without mentioning something related to evolutionism. In contrast, although not the mainstream view in secular environments, yet embraced by many people (some of whom challenge evolutionism) across the globe, creationism defines life as having been created by God. But if so, knowing the plethora of religions and gods or idols in the world, how can one know for sure which God created life, if that was the case? Later in this book, I will delve into all of those viewpoints and many more.

1.2. Why is it so hard to come up with a single definition of life?

Like the Encyclopedia Britannica said, *"no scientist, technician, and other life science expert can give a completely inclusive and concise definition of life"* (Sagan, 2021). As of 2011, more than 120 definitions of life have been assembled (Trifonov, 2011). By now, that number would have climbed. Because of the impact it has been having on the scientific community, Darwinism has influenced the biological sciences so much that even NASA could not define life without introducing it: "*life is a self-sustaining chemical system capable of Darwinian evolution*" (Voytek, 2021). In other words, for these authors, if an organism or a living thing is not evolving, it must not be living. For other people, "*life is a matter that can reproduce itself and evolve as survival dictates*" (Luttermoser, 2012a; Luttermoser, 2012b).

Part of the difficulty in defining life is that some functions present in some

CHAPTER 1: GENERALITIES AND SECULAR DEFINITION OF LIFE

organisms are found with certain clusters of matter or clusters of chemicals not technically labeled as living things. For instance, some abilities found with organisms are found with viruses, yet viruses are not technically considered as living organisms. Certain abilities thought to exist only with one form of life are found with other forms of life and vice versa.

The difficulty that life science experts have been having to come up with a single inclusive definition of life is almost similar to the difficulty that physicists have been having to use a single definition to label all types of celestial bodies and microscopic matter in the universe. In other words, just as the physical and abiotic world consists of celestial bodies and chemical particles (and other things that science has not identified yet), with various characteristics (e.g. size, shape, energy, movement, speed, location, function, inclination, rotation, tilt, mass, volume) difficult to express in one sentence, so also living organisms come in many shapes, organizations, abilities, and other proprieties, which are hard to comprehensively express in a simple sentence. Armed with my discoveries about the mother of all turbulences, I noticed some patterns in the characteristics of living organisms, which I will use to explain how I think life emerged. Considering what I know about turbulence, and that I detailed in my books on the origin of the universe, I felt like the plethora of definitions of life is an encrypted message of how little the scientific community knows about the origin of life and of the features or proprieties of living entities. For if people had for instance understood the law of split-gathering and of intermittence that I spearheaded in *"Turbulent Origin of the Universe"*, they would have already realized that, the way the precursors of living things were split-gathered also allowed the formation of various living entities having different constituents on various scales (e.g. macromolecules, organelles, cells, tissues, organs, apparatus, organisms).

The complexity of life and the difficulty that scientists have to fully understand it and the fact they split life components into distinct categories also affected the inability of the scientific community to fully grasp all aspects of the investigation of life without having disciplines overlapping. And because (even with the growing number of interdisciplinary research groups) most scientists do not cooperate with others in different areas of expertise, the piece of knowledge known in one life science field is not always fully known by the experts in another scientific field. In the end, the little "known" about life (and even about non-living matter that should help explain life) across all scientific disciplines is not fully shared across the board of the scientific fields. Even if they are shared, they are not usually understood by experts of other fields, including abiotic scientists. Consequently, many (sometimes conflicting) approaches are partially given to life problems and challenges, while the authors of these attempts act as if their so-called solutions to the life-origin problem is the ultimate answer, which considers all aspects necessary to give a holistic solution to the challenges related to life. For instance, although many scientific disciplines study life in various areas, biology is the branch of science defined as the scientific study of life. The study of plants is usually referred to as botany although there are various plant science specialties, which address different aspects of plants.

Likewise, the study of animals is referred to as zoology, yet animals are also studied in other scientific disciplines. Bacteriology is the study of bacteria, while mycology is the study of fungi, and phycology the study of algae. Protistology is the study of protists. Finally, virology is the study of viruses.

1.3. All forms of life are made of cells filled with organelles

All forms of life are made of structural and functional units or minimal living units called cells. In other words, just as atoms are perceived as the unit of matter, living things are made of small units called cells. Cells are divided into two types: prokaryotic and eukaryotic. Prokaryotes do not have a nucleus, while eukaryotes have a nucleus. For instance, animals, plants, fungi, and protists are eukaryotes. In contrast, bacteria and archaea are considered prokaryotes. Organisms that are made of a single cell are called unicellular and those containing many cells are called multicellular. For instance, bacteria and yeasts consist of one cell, while other organisms are made up of many cells. All cells have a cytoplasm, which is surrounded by a membrane, which contains various biomolecules (e.g. lipids, carbohydrates, proteins, nucleic acids, vitamins etc.). Another characteristic of most cells is their ability to reproduce themselves through a mechanism called "cell division" during which a parent cell divides into two or more daughter cells.

Organelles are membrane-bound compartments found in organisms. Some of the organelles found in animal cells and in other types of organisms include:

- Cytoplasm, which is referred to as the material or protoplasm within a living cell, excluding the nucleus
- Ribosomes, where proteins are made
- Endoplasmic reticulum, which transports materials and supplies throughout the cell
- Mitochondria, which are the main energy producer in a cell
- Golgi complex, which are complex of vesicles and folded membranes within the cytoplasm involved in secretion and intracellular transport
- Lysosomes, which contain enzymes capable of digesting things in the cell
- Peroxisomes, which comprise enzymes that purify the cell from perilous materials
- Centrosomes, which play a crucial role in mitosis (cell division)
- Microvilli, which act like fingers
- Cilia, which are hair-like structures coming out of the surface of some cells but able to move things
- Nucleus, which consists of nucleolus, nucleoplasm, nuclear pore, nuclear envelope, etc.
- Plastids (e.g. chloroplasts)
- Etc.

CHAPTER 1: GENERALITIES AND SECULAR DEFINITION OF LIFE

1.4. Chemical composition of living things

Some of the chemical elements most needed to biochemically sustain life on Earth are: carbon, hydrogen, nitrogen, oxygen, phosphorus, and sulfur. Consequently, called elemental macronutrients of living things, these chemical elements constitute the primary material with which all organisms are made of. While most people think that life requires oxygen input, it is important to mention that some organisms such as bacteria not only do not require oxygen before living, but they can be poisoned by oxygen. For some are anaerobes. In contrast, no organism can live without water.

The chemical composition varies from one organism of one species to another organism of another species. Likewise, the chemical composition widely changes from one form of life to the other, and some chemical compounds found in some forms of life are absent in others. For instance, the forms of life are not made of the same number and quantity of chemical compounds. For instance, some aminos acids such as lysine and tryptophane are not found in plants, yet they are abundant in animals. Maybe, the processes of the formation of plants did not allow these 2 amino acids to be synthetized and integrated into plant systems.

As I detailed in *"Turbulent Origin of Chemical Particles"*, my book on the origin of chemical particles, the chemical composition of human beings can be summarized as follows. No chemical element is as abundant in human beings as oxygen: 61% (abundance by mass). More than 97.9% of the mass of human beings is made of nonmetals. In fact, about 94% of the mass of human beings is made of 3 nonmetals: oxygen, carbon, and hydrogen. No synthetic element, lanthanoid, or noble gas was found in human beings.

The abundance of the chemical elements in human beings varies between 0 and 61%, with Oxygen (O, Z=8) being the most abundant (61%); here Z is the atomic number or the number of protons in the nucleus of the atom. More than 97.9% of the mass of human beings is made of nonmetals (Fig. 1), and 3 of these nonmetals account for about 94% of the mass of human beings: Oxygen (O, Z=8), Carbon (C, Z=6), and Hydrogen (H, Z=1). No synthetic element, lanthanoid, or noble gas was found in human beings.

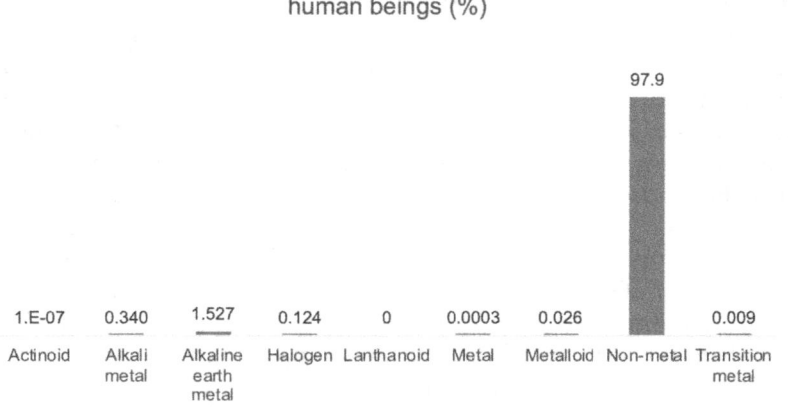

Fig. 1: Sum of the Abundance of chemical elements in human beings (%)

By decreasing abundance order, the 15 most abundant chemical elements in human beings are (Periodictable.com, 2014):
1. Oxygen (O, Z=8), a nonmetal: 61%
2. Carbon (C, Z=6), a nonmetal: 23%
3. Hydrogen (H, Z=1), a nonmetal: 10%
4. Nitrogen (N, Z=7), a nonmetal: 2.6%
5. Calcium (Ca, Z=20): 1.5%
6. Phosphorus (P, Z=15), a nonmetal: 1.1%
7. Sulfur (S, Z=16), a nonmetal: 0.2%
8. Potassium (K, Z=19): 0.2%
9. Sodium (Na, Z=11): 0.14%
10. Chlorine (Cl, Z=17): 0.12%
11. Magnesium (Mg, Z=12): 0.027%
12. Silicon (Si, Z=14): 0.026%
13. Iron (Fe, Z=26): 0.006%
14. Fluorine (F, Z=9): 0.0037%
15. Zinc (Zn, Z=30): 0.0033%

The abundance of any other chemical element found in human beings is not higher than 4.6E-04%. The most abundant actinoid in human beings is Uranium (U, Z=92), contributing to 1E-07% of the human mass. I was a little surprised that our bodies contain Uranium, which is known as very radioactive.

As far as alkali metals are concerned, the least abundant is Francium (Fr, Z=87), with an abundance reported as 0%, while the most abundant are:
- Potassium (K, Z=19): 0.2%
- Sodium (Na, Z=11): 0.14%

The alkaline earth metals in human beings are dominated by:

CHAPTER 1: GENERALITIES AND SECULAR DEFINITION OF LIFE

- Calcium (Ca, Z=20): 1.5% followed by
- Magnesium (Mg, Z=12): 0.027%

The most abundant halogens in human beings are:
- Chlorine (Cl, Z=17): 0.12% and
- Fluorine (F, Z=9): 0.0037%

Concerning the other 3 halogens, while Bromine (Br, Z=35) and Iodine (I, Z=53) were recorded at a concentration smaller than 0.0003%, Astatine (At, Z=85) was not found at all.

Of the typical metals, the most abundant are:
- Lead (Pb, Z=82): 1.7E-04%,
- Aluminum (Al, Z=13): 9E-05%, and
- Tin (Sn, Z=50): 2E-05%

The most abundant metalloids are:
- Silicon (Si, Z=14): 0.026%,
- Boron (B, Z=5): 7E-05%, and a trace of
- Arsenic (As, Z=33): 5E-06%

The most abundant transition metals in human beings are:
- Iron (Fe, Z=26): 0.006% and
- Zinc (Zn, Z=30): 0.0033%.

Among the heaviest chemical elements (which atomic numbers > 38) found in human beings, Lead (Pb, Z=82) is the most abundant: 0.00017 %. This may explain why lead intoxication is a serious issue, for it can have a detrimental effect on human beings. Human beings do not passively incorporate into their bodies all of the most abundant chemical elements in nature, but elements that align with their physiological needs.

Before I close this segment on the abundance of chemical elements in human beings, I would like to quickly address their location in the human body. Indeed, chemical elements are found in different concentrations in different parts of the human body. While some chemical elements like Carbon (C, Z=6) are found everywhere in human bodies, others are found in specific locations:
- bones and teeth (e.g. Fluorine (F, Z=9)),
- enzymes (e.g. Potassium (K, Z=19), Manganese (Mn, Z=25), and Iron (Fe, Z=26)),
- body liquids (e.g. Chlorine (Cl, Z=17)),
- proteins (e.g. Sulfur (S, Z=16)),
- urine and bones (e.g. Phosphorus (P, Z=15)),

- all liquids and tissues: Sodium (Na, Z=11),
- all liquids, tissues, bones, and proteins: Hydrogen (H, Z=1) and Oxygen (O, Z=8),
- all liquids, tissues, and proteins: Nitrogen (N, Z=7),
- lungs, kidney, liver, thyroid, brain, muscles, and heart: Magnesium (Mg, Z=12). In addition to all of the places where Magnesium is found, Calcium (Ca, Z=20) is also found in bones.

The types of chemicals present in specific locations and the characterization of these elements could help to understand some of the functionalities of the organs present at these locations.

1.5. Energy consumption by organisms

One of the key chemical requirements by organisms is the capability to store and release energy to meet their needs. Most organisms get their energy through light, chemical oxidation, or other means. For a long time, scientists have defined life to fit only organisms that depend of photosynthesis, which is a chain reaction that allow plants to capture solar energy and convert it into biochemical energy, which, at its turn, can be consumed by animals and other organisms. The discovery of organisms which energy source is based on chemical oxidation is more recent.

Organisms that are primary producers are called autotrophs. Green plants are photoautotrophs, while organisms that get their energy from the oxidation of chemical reactions are called chemoautotrophs. Many chemoautotrophic organisms are oxidizers of ammonia, hydrogen, methane, and sulfur. In general, most autotrophs are algae, bacteria, and plants. In other words, plants are not the only organisms that depend on no other organism to produce their biomass.

Because human beings and most animals depend on the autotrophs, they are called heterotrophs. Except the autotrophs (which can produce their own food using light and/or chemical reactions as I already explained), most organisms need food consumption and respiration (which involves reactions of oxidation and reduction allowing the release and capture of various chemical elements needed) for their subsistence, growth, and maintenance.

ATP (adenosine triphosphate) and other similar molecules such as guanosine triphosphate (GTP) are currently known as the main or exclusive energy currency or energy-exchange molecules of most living organisms found on Earth. Mitochondria are the energy storehouse (you could consider them like the power plant that has the power for your home) of the cells. Because the sequence of its DNA is different from that of the nucleus, but similar to that of a bacterium, some people have postulated that organisms which have mitochondria must have descended from bacteria. What a linear way of thinking. As you will learn later in this book, mitochondria did not descend bacteria. Likewise, because the adenosine triphosphate (ATP) is "the" main energy carrier does not mean that all organisms descend from the same ancestor, but that the processes of their formation and functioning were built on similar mechanisms. Later, I will explain why the

CHAPTER 1: GENERALITIES AND SECULAR DEFINITION OF LIFE

mitochondrial DNA is similar to that of bacteria.

Throughout my writing, wherever you see "universe-origin", please know that I meant "origin of the universe" or "the origin of the universe". Likewise, wherever you see "life-origin", please understand that I meant "origin of life" or "the origin of life". In the same manner, wherever I mention "chemicals-origin", please know that I am referring to "origin of chemicals" or "the origin of chemicals".

Another Book by Nathanael-Israel Israel:
TURBULENT ORIGIN OF THE UNIVERSE

THE FIRST AND ONLY SCIENTIFIC BOOK THAT ACCURATELY EXPLAINS EVERYTHING YOU NEED TO UNCONVENTIONALLY, EASILY, AFFORDABLY, AND ENJOYABLY DECODE THE UNIVERSE FORMATION

In *"Turbulent Origin of the Universe"*, filled with great diagrams and digestible scientific facts, you will discover, learn, or get:

- The all-in-one, proven & uncomplicated scientific formula that accurately decoded the formation of the universe, and that explained the birthdate of the stars, planets, satellites, asteroids, and all other celestial bodies in the universe, so you can position yourself to stay on top of your competitors, avoid repeating crucial mistakes that many people have ignorantly made at their own perils
- Extraordinary, unprecedented, accurate insights into the first factors (e.g. early universe physics) that defined the history and formation of the universe so you can tap into deep scientific secrets you ignore, and set yourself apart from others
- The new physics that will revolutionize science forever and land you into a zone of original ideas that improve lives nonstop regardless of your expertise
- The 4 simple things without which it is impossible for anyone to ever understand the formation of the universe, think accurately, work differently, achieve, or perform better for superior results
- The verified key to move the cosmological mountains of misunderstanding, so you can confidently free your mind from doubts, improve your health, and prevent you from any danger connected with sticking with wrong assumptions
- Save time and money, and enjoy your life once you remove errors holding your true understanding of the universe-origin captive

- Historic scientific proof of whether a planet was formed in 2.82 days, whether a satellite was formed in 3.32 days, and whether a star was formed in 3.69 days after the beginning of the universe; so you can creatively produce and address a broader work spectrum by learning how to effectively communicate with and establish unusual connections between otherwise disconnected and disparate scientific data
- The scientific formula that successfully tested the existence of God in a way that shocked believers, skeptics, and all other freethinkers
- Why the scientific community has failed to sufficiently explain the origin of the universe; and understand how existing theories have missed and undefined central ideas, and imposed limits on the vision of scientists
- Specific in-depth knowledge, up-to-the-minute information, and ideas so you can expand your market, cut useless costs, stop wasting time on inadequate projects, and start focusing on the profitable solutions (Science180.com/scientific)
- How Science180 Academy can strategically enlighten you, guide you to navigate and filter the massive data collected on the universe, so you can answer the world's most challenging questions, remove any scientific and philosophical cataracts that may be blocking you, and bring you many steps closer to your best life
- How to better resonate with your target market that is craving something original that breaks wrong explanations of the universe-origin

Get *"Turbulent Origin of the Universe"* today to begin an incredible journey of accurately decoding the universe and change your life forever!

Dr. Nathanael-Israel Israel is told by people that he is the #1 Universe-origin, Life-origin, and Chemicals-origin Expert. He is the founder of Science180 and the author of many books on the origin of the universe and its content. To learn more about how he may help you, visit Israel120.com.

CHAPTER 2

CAN YOU IGNORE THE TYPES OF ORGANISMS ON EARTH AND STILL BE ABLE TO DECODE THEIR ORIGIN ACCURATELY?

To understand the origin of life, we need to first know what kind of living organisms exist in nature. Unless you are a well-informed expert in life sciences, this chapter will help you to get accustomed to the forms of life that I will later use to present how they were formed. The understanding of the similarities and differences between the types of organisms will prepare you to comprehend the processes that molded them.

2.1. Systematics or classification of organisms

Throughout the ages, human beings have tried to explore, inventory, and classify things in nature. They did so for celestial bodies, chemical particles, and living things. Some of those efforts have been systematized into scientific classifications used as references for most works on living things. For instance, since the publication of the historic book on the hierarchical classification of organisms by the Swedish botanist Carl Linnaeus in 1758 (Linnaeus, 1758), many modifications of the classification of living things have been done. As of 2021, taxonomists have tried to classify living organisms using about 8 major taxonomic ranks: domain, kingdom, phylum, class, order, family, genus, and species. My goal in this book is not to argue about those classifications or to build a new one, but to mainly focus on some of them to explain the origin of life.

As of 2025, species are considered as the lowest level of classification of organisms. Although many characteristics are used to define species, one of the most used is the ability of individuals of the same species to reproduce. But that definition does not always stand, for some organisms belonging to different species are able to cross reproduce. For instance, a male donkey and a female horse can mate to birth a mule as offspring, yet donkeys and horses belong to different species. Furthermore, for cultivated plants, some differences can be seen in cultivars or varieties of the same species. Even within the same variety or cultivar, the individuals in a population can look different.

Instead of the scientific classification mentioned above, the Bible (one of the books that also deals with the origin of life) used the term "kind", which some people refer to as the Bible's highest hierarchical classification, or God's classification. The Biblical "kind" used to recount the formation of organisms in the Bible's Book of Genesis is said to be most of the time close to the taxonomic rank termed "family". In other words, the term "kind" used by the Bible can refer to many genera containing many species. For, beneath the taxonomic rank called family are the genus followed by the species. At this point, I will review each of the 6 forms of life I mentioned above.

2.2. Plants

Until recently, plants were considered as photosynthetic organisms that include all algae and fungi. But today, fungi, some algae (e.g. red and brown algae), the prokaryotes (the archaea and bacteria) are no longer classified as plants. The green algae are considered plants. Based on their molecular characteristics, fungi are said to resemble animals more than plants (Deacon, 2005), yet based on their physical appearances, they have been classified as plants for centuries. For instance, when the Swedish botanist Carl Linnaeus classified organisms, he put fungi within the Plantae. However, unlike plants, which are able to make carbon-based compounds through photosynthesis, fungi lack chloroplasts and get carbon by breaking down and absorbing materials in their environments.

As the Merriam-Webster Online Dictionary and the Britannica Online Encyclopedia reported, plants are commonly defined as multicellular organisms having cell walls (made of cellulose), and capable of photosynthesizing by primarily using their chloroplasts. Some plants are found in water, on the ground, underground, and in the air. The term embryophytes is usually used to refer to land plants. Based on a recent count by "World Flora Online" (World Flora Online, 2020) and by "World Plants" (Hassler, 2020), by 2020, about 350,000 plant species have been identified in the world. Some of these plants produce seeds, while others do not. For instance, of the about 320,000 plant species acknowledged by 2010, about 260–290 thousand produce seeds (IUCN, 2010), meaning that, according to those statistics, about 90% of the known plants produce seeds.

2.3. Animals

Also called metazoa, animals are composed of many cells known for feeding on organic materials, breathing oxygen (O_2), capable of moving and reproducing sexually. The word "animal" derives from the Latin "animalis", which means 'having breath', 'having soul' or 'living being' (Cresswell, (2010). Animals (Fig. 2) can be classified into 5 groups, which I will detail later in this book:

CHAPTER 2: TYPES OF ORGANISMS

- Fish
- Bird
- Insect
- Reptile
- Mammal

To avoid redundancy of information while dealing with these groups of animals, I decided to provide more information about them in the chapters related to their formation. Furthermore, because this book is not just about anatomy, physiology, biochemistry, or other general studies of organisms, I refrained from delving into certain general descriptions already furnished by other research characterizing life. Although it is speculated that the Earth contains millions of animal species, only about 1.5 million animal species have been identified, and one third of them are insects. However, the "total biomass of humans and livestock account for more than 90% of all terrestrial vertebrates and nearly all insects combined" (Eggleton, 2020).

Fig. 2: Examples of animals

2.4. Fungi

Also called funguses, fungi (singular: fungus) are eukaryotic organisms such as mushrooms (Fig. 3), yeasts, and molds. Unlike plants, but just like animals, fungi are heterotrophs, meaning that they are unable to primarily produce their own food, but must rely on absorbing dissolved molecules (e.g. from chemical and organic compounds) that they decompose in their environment. In other words, unlike plants, fungi are unable to photosynthesize. They are perceived as the main decomposers of organic matters. Just like plants, fungi are mostly immobile. Their morphology and growth habitat look the same as that of plants. Fungi are able to reproduce both sexually and asexually.

Just like plants, fungi have a cell wall, but this cell wall is characterized among other things by the presence of chitin in it. That specific trait distinguishes fungi from plants, bacteria, and some protists. Fungal cell wall also contains glucans (a

CHAPTER 2: TYPES OF ORGANISMS

type of carbohydrate), which are also found in plants, while chitin is found in the exoskeleton of arthropods (Alexopoulos et al., 1996). Unlike most fungi, the water molds (oomycetes), which were classified as fungi, contain cellulose in their cell wall, but lack chitin. By the way, chitin is the second most abundant polysaccharide found in nature. According to the previous author and the following, fungi are the only organisms that combine both glucan and chitin in their cell wall, which, unlike that of plants, does not contain cellulose (Gow, et al., 2017). Unlike plants, most fungi do not have an effective xylem and phloem, which plants use to transport water and nutrients.

The cells of most fungi usually grow and look like elongated, filamentous structures called hyphae. However, yeasts are single-celled fungi that do not form hyphae. Many hyphae interconnect and form a network called a mycelium. The hypha is also able to branch or fork as plants do. Just as plants grow at their meristems, so also fungi grow at their tips (apices). Like plants, fungi usually grow on soil and some of them (e.g. mushrooms) form noticeable fruit bodies, similar to that of moss, a type of plant. As of 2021, more than 148,000 fungi species have been described (Cheek et al., 2020).

As that of most eukaryotes, fungal cells have membrane-bound nuclei containing chromosomes, membrane-bound cytoplasmic organelles (e.g. mitochondria, membranes, ribosomes), and they contain carbohydrates (Deacon, 2005). Just like animals, they lack chloroplasts and are heterotrophic organisms (Alexopoulos et al., 1996).

Due to some of their differences with respect to the so-called true fungi, some organisms originally classified as fungi or "fungus-like" such as hyphochytrids (which have both chitin and cellulose), water molds, and slime molds, are no longer considered as real or true fungi. In other words, most of the time, as more details emerge about the characteristics of organisms, scientists try to adapt the frontiers of their classification and theories accordingly. This also implies that many concepts "accepted" today will be reversed sometime in the future.

Fig. 3: Mushroom, an example of fungi (Photo credit © Nathanael-Israel Israel)

2.5. Protists

Although people usually define protists as organisms that are neither plant, animal, or fungus, the nuance is more than that. Algae are the main dominant example of protist. Other examples of protists are water molds and slime molds. *Plasmodium falciparum*, the human parasite that causes malaria (also called paludism), is a protist. That unicellular protist has caused so many deaths that it is considered the deadliest human parasite. Most of the victims of malaria are found in Sub Saharan Africa. Those who are not informed may think that mosquitos are the organisms that kill people through malaria, but they are just the carrier of the aforementioned protist, which feeds on human beings.

Likewise, *Trichomonas vaginalis*, a flagellated parasite that causes trichomoniasis (a human sexually transmitted infection which symptoms manifest in women as vaginitis or vulvovaginitis, an inflammation of the vagina and vulva), is also a protist.

While most algae are autotrophic, meaning capable of doing photosynthesis, other protists do not have chloroplasts, which contain the chlorophyll, the pigment required for plants photosynthesis. Instead of understanding that some protists were originally formed without their own chloroplasts but endowed with other means to live, some scientists think that they have lost the pigments. I will elaborate on that later.

2.6. Archaea

Archaea (singular: archaeon) are single-celled prokaryotes, meaning organisms lacking cell nuclei. As of 2025, no clear definition has been given to archaea and I expect them to be reclassified very soon once more is known about them in this era of molecular biology during which more details are coming out more often than

CHAPTER 2: TYPES OF ORGANISMS

ever before. To some extent, the archaea have characteristics intermediate between bacteria and eukaryotic. Unlike in the past, they are no longer classified as bacteria. For instance, although their size or shape is similar to that of bacteria, they are said to contain genes and metabolic pathways similar to those of eukaryotes. Archaea are said to range from 0.1 micrometers (μm) to over 15 μm in diameter and usually look like "spheres, rods, spirals or plates" (Krieg, 2005). The cellular membranes of archaea are made of molecules different from those of other organisms. Archaea are said to reproduce asexually.

Many of them are found in the gut, mouth, skin, and other locations of human microbiome (Bang and Schmitz, 2015). Some of them play crucial roles in the fixation of carbon, the cycling of nitrogen, the decomposition of organic compounds, and the maintenance of microbial communities (Moissl-Eichinger et al., 2018). As of today, they have not been classified as clear pathogens or parasites of other organisms. Some of them called methanogens are found in the gastrointestinal tract in humans and ruminants, where they produce methane, and facilitate digestion. Some of them are involved in the production of biogas and in the treatment and management of sewage wastes. Some extremophile archaea are able to live in extremely hot environments, while others thrill well in extremely cold environments.

2.7. Bacteria

Bacteria (singular: bacterium) are generally free-living organisms habitually consisting of one cell. They used to be classified as plants. These prokaryotes (meaning their cells lack a nucleus) are present almost everywhere on the planet. They are usually a few micrometers long and their organelles usually lack membranes. The cell wall of bacteria is made of peptidoglycan, which is a polysaccharide. The size of the cell wall of all bacteria is not the same. For instance, the cell wall of some bacteria (termed Gram-positive bacteria) is very thick, while that of others (called Gram-negative bacteria) is very thin and that of others yet is neither thick or thin. For instance, the "clinically important bacteria called mycobacteria have a thick peptidoglycan cell wall like Gram-positive bacteria, but they also have a second outer layer of lipids (Alderwick et al., 2015).

Although many bacteria are harmless and some even live in the human gut, others are very pathogenic, meaning capable of causing infectious diseases such as anthrax, cholera, leprosy, syphilis, and respiratory infections. Other bacteria are responsible for bacterial meningitis, pneumonia, urinary tract infections, gastritis, respiratory tract infections, sinusitis, eye infections, and many sexually transmitted diseases. Example of pathogenic bacteria include (Clark, 2010):
- *Clostridium tetani,* responsible of tetanus
- *Mycobacterium leprae,* the causing agent of leprosy
- *Mycobacterium tuberculosis,* responsible of tuberculosis

Bacteria do not grow and reproduce like most multicellular organisms do. Under some growth and reproduction conditions, bacteria can double every 17 minutes

(Pommerville, 2014).

2.8. Viruses

Although not usually identified as living organisms to the same extent as the 6 above-highlighted forms of life, viruses can perform some functions seen with conventional organisms. Hence, some scientific studies classified them as organisms. Because they don't meet all of the characteristics of most conventional living things, viruses are considered by some scientists as "organisms at the edge of life" (Rybicki, 1990). My goal here is not to argue in favor or against such a classification, but to just present a few facts, which can help explain the origin or viruses.

By 2020, the International Committee on Taxonomy of Viruses described more than 9000 virus species in detail (ICTV online, 2021). As of 2017, the National Center for Biotechnology Information (NCBI), a United States agency, has reported in their virus genome database more than 193,000 complete genome sequences (NCBI, 2017). Viruses are said to be found in nearly every ecosystem on Earth. Viruses usually exist in the form of DNA and/or RNA surrounded or protected by a coat protein (also called a capsid) and sometimes also by an external lipid envelope. Although some people consider bacteria as the smallest organisms, viruses are even much smaller than bacteria, sometimes even less than one-hundredth the dimension of most bacteria. Outside of their host organisms, viruses exist as independent particles called virions. Some RNA molecules called viroids (not considered as viruses) do not have the aforementioned protein coat although they share some features with numerous viruses (Collier et al., 1998).

Viruses are mostly known for their ability to infect all kinds of organisms and to replicate or reproduce themselves only when using or depending on the machinery of their host organisms. Viruses are responsible for diseases such as: avian influenza, some cancer viruses, chickenpox, cold sores, common cold, Covid-19 (which shook the world in 2020-2022) and other forms of coronaviruses, Ebola virus disease, hepatitis viruses, herpes virus, HIV (AIDS), influenza, lymphotropic virus, papilloma virus, polyomavirus, SARS, smallpox, etc.

2.9. Neglected spiritual lives

At this point, I should have talked about spiritual beings, but like I previously said, I need to wait a little longer until some few more chapters are introduced before I do. However, I would like to say a few things about how some people deliberately neglect to address spiritual beings in major discussions about life. Indeed, in all secular scientific books about life that I have consulted from the elementary schools all the way to the doctoral schools, I have never seen any that has a full chapter solely dedicated to spiritual beings, yet the spiritual world is real. When the authors of most of those books talked about spirituality, they usually attack some religious people (e.g. creationists) who oppose their evolutionist views. Likewise, most philosophical books that approach the spiritual aspects of life usually ignore key

CHAPTER 2: TYPES OF ORGANISMS

scientific details presented in scientific books. In other words, philosophical efforts to understand life are not synchronized with scientific efforts, and vice versa, and in the end, the holistic understanding of life is missed. Therefore, beings such as angels, demons, and God are usually removed from most secular biological discussions. Later in this book, I will handle those forms of neglected lives.

'Science180 Academy' Success Strategy:
SCIENCE180 ACADEMY OVERVIEW

Science180 Academy is a training, speaking, consulting, and mentoring program designed to groom and empower people of all backgrounds in the truth about the origin of the universe, life, and chemicals. According to their background and interest, trainees are taught different levels of scientific facts to grasp a deeper understanding of the origin of the universe, how to properly think to unearth mysteries hidden in the massive scientific data collected across the globe, but which is unfortunately less analyzed. If you want to be enlightened and equipped so you can cause positive changes in your respective field of expertise, then Science180 Academy program is for you.

Science180 Academy does not confer college credit, grant degrees, or grade its attendants, participants, or students. It is not an accredited university or college, but is the one-stop-destination for universe-origin, life-origin, and chemicals-origin experts. It is where scientists and laypeople get all their origin-related questions properly answered. It is the only place where the accurate interpretation of the universe-origin, life-origin, and chemicals-origin data matters a lot.

Science180 Academy brings together Dr. Nathanael-Israel Israel (the Founder of Science180) and other experts to deliver outstanding value, insight, and lessons to assist you to accurately understand the true origin of the universe, chemicals, and life, so you can tap into that knowledge to improve lives perpetually. Nathanael-Israel's goal is to give you practicable and undeniable proofs of the formation of the universe so you can be fired up to become the best version of you, and to cause positive changes to your initiatives that will profit you today and forever. For Nathanael-Israel, decoding the origin of the universe and everything in it is not a job, but his life mission, and helping others to fully understand that is his mission. Visit Science180Academy.com today to start.

Science180's clients and prospects have a profound technical knowledge and background in science, while others don't. Some are creationists (e.g. Science180 creationism, Young Earth creationists, Old Earth creationists, Intelligent design proponents), others are anti-creationists. Some are believers, others are freethinkers (including atheists, humanists, rationalists, agnostics, nontheists, nonreligious, skeptics, nonbelievers, religiously unaffiliated, spiritual-not-religious, ex-believers, and doubters).

TURBULENT ORIGIN OF LIFE

Regardless of their background, belief, or disbelief, Science180 works with each of these people to figure out their needs, priorities, and the products and services that best fit them. Science180 improves their knowledge, experience, performance, and answer their questions (related to the universe-origin, life-origin, and chemicals-origin) by crafting a personalized program that perfectly matches their interests, needs, and things that are dear and meaningful to them whether it is to:

- Bypass technical knowledge that restricts non-experts from accessing the life-origin truth contained in the massive scientific data, and get to the bottom of scientifically-locked origin-related secrets regardless of your background.
- Challenge the cosmological status quo and embrace the real change that will disrupt the cages that were holding you
- Connect with practical tips about how to decode the origin of the universe, life, and chemicals and protect yourself from wrong theories in the literature and the media
- Empower yourself to leave unforgettable marks and to stand tall as a symbol of freedom, power, creativity, and originality in your field of expertise
- Fearlessly push the boundaries of the human abilities to properly understand what is perceived as un-understandable, mysterious, supernatural, , impossible, and unthinkable that holds you back
- Get inside secrets about how to locate flaws in life-origin related theories so you can save time, money, and other resources to improve lives
- Have a reliable access to the world's authority on origin-related matters and get your origin questions professionally answered with the truth step-by-step
- Protect yourself and loved ones by keeping all of you secured and empowered with the true knowledge of the origin of life and the universe
- Satisfy your burning desire for freedom from beliefs and scientific theories about the universe-origin and life-origin that suffocate you and bind your mind, faith, unbelief, heart, and education
- Scientifically test and know whether there is a God that created the universe or not, and which God it is
- Ultimately boost your confidence in detecting, confronting, and avoiding wrong theories by knowing the facts and processes involved in the formation of life and the universe

To register or to learn more, visit Science180Academy.com today.

CHAPTER 3

FROM CRACKING THE DNA OF THE UNIVERSE-ORIGIN TO UNLOCKING THE CODE OF THE FORMATION OF LIFE AND OF CHEMICAL PARTICLES, THIS SCIENTIST AND MATHEMATICIAN HAS DONE IT ALL

Do you really want to know why and how I wrote this book on the origin of life? This book on the origin of life is part of a series of books I wrote about the origin of the universe. Like I detailed in *"Turbulent Origin of the Universe"*, my journey of researching and writing about such a topic started in 2013. In fact, after graduating with my PhD in plant, insect, and microbial sciences from the University of Missouri, where I was the Doctoral Marshall (meaning first of my class of hundreds of doctorates), I practiced in the field on biotechnology and other aspects of plant science research before a concourse of events branched out my investigation of the origin of many things including the universe and everything in it. Indeed, I wrote many things you will see in this book in 2013 long before I knew that I was going to write books on the origin of the universe. In fact, it was the curiosity of double checking what I was writing on life that partially led me to examine what is scientifically known about nonliving things (e.g. celestial bodies) in the universe. In those days, because, in spite of my achievements in academics, certain things were missing in my life, which I was trying to address, I was searching for the truth and ways to get out of some struggles to succeed. In other words, my efforts to comprehend the meaning, mission, fate, and struggles in life preluded my research on the formation of the universe.

When I started learning about celestial bodies in 2013, I noticed that the scientific community collected a wealth of information, which I quickly understood were improperly analyzed. Because a lot of ideas were already running through my head by then, I quickly noticed how some of the data I was finding in the scientific literature could make a sense different from what they had been framed for. Without delay, I realized that some of those scientific data could help me to back some philosophical ideas I was having. By 2013, I already wrote many pages on life and the universe and I was planning to write about them in a book someday. In those days, the information I wrote down concerned also philosophical explanations of the origin of the universe, but in a way that I never heard of before. In fact, the 4 people with whom I shared the primer of those ideas in 2013 and even later, thought I was really crazy. Part of my motivation of learning about what the scientific data said about the philosophical ideas that were racing through my mind by 2013 was to compile a few scientific facts to write a chapter in that book. In those days, I did not know that what I thought could be a small book chapter would become a data mine that would occupy me for 9 years. Beforehand, due to the scientific data I analyzed, I went back to reinterpret the aforementioned philosophical intuitions that started in 2013 and never ended until today.

Due to the immensity of the scientific data and the time needed to properly analyze them, I decided to put aside for a moment the life-origin data and work on the universe-origin before I resumed my efforts to explain the formation of living things. Hence, I ended up writing my books on the origin of the universe before the one you are reading now on the origin of life. Because of the size of the information I gathered and that I needed to share, I realized that I could not combine all of it into one book. As I split the content according to the audience, I ended up writing many books on the origin of the universe including:

- *"Turbulent Origin of the Universe"*
- *"Turbulent Origin of Chemical Particles"*
- *"From Science to Bible's Conclusions"*
- *"How Baby Universe Was Born"*
- *"Science180 Accurate Scientific Proof of God"*

Along the years, I always took note of every idea and inspiration I had about life and how my understanding of the formation of the celestial bodies shaped my perception on the origin of life, which is completely different from any scientific theory I had ever heard of. By the time I finished writing my books on the origin of the universe, I had tons of notes about the formation of life. But to my surprise, I was unsatisfied with the theories elaborated about the origin of the universe and life not only by the secular people in the academia, but also by many people who framed themselves as believers, including some creationists. For humankind has failed to properly understand how life and the universe that contains it originated. Hence, I felt the need to also write a book on the origin of life. But as I was finishing writing the books on the origin of the universe, I felt like combining them into one book and adding details about the origin of life would have made those books very heavy. To avoid diluting my understanding about the origin of life, I decided to devote a different book to life and I called it the biological version of my books on the origin of the universe. In the end, I realized that the ideas that raced through my mind and

CHAPTER 3: WHY AND HOW I WROTE THIS BOOK

that I captured on paper and digital recordings in 2013 were a foundation of all my books about the universe.

After "finishing" the writing of the books on the origin of the universe, I initially decided to launch them first and then slowly write the book on the origin of life. Then, in September 2021, I strongly felt like I should write the book on the origin of life before I go live with any other book on the universe. Doing so allowed me to put my original thoughts on paper before people could try to influence my way of thinking about life. I felt like this strategy could also allow me to "prevent" some people from using my original discovery on turbulence and the origin of the universe to introduce wrong theories or doctrines about the origin of life, which may take a long time to correct. For it will be harder to change some minds after they swallow wrong doctrines based on people's wrong interpretations of my discoveries on turbulence, which I know are very original. Finally, writing this book will also help me to feed the thought of the countless biologists (and other life science specialists) who will be interested in knowing how they can apply my discovery on turbulence to their fields.

My goal for this book is to briefly introduce the main cornerstones and foundational facts about the true origin of life. Before I start handling the process involved in the formation of living things, I will first present some key facts I discovered about the origin of the universe. Then, after showing how those discoveries affected my explanation of livings things, I will elaborate on how all organisms were formed.

CHAPTER 4

TO NEVER GET THE ORIGIN OF LIFE WRONG, PAY ATTENTION TO THIS SCIENTIFIC BREAKTHROUGH ON THE STUDY OF TURBULENCE IN THE EARLY UNIVERSE

To explain the origin of life, I will be using some key lessons I learned from my investigation of the origin of the universe. This chapter and the next one will give you a quick summary of some keys I discovered on celestial bodies and that will help you to decode the formation of life.

Before any living thing was made, the matters in the universe had to be made first. This does not mean that all kinds of matter present in the universe today were all completely formed before life originated. But for each form of life, the corresponding matter was made first. In other words, the first step in the formation of life was the formation of the matter it is made of. Hence, in this book, I felt compelled to first handle the origin of celestial bodies and chemical particles before I delve into the origin of life. In other words, because living things are made of chemical elements and spiritual (invisible) entities, it is important at this point that I summarize my discoveries about the formation of the universe and chemical particles, which I covered on thousands of pages in other books.

This book on the origin of life contains some key information thoroughly handled in *"Turbulent Origin of the Universe"*, but here, I broke down the raw scientific data that can annoy, scare, or confuse the profane, who may be unfamiliar with them. So the readers may know a little bit about the philosophical (including religious) implications of the explanation of the origin of life and other things in the universe, this book also contains a few information from *"Reconciling Science and Creation Accurately"*. For it will be unjust not to clearly state my religious viewpoint as I dealt with the origin of the universe, which cannot be completely understood

CHAPTER 4: WHAT IS TURBULENCE AND HOW I STUDIED IT

without referring to some religious facts revealed in some religious book, which science can likely never be able to explain using conventional or traditional methods. Therefore, here, instead of detailing things that I already approached in those books, I will just present a limited take home message concerning selected crucial chapters of those books, knowing that those who want to know more, can consult those records.

Because, in this book, you will come across the term "biological turbulence" or "turbulence" a lot, to ensure everybody understands what turbulence means, I decided to define it first. Later in this book, after presenting a few more data, I will also explain biological turbulence.

To illustrate what turbulence means, I would say that, if you have ever seen the movement of cream poured into coffee or the movement of clouds in the atmosphere, you have already experienced turbulence. Turbulence can occur when fluids such as gases (e.g. air) or liquids (e.g. water) flow around a moving vehicle, airplane, and boats. Wherever there is turbulence, fluids formed multi-dimensional structures (some of which are termed vortices, vortexes, or eddies) which are moved in a complex fashion that some people defined as random and chaotic. Therefore, for centuries, people have usually associated turbulence with chaos, a state of disorder, randomness, trouble, or a disordered motion of things not just fluids but even organisms.

Indeed, considered as "the last great unsolved problem of classical physics", turbulence is a phenomenon that has no clear definition despite it being investigated for more than 500 years. However, turbulence is not just a problem found only in physics, but also in many other disciplines, which unfortunately ignore it, not because the experts in those scientific fields want to, but mainly because most of them are not aware that some of the complex data they deal with are encoded in turbulence. Not much progress has been made in the field of turbulence. Due to the significance and the difficulties of studying turbulence, when asked what he would ask God if given the opportunity, Quantum physicist Werner Heisenberg said *"I'll ask him two questions. Why relativity? And why turbulence? I really believe he will have an answer to the first."*

In my books on the origin of the universe (i.e. *"Turbulent Origin of the Universe"* and *"From Science to Bible's Conclusions"*), I lengthily explained why turbulence is not really a chaos or random phenomenon, but a high level of order that people have failed to distinguish. Therefore, after years of investigating the formation of the universe through the glance of turbulence, I ended up crafting my own definition of turbulence on October 28, 2018:

"Turbulence is a complicated phenomenon that can lead to an ordered state of matter that did not always allow scientists to comprehend it because they were approaching it with a limited and corrupted mindset, which does not even know that every kind of matter in the universe is a product of the original turbulence, the mother of all turbulences, which occurred at the beginning of the universe. I saw order everywhere in turbulence, but the problem is that the order in turbulence is so complex and filled with details that a mere human being using the current limited computational machine cannot easily understand. In other words, turbulence contains too many data and grabbing the full picture of its information is impossible to those who like to think linearly. To see some details in turbulence, scientists will have to magnify their data and the scale of their experiments as well as the power of the equipment they use to visualize these data. Unfortunately, they are limited in what they can see and when they see things, they cannot properly analyze them because they think turbulence is a disorder. They also tend to focus on the bigger things, while neglecting the smaller ones, which should help them to have a complete understanding of how things were formed. Turbulence in the universe occurred at many scales and locations in space. Although many things in the universe today may no longer be going through the kind of strong turbulence at the beginning of the universe, they are still "feeling" or bearing the consequences of that turbulence, which birthed them. To put it in another way, many kinds of turbulences have been occurring in the universe today, but they all can be linked to the initial turbulence that I called the "mother of all turbulences", the strongest and largest turbulence of all time. Before that turbulence, no other turbulence has ever existed and since it occurred so the universe can be formed, no other turbulence as strong has occurred yet although many strong turbulences are still happening in the universe on many scales" (Israel, 2025a).

Likewise, I had to design a methodology to study turbulence based on some of the following aspects usually handled when turbulence is addressed as I described in *"Turbulent Origin of the Universe"*:

- description of the various scales of motion of a fluid in turbulence and if possible, the equations that explain them
- source and the transfer of energy to these scales and from one scale to the others
- interactions between the structures formed by the turbulence and/or their "evolution" from a certain precursor
- whether or not a fluid jet under the influence of turbulence will break, and if so, how long will it take
- sensitivity of a turbulent fluid to the density, viscosity of its environment
- processes that can explain the separation between the fluid fragments if they break
- size dispersion of the fluid fragments or scales of motion after a fluid breakup
- factors controlling the size distribution of the structures formed in a turbulent fluid, etc.

For instance, on the scale of celestial bodies, I explained the energy transfer in the precursor of the Solar System by clarifying how the precursors of the planetary systems and the Sun got their energy from the precursor of the Solar System, how a planet and its satellites got their energy from the precursor of their planetary system. I showed that, structures found in turbulent fluids get their energy from

CHAPTER 4: WHAT IS TURBULENCE AND HOW I STUDIED IT

their precursors, not just from the interactions with one another.

The summary you will see in the next chapter and across this book about my findings on the origin and formation of the universe is based on my study of hundreds of variables collected on:

- Hundreds of celestial bodies in the Solar System including the Sun, the planets (Mercury, Venus, Earth, Mars, Jupiter, Saturn, Uranus, Neptune, and Pluto), asteroids, satellites, rings systems, as well as the atmosphere and crust of some celestial bodies
- Stars in the universe and their clusters into galaxies, globular clusters, galaxy clusters of various scales
- Every chemical element and most subatomic particles known as of 2025
- Etc.

For the sake of space, I will not detail every variable I studied on the celestial bodies here, but I would like to list a few:

- Angular momentum (related to the quantity of rotation or the ability or propension of a body to move around an orbit or a rotational axis)
- Axial tilt (inclination of the rotational axis with respect to a reference plane or axis)
- Density (which expresses the ratio between the mass and volume of a body)
- Eccentricity (related to the elongation of the orbit of a celestial body)
- Escape velocity (related to the minimum speed generally required to escape the gravity of a celestial body)
- Gravity (which is used to explain why things fall back on earth after they are thrown in the air with a speed inferior to the escape speed)
- Mass (related to the amount of matter in a body)
- Orbital inclination (the inclination of the orbit of a body with respect to a referential plane, which for the planets in the Solar System is the elliptic, the place on the earth's orbit around the Sun)
- Orbital speed (the speed of the motion of a body on its trajectory around the primary body it orbits)
- Radius (an expression of size of the straight line from the center to the circumference of a circular or spherical body)
- Rotation period (the time needed for a body to complete one rotation around its rotational axis)
- Rotation speed (the speed of rotation of a body at its equator)
- Rotational kinetic energy (energy related to the rotational motion of a body)
- Semi major axis (the average distance of a body from its primary body; for the planets in the Solar System, the semi major axis is their average distance from the Sun)

TURBULENT ORIGIN OF LIFE

- Time scale (the average time it took for a process or event to occur)
- Total kinetic energy (the sum of the translational kinetic energy and the rotational kinetic energy of a body)
- Translational kinetic energy (kinetic energy related to the orbital speed and mass of a body)
- Volume (the product of the mass and the density of a body, meaning an expression of the 3-D amount of space that a body occupies)
- Hundreds of other variables and ratios.

Likewise, for the chemical particles (atoms, molecules, minerals, chemical compounds, and rocks), I studied many variables including the following:

- Abundance (e.g. abundance in the universe, in the Sun, in humans, in oceans, in the Earth's crust, in the atmosphere of the planets, in the atmosphere of the Moon, and in the surface and crust of the Moon)
- Composition, density, electric charge, energy, hardness, mass, melting point, radius, rotation, state at the standard temperature and pressure, stability, and dozens of other variables.

For most of the aforementioned variables, I did some descriptive and analytical statistics, regression, and other mathematical analysis before finalizing my conclusion about the formation of the universe, which I summarized next.

CHAPTER 5: SUMMARY OF MY DISCOVERIES ON THE TURBULENT ORIGIN OF THE UNIVERSE

CHAPTER 5

WHY DON'T PEOPLE TAKE SERIOUSLY THIS SUMMARY OF THE TURBULENT ORIGIN OF THE UNIVERSE THAT SHOOK THE FOUNDATION OF ALL EXISTING COSMOLOGICAL THEORIES ... SO THEY CAN LIVE HAPPILY AND THINK MORE CLEARLY?

Because I decoded the formation of life using some keys I discovered while studying the origin of the universe, I felt like you will benefit from me first sharing some of the take home messages about the universe formation. As I will start presenting the formation of life later, this summary will help you to better visualize and easily relate to how living things were formed. Here, due to space limitations and what is relevant for this book, I will briefly outline just some key findings, hoping that those who are interested to know the full story of the universe-origin can consult the books I wrote about that. Indeed, I scientifically showed that (Israel 2025a), in the beginning of the formation of the universe, a certain kind of matter, which I call the "turbulent prima materia", which is different from any matter known today, mysteriously appeared in the universe out of nothing..., and through very complex, dynamic, and turbulent processes, that it was progressively molded into all types of matter, bodies, systems or clusters of matters and of bodies known in the universe today. The initial matter in the universe was formless and could have occupied a very huge portion of space. The state of the matter of the "turbulent prima materia" was none of the 4 states of matter known today (solid, liquid, gas, and plasma). The initial state of the original matter went through complex changes and processes so that each state of matter could be formed. Quickly after its appearance, the turbulent prima materia was broken open into giant and small pieces by a violent event accompanied by a huge noise as that of an "explosion". The original mysterious scattering is a term I coined to explain how the turbulent prima materia was divided into blocks of matter of various scales. Some of the aforementioned "broken pieces" of the turbulent prima materia quickly became the precursors of the systems of bodies (galaxy clusters, galaxies, and all other celestial bodies and particles found in the universe). Some of the processes that molded the original matter in the universe and which led to the formation of the bodies in the universe involved:

- Fluid Instability
- Split of fluid bodies and formation or precursors of bodies
- Birth, split, and transfer of energy
- Initiation of movement
- Fluid flow
- Organization and separation of fluids into layers, formation of fluid ligaments, pinch off, mixing, breakup, relaxation, recoiling, tilting, or overturning of fluid bodies
- Acquisition and transfer of momentum
- Initiation and development of turbulence
- Positioning, sizing, and spacing of bodies
- Gathering together of fluid layers
- Spiraling or spinning of layers of fluids to birth bodies of various shapes
- Formation, squeezing, and loosening of vortical structures
- Birth and strengthening of various forces (which compressed matter to increase its density or loosen it to decrease density)
- Wrapping, tilting, stretching, squeezing of vortical structures
- Coalescence of fluid bodies, etc.

Most of the aforementioned processes of the formation of the universe also exist in some biological systems, reactions, and pathways. In the chapter on developmental biology, I learned lessons from morphogenesis and biology of development, I addressed some of these similarities. For instance, the first or initial cells (stem cells) of living organisms are undetermined, but as the body grows and goes through developmental stages, biochemical changes occurred to modify or mature the initially undetermined cells into determined cells, which are generally unable to reverse back to other kinds of cells, including the undetermined ones. Likewise, not only was the turbulent prima materia (the original matter in the universe) undetermined, but it was also transformed into specific daughter bodies most of which are unable to reverse back to their precursors. I demonstrated that the split-gathering of the initial matter led to the formation of various precursors of bodies on various scales:

- Invisible or spiritual bodies
- Chemical particles (subatomic particles, atoms, molecules, minerals, and rocks)

CHAPTER 5: SUMMARY OF MY DISCOVERIES ON THE TURBULENT ORIGIN OF THE UNIVERSE

- Celestial bodies (e.g. stars, planets, asteroids, and satellites)
- Stellar systems
- Globular clusters
- Galaxies and clusters of galaxies
- The whole universe

In the rest of this chapter, I will quickly explain how turbulence led to the formation of specific celestial bodies and the particles they contain. I have another book called *"Turbulent Origin of Chemical Particles"*, which is devoted to the origin of chemical particles and another called *"Turbulent Origin of the Universe"*, which is devoted to the origin of the universe. In the later, I showed that the fluids in the precursors were characterized by the:

- Formation and shearing of layers of fluids in them
- Intermittence and mixing of the fluids in the layers
- Top layers leading to the formation of innermost secondary bodies, whereas the bottom layers birthed the outermost secondary bodies
- Stratification of the fluid layers of the precursors is responsible for the stratification of the crust and atmosphere of celestial bodies
- Speed of the fluid layers was later translated into orbital and rotational speed of the daughter bodies
- Direction of the flow of the fluids in the layers can explain the counterclockwise movement of some bodies as seen from the north pole
- Topological and structural changes of the fluids of the precursors of bodies account for some of the variations in the composition of the bodies.

The fluid layers at the top and at the bottom of the secondary flows for instance were not as turbulent as those which were somewhere "between" them. The layers of fluids in the precursors of bodies can be divided generally into 7 regions that I called the turbulence zones, which host bodies having different characteristics. Examples of these turbulence zones include the following 5, which were my focus when I studied planets and satellites in the Solar System:

- Turbulence Zone 1: where in planetary systems, the innermost and fastest satellites are found.
- Turbulence Zone 2: a transition zone from the "laminar" turbulence of Zone 1 to the fully developed turbulence which will be reached in Zone 3.
- Turbulence Zone 3: where the highest turbulence leading to the formation of the biggest secondary bodies in a fluid flow are usually found.
- Turbulence Zone 4: a transition zone out of turbulence, but, unlike Zone 2, which is a transition zone into turbulence.
- Turbulence Zone 5: the outermost zone where the remote and the slowest and smallest celestial bodies in each system are found.

During the formation of the universe, cascades of split-gathering were observed and precursors of bodies were borne and reorganized into their daughter bodies. I

TURBULENT ORIGIN OF LIFE

demonstrated that the cascades of split-gathering led to the formation of the:
- precursors of clusters of galaxies and clusters of globular clusters
- precursors of stellar systems
- precursors of primary stars and the precursors of the bodies orbiting them (e.g. planetary systems, planets without satellites, asteroid systems, and asteroids without satellites)
- precursors of their primary planets and of their satellite system
- precursors of asteroid systems and their satellite system
- precursors of minerals, mineraloids, and rocks
- precursors of atoms and their clustering into molecules and chemical compounds
- precursors of subatomic particles
- precursors of the smallest particles that will never be scientifically discovered

One of the key features in turbulence is what is called intermittence, a phenomenon that caused the formation of smaller bodies between larger ones. Due to the importance of this process, I devoted a special chapter to it later in this book. It is not by chance that smaller organisms abound wherever larger ones are found in nature. In other words, regardless of the form of life, smaller organisms are usually located between larger ones. Regardless of the system of bodies, the number of the bigger bodies is usually smaller than that of smaller bodies. The way that the precursors of the celestial bodies were split-gathered explains why celestial bodies are usually organized as primary bodies "orbited" or surrounded by secondary bodies.

In *"Turbulent Origin of the Universe"* I justified how and why, during the turbulence of a body of fluids, more than 99% of what I called the 9 "system-additive" variables (e.g. mass, energy, volume, momentum) went into the precursor of the primary body, while less than 1% went into the precursor of the secondary bodies. In other words, when a body of fluids breaks up under the influence of turbulence, a very few bodies amass most of the energy, mass, volume, momentum, but several small bodies amass just a tiny amount. I also showed that physical laws are not the same everywhere in nature. I also showed that, in each system of bodies (e.g. Solar System, planetary systems), the largest body is the primary body, while the largest secondary bodies are usually located in the most turbulent zone, usually Zone 3. Some of the factors that affected the breakup of the precursors of the celestial bodies include viscosity of the fluids, their energy, and their position in the stack of fluid of precursors. The size of the precursors of the celestial bodies affected the destiny of their internal particles and consequently their overall state.

Some of the processes involved in the split-gathering of the precursors of celestial bodies are similar to the processes which led to the branching and organization of some organisms even until today. I demonstrated that the celestial bodies escaped the precursor of their primary body with a speed almost equal to the

CHAPTER 5: SUMMARY OF MY DISCOVERIES ON THE TURBULENT ORIGIN OF THE UNIVERSE

escape velocity of the primary body. For instance, after the precursor of the bodies orbiting the Sun escaped the precursor of the Sun at a speed equal to the escape velocity of the Sun, its stack of fluids flowed until reaching a point where the top fluid layers split from the rest, which, at its turn, continued its journey until another layer split and so on and so forth until the last fluid layer was gathered into a body. At the planetary levels, after the precursor of a planetary system split into the precursor of the primary planet and the precursor of the satellites and rings around it, the precursor of the bodies orbiting the primary body started flowing and its fluid layers started separating from one another until the bottom fluid layers were reached.

The cascade of split-gathering of precursors of celestial bodies is similar to the processes of the growth of plants and other organisms: a seed germinates and after yielding a precursor of roots and shoots, it continues growing, leading to the emergence of branches from a trunk, which continues growing until another branch emerges again and so on and so forth until the tree reaches its maximum height and eventually dies. Each branch continues growing at its own rate which can be different from that of the trunk. Based on a formula I discovered for the birthdate of the celestial bodies using scientific data, and that I detailed in my books on the origin of the universe, the first body to be formed in the Solar System (according to the celestial bodies known as of 2020) was Mercury, the innermost body orbiting the Sun, while the Earth was formed about 2.82 days after the beginning, the Moon was born 12.14 hours later, meaning 3.32 days after the beginning, and the Sun 3.693 days after the beginning of the formation of the Solar System (Israel, 2025a).In other words, the scientific evidences support that the Earth was formed on the 3rd day of creation, while the Moon and Sun were formed on the 4th day just like the Bible's Book of Genesis said more than 3500 years ago (Genesis 1:14-19). Using the scientific data, I showed that, among the creation narratives of the major religions and worldviews in the world (e.g. Animism, Buddhism, Confucianism, Hinduism, Islam, Judeo-Christian religions, and Evolutionism), only the Biblical account of creation (Bible's Book of Genesis 1:1-19) perfectly matches the scientific data (Israel 2025a) concerning the formation of the universe according to which Moses revealed that, on the first day of creation, God created the heaven and Earth, but on that day, the Earth was still not fully formed yet and it was on the 3rd day that the formation of the Earth was completed, while it was on the 4th day that the formation of the Moon and the Sun were completed. Before moving to the next chapter, I want to emphasize that this chapter is just a glimpse at the details of the formation of the universe and, if you want to have the full picture of how the entire universe and everything in it (e.g. celestial bodies and chemical elements) were formed, please consult the books I wrote on those topics: *"Turbulent Origin of the Universe"*, *"From Science to Bible's Conclusions"*, *"How Baby Universe Was Born"*

You can find more information at www.Science180.com/books.

Another Book by Nathanael-Israel Israel:
TURBULENT ORIGIN OF CHEMICAL PARTICLES

FIND ALL THE RELIABLE, CONVINCING, SCIENTIFIC ANSWERS YOU NEED TO SUCCESSFULLY DECODE THE ORIGIN OF CHEMICAL PARTICLES SAFELY

Where did all elementary particles and composite particles including atoms, molecules, minerals, and rocks come from? What are the fundamental factors, the machinery, and the generic processes that defined their formation and proprieties? What was the nature of their precursors at the beginning of the universe and what underlying processes shaped or molded them into the chemicals we know today? What was the primary cause of the abundance and diversity of chemicals in the celestial bodies in the universe? What is the accurate link between the formation of chemical particles and the formation of galaxies, stars, planets, asteroids, and satellites? What light can the origin of chemicals shed on the real cause and meaning of gravity and the other so-called fundamental forces in nature? How does the formation of the chemical particles fit into the big picture of the formation of the universe?

After studying these questions for more than 12 years, Dr. Nathanael-Israel Israel discovered that the proper understanding of the origin of chemical particles is a very challenging but profitable task that requires original, scientific, mathematic, and philosophic efforts beyond the current state of modern science—until recently. The solution for all of these puzzling problems: *"Turbulent Origin of Chemical Particles"*, the straightforward and trustworthy book that will help you to quickly, cheaply, easily, and efficiently navigate everything you need to know to finally solve the hard problems about the origin, the formation, and the functioning of all chemical particles. Whether you are a chemist, a biochemist, any other scientist, an engineer, as long as you have a reasonable background in chemistry but ignore how to scientifically demonstrate the origin of all chemical particles, this marvelous book is for you!

Amazingly packed with eye-popping analysis, fantastic graphs, tables, and the historic formula that broke the universe-origin code, *"Turbulent Origin of Chemical Particles"* will:

- Make it easier than ever for you to properly understand, decrypt, and articulate the real origin of natural chemical particles in the universe, therefore freeing you from false and boring explanations of the origin of all matters, and embrace the proven theory that opens doors to unparallel opportunities

CHAPTER 5: SUMMARY OF MY DISCOVERIES ON THE TURBULENT ORIGIN OF THE UNIVERSE

- Professionally teach you how to transform the true knowledge of the origin of chemical particles into insights that significantly add value to your life in less time, and successfully establish you as a symbol of freedom, power, creativity, and originality in your field of expertise
- Fire you up to become the best version of you, and to cause positive changes to your initiatives that will profit you nonstop
- Discover thrilling illustrations and unconventional explanations of the formation of all matter in the universe, written in a simple language that brings humankind much closer to the complete deciphering of the mysteries at the very heart of chemistry, and open the way to a future of technology, innovation, discoveries, and breakthroughs
- Equip you to bypass technical knowledge that restricts non-experts from accessing the origin-related secrets contained in the massive scientific data, and get to the bottom of origin-related mysteries regardless of your background so you can empower yourself to leave unforgettable marks in your field of expertise
- Learn more at Science180.com/chemical

With *"Turbulent Origin of Chemical Particles"*, the accurate decrypting and understanding of the formation of chemicals has never been profitable and easy. Hence this great book is THE ultimate how-to guide for great people wanting to correctly decode the origin of the chemicals and positively transform their lives. Get this celebrated book today. Don't wait!

Known as the nonconformist, rule-breaker, and accurate demonstrator of the universe-origin, **Dr. Nathanael-Israel Israel** is the founder of Science180, the one-stop for answering the most crucial universe and life's origin questions. He has had the honor to be acknowledged as the fearless universe-origin decryption trailblazer. Learn more at Israel120.com.

TURBULENT ORIGIN OF LIFE

CHAPTER 6: TURBULENT TREES, BRANCHES, AND LEAVES OF THE PRECURSORS OF CELESTIAL BODIES

CHAPTER 6

HOW NATHANAEL-ISRAEL ISRAEL DISCOVERED TURBULENT TREES, TURBULENT BRANCHES, AND TURBULENT LEAVES IN THE PRECURSORS OF CELESTIAL BODIES ... (AND HOW THEY WILL ENLIGHTEN YOUR PATH TO FINALLY DECODE THE ORIGIN OF LIFE)

Based on simple but deep concepts borrowed from plants (tree, branches, and leaves), this chapter is crafted to strategically start opening your eyes on significant similarities and differences between living and nonliving things that helped me and will also help you to crack the code of the origin of life.

6.1. Turbulent tree of the precursor of a system of bodies
As I was nearing the end of the writing of *"Turbulent Origin of the Universe"*, I noticed that the tracing of the organization of the paths taken by the fluid layers of the precursors of the celestial bodies in the universe can be summarized using a graph looking like a tree. Indeed, when I was working on the timescale of the split-gathering of the fluid layers according to their position in the stack of fluids of their precursors, I realized that how the fluid layers separated from the stack, each at its turn, is similar to how a plant germinates, grows, produces a trunk from which branches and leaves emerge with time. When I investigated the organization of bodies formed in turbulence, I also identified features looking like the trunk, branches, leaves, and roots of a tree. What has prevented people from discerning these patterns was that they did not know the history of the formation of those bodies and also, unlike real tree parts (roots, trunk, branches, leaves, and inflorescence), which are all connected, in the case of the celestial bodies, the connection between the bodies were broken during the split-gathering of their

precursors, leaving behind just systems of bodies that seem independent, yet some of them descended from common precursors, which, altogether can be traced back to the original matter or first matter in the universe that I term "turbulent prima materia".

Because, in addition to the formation of the celestial bodies, I also needed to decode the formation of the living things, I needed to find some terminologies that can ease the explanation of the processes and allow people to quickly grasp the similarities and differences between the formation of living things and nonliving things. Therefore, I coined the terminologies related to turbulent tree, turbulent trunk, turbulent branches, and turbulent leaves. Because details about the timing or the processes involved in the formation of the body parts of the turbulent tree of the precursors of the Solar System were presented in *"Turbulent Origin of the Universe"*, in this book, I just pinpointed a few things that can help you to understand the origin of living things, namely organisms.

The precursor-tree of a system of bodies is a graphical description of what the fluid layers of the precursor of the system could have looked like during the formation of its daughter bodies, with the root of that tree positioned as the precursor of the primary body of that system, the trunk of the tree being the fluid layers (or the leftover, trajectory or track of the fluid layers) of the precursor of secondary bodies (which escaped the precursor of the primary body) before it branched out, and the branches of that tree being the systems of bodies that emerged from that trunk. Just as all parts of a tree are connected together from the roots all the way to the apex of the trunk or main stalk passing by all the branches, with the leaves attached to them, so also, when I talk about the precursor-tree of the celestial bodies, the trunk of the precursor-tree is a virtual or apparent connection between the components or parts of that tree, although in reality, the fluid layers of the precursors of the systems of bodies were not connected altogether anymore once their split gathering started. For instance, on the scale of the Solar System (which is a stellar system), the precursor-tree consists of a tree which roots are the precursor of the Sun, the trunk being the trajectory or the track left by precursor of the bodies orbiting the Sun, the systems of primary branches connected to the trunk being the precursor the planetary systems or asteroid systems, while secondary branches are like the satellites of the planets and asteroids. Leaves are like some of the smallest levels of organization of matter on the scale of these celestial bodies, meaning they are like satellites and asteroids which have no satellite. Yet, at the level of the leaves, systems of primary and secondary bodies can be defined. Later in this book, I will detail that in the chapter on the formation of plants. Later on in this chapter, I have a graph illustrating this turbulent tree. The turbulence in the precursor of the trunk, which was partially transferred to the turbulence in the branches, birthed many orders of branching such as primary branches, secondary branches, tertiary branches, and so on and so forth until no branching could occur no more downstream of the flow in its veins. Hence, leaves can be found even attached to the main trunk or attached to a branch. In other words, if all the

CHAPTER 6: TURBULENT TREES, BRANCHES, AND LEAVES OF THE PRECURSORS OF CELESTIAL BODIES

branches and the trunk of a tree or plant can be removed without displacing the leaves, what would be left would be a system of leaves which organization may look weird, but a careful study of the position and distribution of those leaves could allow to pinpoint the location of the position, track, or trajectory of the trunk and branches of their precursors. To put it another way, by carefully investigating the organization of leaves with respect to one another and with respect to the different clusters of leaves on a tree, it is possible to describe the position of the branches and trunk and even describe how the entire tree could have grown from its roots all the way to the top of the tree. On the scale of the entire universe, all of the celestial bodies formed a tree that I called the "Turbulent Tree of the Universe" or the "Tree of the Universe" in short. Using the turbulent tree of the precursors allowed me to demonstrate how the universe was formed through a process of split-gathering of fluid layers of precursors on various scales. After a presentation of the parts of the turbulent tree of the precursor of the Solar System, I will show its graphical representation.

6.2. Turbulent branches of the precursors of celestial bodies

Just like most trees or plants are made of branches connected to a trunk or main stalk, I also found that the paths taken by the precursors of the celestial bodies look like branches. In *"Turbulent Origin of the Universe"*, I showed that the fluid that escaped the precursor of the Sun branched to yield the precursor of the planetary systems, and then the latter flowed and branched before yielding the precursor of the planets and their satellites. Hence, satellites do not directly orbit the Sun, but they orbit their primary planet, which at their turn, orbit the Sun. I will show you a graphical illustration of this process very soon. I also demonstrated that the formation of fluid layers is a form of separation or splitting of a fluid. The branching site is the point or position where a fluid layer leaves, or splits from, or is removed from a stack of fluid layers of its precursor so it can continue its own journey of gathering or split-gathering. As a layer branched out, the remaining of the stack it escaped continued its journey and released other layers with time according to their position, the top layer first and the bottom layer last. In some cases, from the branching site or branching point until the completion of the formation of the body, the gathering of the fluid layers dominated the rest of its split-gathering. For from this point, no more split may be needed, but smaller layers located inside the major one can be found inside the bodies being gathered. Hence, the presence of strata or layers of materials in some celestial bodies. An example is the layer of some rocks (that are not sedimentary) found in the Earth's crust. Unfortunately, some people think that all of the layers in the crust are deposited by sedimentation.

For instance, when I considered the stack of fluid layers of the precursors of the planets in the Solar System, the branching site of their precursor is the site where each of them splits from the rest of the stack of the precursor of the bodies orbiting the Sun. More specifically, the site at which the precursor of the Earth-Moon system split from the rest of the stack of fluid layers beneath it is its branching site. The position of that branching site is about 1 AU from the precursor of the Sun. In other words, the branching site of the precursor of each planetary system is about the semi major of their primary planet. The distance between 2 consecutive branching sites is the semi major increment separating them. Likewise, the fluid layer of the precursor of the satellites split from the stack of fluid layers of the satellites of their system at different branching sites corresponding to the semi major axis of those satellites. The distance separating two branching sites of satellites is the semi major increment of these satellites.

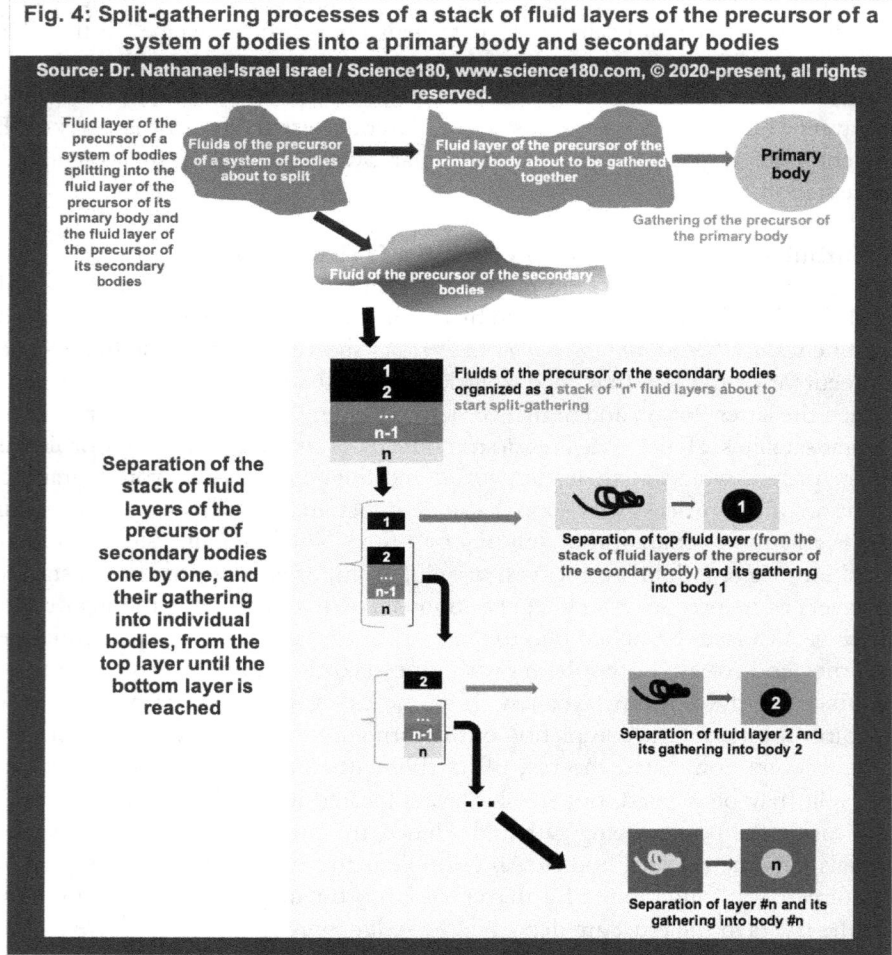

Fig. 4: Split-gathering processes of a stack of fluid layers of the precursor of a system of bodies into a primary body and secondary bodies
Source: Dr. Nathanael-Israel Israel / Science180, www.science180.com, © 2020-present, all rights reserved.

Fig. 4 illustrates the split-gathering of the fluid layers of the precursor of a system of bodies into a primary body and secondary bodies. First, the precursor of the system of bodies split into the precursor of a primary body and the precursor of the secondary bodies. The stack of fluid layers of the precursor of the secondary bodies

CHAPTER 6: TURBULENT TREES, BRANCHES, AND LEAVES OF THE PRECURSORS OF CELESTIAL BODIES

moved away from that of the precursor of the primary body and split into individual layers. After a layer splits, the rest move for a certain distance before another one splits, and so on and so forth until all of the layers are separated. Each of them goes through similar processes before yielding its daughter bodies and being collected into a special body after splitting from the rest. I coined the term "turbulent branching" to express the process by which turbulent branches appear on a turbulent tree.

6.3. Turbulent leaves and leaf-like shape of systems of celestial bodies

Armed with the turbulent understanding of the formation of the Solar System and other systems of bodies in the universe, I realized that many variables can be used to illustrate the leaf-shape of the precursors of the systems of celestial bodies. But in this chapter, I will focus only on one of the 5 variables encrypting the leaf-like shape of turbulent processes: the relationship between the semi major axis of the planets and the semi major axis of their outermost satellite.

These leaf-like shape patterns helped me to explain the biological turbulence that took place during the formation of living things. For the turbulent leaf-like shape based on the relationship between semi major axis of the primary planets and semi major axis of their outermost satellite, I plotted the semi major axis of the primary planets on the X axis, and on the Y axis is the semi major axis of their outermost satellite (Fig. 5). The goal here is to graphically show how the semi major of the primary bodies can be related to the semi major of their secondary bodies to yield a shape looking like a leaf. In the turbulent leaf-like graph below (Fig. 5 and Fig. 6), the widest area corresponds to the location of the semi major axis of the outermost Neptunian satellite, which has the highest semi major axis recorded on the satellites in the Solar System. The area covered by Jupiter and Saturn on the leaf-like shape below is not as wide as that covered by Neptune and its outermost satellite, yet turbulence was more developed in the Jovian and Saturnian planetary systems than in the Neptunian planetary system. Because the length of a secondary vein on a leaf is like the furthermost distance its precursor could have elongated to, the widest area of a leaf is not necessarily where the turbulence of the precursor of the leaf was most developed, but rather the zone where the precursor of the leaf had to travel further before reaching the end of its turbulence.

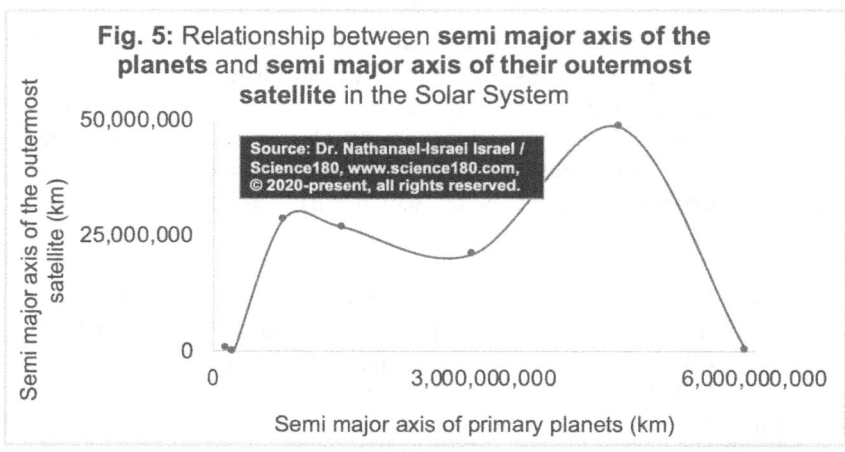

Fig. 5: Relationship between **semi major axis of the planets** and **semi major axis of their outermost satellite** in the Solar System

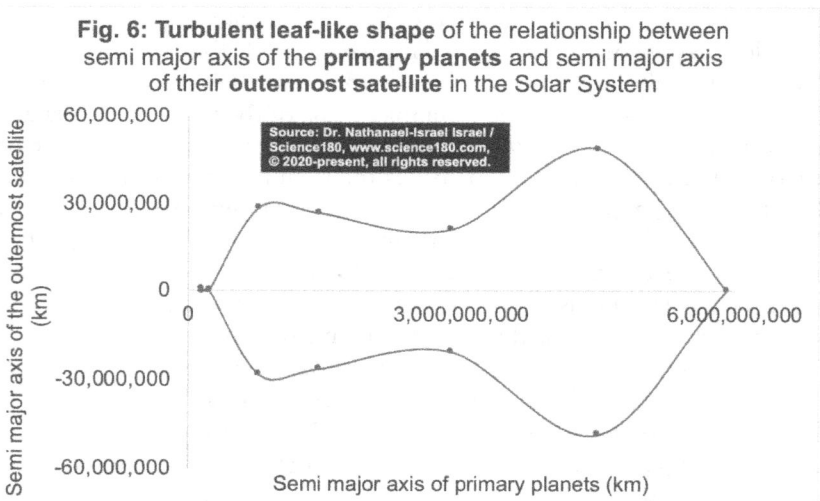

Fig. 6: **Turbulent leaf-like shape** of the relationship between semi major axis of the **primary planets** and semi major axis of their **outermost satellite** in the Solar System

6.4. Turbulent tree of the precursor of the bodies in the Solar System

The turbulent tree of the precursor of the bodies in the Solar System illustrates the distribution of the celestial bodies according to the order of the branching of their precursor in the stack of fluid of the bodies forming the Solar System (Fig. 7). I presented in parenthesis the escape velocity of the primary body in each system— by the way, the escape velocity represents the speed with which the precursor of a secondary body escaped the precursor of its primary body. For instance, as illustrated in the aforementioned graph (Fig. 7), the precursor of all the bodies orbiting the Sun escaped the precursor of the Sun at about 617.6 km/s, while the precursor of the Moon escaped the precursor of the Earth at about 11.19 km/s.

CHAPTER 6: TURBULENT TREES, BRANCHES, AND LEAVES OF THE PRECURSORS OF CELESTIAL BODIES

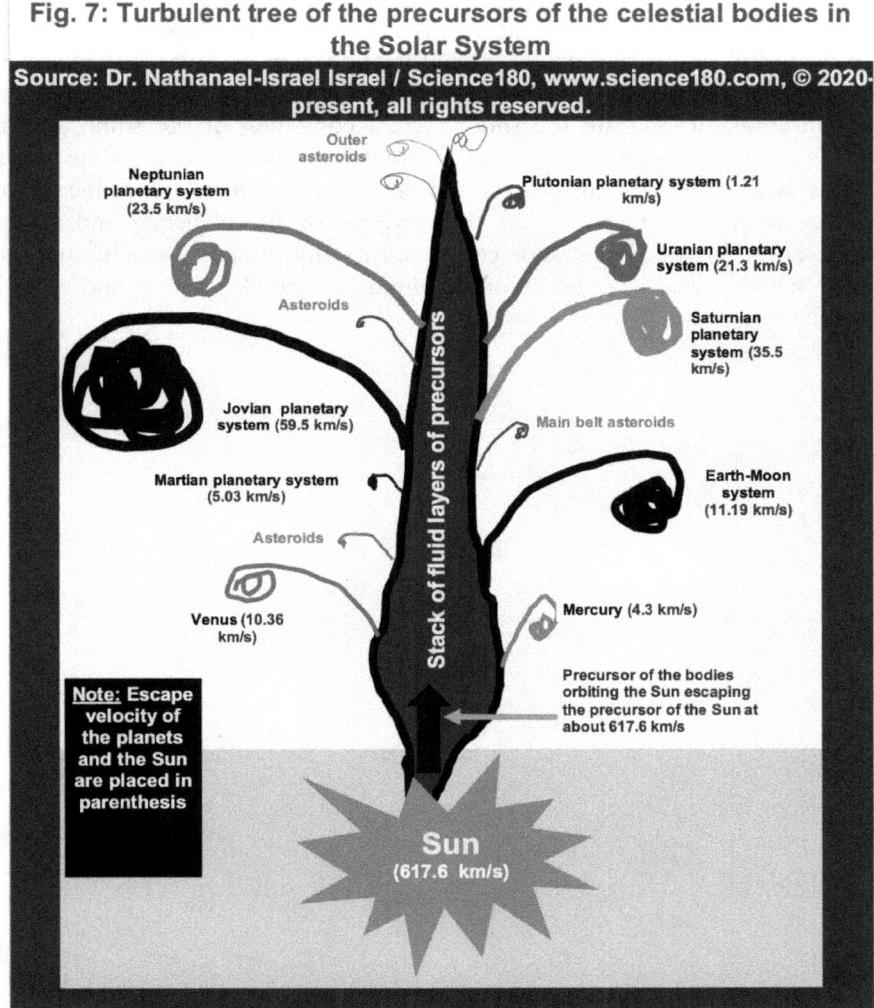

Fig. 7: Turbulent tree of the precursors of the celestial bodies in the Solar System
Source: Dr. Nathanael-Israel Israel / Science180, www.science180.com, © 2020-present, all rights reserved.

As a general rule, the speed of the flow in the trunk is about the escape velocity of the fluids from the root (here represented as the Sun). the speed of the flow through the primary branches is about the escape velocity of the fluids from the trunk (which here is represented by the escape velocity of the planets). Finally, the speed of the fluid flow through the secondary branches is about the escape velocity of the fluid flow from the primary branches into the secondary branches (which here is represented by the escape velocity of the satellites). The same process can explain the arrangement of the trunk, branches, and leaves on plants. For matters leaving the roots go to the trunk with a certain speed, then leave the trunk going to the branches with another speed, and then from the primary branches to the secondary ones with another speed. All of those speeds or flow of fluids may not be the same but decreased slightly as the distance away from the source increased.

In the turbulent tree of the precursor of the bodies in the Solar System, the Sun (being the main left over of the precursor of the Solar System) is the root, while the bodies orbiting the Sun are the shoot system consisting of the trunk, and the branches are the planetary systems and asteroids systems. The stem of the primary branches is like the primary planets and asteroids to which are connected the secondary branches which are like the satellites of the planetary and asteroid systems. Some isolated leaves not connected to any primary branch, and some primary branches that have no secondary branches are like planets and asteroids that have no satellite.

CHAPTER 7

UNCOVER HOW THE FEATURES AND PATTERNS ON LIVING THINGS POINT AT THE BIOLOGICAL TURBULENCE THAT SHAPED THEM ... BUT TOP SCIENTISTS HAVE STRUGGLED TO DISCOVER THEM

Based on my discoveries on the turbulent origin of celestial bodies, I understood that it is and it will be impossible for scientists to explain the origin of life without properly understanding turbulence. In this introductory chapter, you will learn how some patterns seen with living things are signatures or footprints of the biological turbulence that took place during the formation of life.

7.1. Shapes and patterns of biological components are marks of the turbulence of their precursors

Nature is filled with patterns, some of which are biological and others are not. Some biological patterns are seen with animal markings, animal segmentation, arrangement of leaves on the stems of plants, and many other anatomical and morphological characteristics of organisms. Although usually overlooked by most people until today, some of these patterns provide key information about the processes involved in the formation of the bodies that bear them.

In these following segments, I will present some evidences of how I discovered that a biological turbulence took place during the formation of living things. Indeed, unlike celestial bodies, which are characterized using variables such as semi major axis, inclination, eccentricity, radius, density, kinetic energy, etc., living things are characterized using variables such as their form, shape, size, weight, and height. Therefore, I used similar variables to present the story on their origin. As far as radius is concerned, one thing I learned about the turbulence of the precursor of bodies is that, in each system of bodies (e.g. stellar system, planetary system, and atomic system), the largest body is the primary body and the next largest bodies are usually found in the most turbulent zone of the precursor of the secondary bodies, usually Zone 3. In the Solar System for instance, I discovered and detailed in *"Turbulent Origin of the Universe"* that, turbulences occurred at different locations and with different intensities and other characteristics, leading to the formation of various systems of bodies including planetary systems, asteroids systems, without forgetting some planets and asteroids that have no satellite, etc. Likewise, when I looked at living things, I saw that some are very big and others are small. Within each family of animals or plants, I made similar observations. Like I demonstrated in *"Turbulent Origin of the Universe"* that the size of the celestial bodies was defined by the nature of the turbulence that their precursors went through, I also felt like the size of living things was defined by the original biological turbulence that birthed them. In fact, when I looked at each living thing, I saw turbulence everywhere. For instance, the organization of the root system, shoot system, and their components (e.g. leaves, fruits, branches, and roots) bear marks of turbulence. I also noticed that the shape and the organization of animal body parts were impacted by the turbulence of the matters used to form them. Whether or not a turbulence took place during the formation of living things is not an issue (for all the features I studied pointed to turbulence), but the challenging problem is the explanation of how that type of turbulence (which I termed "biological turbulence") took place for the first time. What were its characteristics? How did it yield the variety of beings seen in the universe, even beings beyond the reach of human beings?

My goal in this book is to help the readers perceive the biological world through the glance of the turbulence that birthed life. Because they were made of chemical particles and are living on Earth, I perceived that, as most ecologists and biologists will agree with me, living things are influenced by the nature of the turbulence that birthed their constitutive matters. The turbulence that formed the universe affected how the components of living things were assembled and how living organisms live today. Nevertheless, I realized that the type of turbulence that birthed the types of living things had specific characteristics different from those that birthed celestial bodies. Throughout this book, I will walk you through how I discovered the traits of the biological turbulence and how it shaped the living world.

7.2. Shape of plants

Like I presented earlier, most plants have:
- roots
- trunk, or principal stalk or stem
- branches
- leaves
- inflorescence (if they are flowering plants)

The shape of each of these components reflects the nature of the turbulence that formed them. For instance, the shape of a tree as a whole, that of a branch, or that

CHAPTER 7: FEATURES AND PATTERNS POINTING AT A BIOLOGICAL TURBULENCE

of a leaf can be presented as having 5 turbulence zones, whereas the largest bodies of the components are usually located around turbulence Zone 3.

7.3. Shapes of animals

When I looked at most animals, I felt like the head is the primary body, the rest of the body is the secondary bodies, while arms are like branches coming out of Zones 1 and 2, and legs are like branches emerging from Zones 4 and 5. The belly belongs to turbulence Zone 3. In general, the abdomen belongs to Zone 3. Although some people may place the tail in Zone 5 (the outermost zone), it can also be considered as a branch or ramification from Zones 3 or 4.

During my study on the origin of life, I realized that the shape of animals matches the turbulence in their environment today and that which shaped the first specimen of their kind. The body plan of most animals is bilaterally symmetric. Examples of non-bilaterians animals include sponges and corals. Although some animals have shells, bones, and spicules, which give them some structural support, their cells are not strongly held by cell walls as those of algae, plants, and fungi. The shape of all animals reflects the turbulence that took place during their formation. Although current living things are surrounded by various forms of turbulence throughout their lifespan, those turbulences are different from the one that birthed them. For instance, the water in which aquatic organisms live and the air that surrounds most terrestrial or land animals are all subject to turbulence. The shape of most organisms follows the characteristics of the turbulence zones (more details can be found in *"Turbulent Origin of the Universe"*. The size and the position of the belly in fish respect the turbulence which follows them as they move in water. Likewise, even on land, terrestrial animals are shaped after the turbulence that took place during their formation, and which also respects the turbulence that the atmosphere can bring on them during their lifespan. Looking at the shape of some massive marine animals (e.g. whales) made me think of the shape of birds and other animals that are in close contact with turbulence most of their lives. This is not a matter of evolutionism, but of features that considered the environment where organisms were initially formed and also in which reproduction will follow as a way to synchronize organisms with the environments they were formed, and in which they (and their descendants) will spend the rest of their lives.

The turbulent tree I presented in the previous chapter, and which is blatant with plants, is also present with animals although the nature of animal cells does not allow the turbulent trees of animals to be easily noticed. For with animals, most body parts are not organized with a vertical stature as is the case for plants where the trunk can support most of the weight. Although the skeleton of some animals helps to hold the bodies in a certain position (e.g. vertical for human beings), the internal components which are not supported by bones are still relatively loose in such a way that body parts belonging to the same system can seem mixed with others of different systems.

Even the components of the digestive tract look like a system of bodies born

TURBULENT ORIGIN OF LIFE

through a turbulence:
- mouth (entrance or beginning of the turbulence),
- esophagus (is like the turbulent neck which is like Zones 1 and 2),
- stomach (Zone 3),
- intestines (Zones 4 and 5) – which can also explain why intestines are smaller and long and located at the end
- anus (end of the turbulence of the digestive tract) – Zone 5.

The respiratory system also shows signs of turbulence:
- Thoracic canal = turbulent neck
- Lung = Zone 3
- Other respiratory parts = Zones 4 and 5.

CHAPTER 8

CAN THE LENGTH OF THE NECK HOLD A KEY TO DECODING HOW AN ANIMAL WAS FORMED? THIS EXPERT SAYS YES ... AND HERE'S HOW TO PROPERLY UNDERSTAND WHAT HE TERMED "TURBULENT NECK"

The decrypting of the origin of life requires the understanding of some fundamental facts; and many of these are open secrets that people have failed to grasp. This crucial chapter will introduce you to a code related to the neck of living things that I will later use to crack the formation of life.

8.1. Definition of a turbulent neck
Also called cervix or column, the neck of an organism is the body part connecting the head with the torso (the central part or trunk of most animals). The word cervix is also used to refer to the uterine cervix, which is the neck of the uterus (Whitmore, 1999). The neck of some animals contains some visceral organs such as the larynx, pharynx, trachea, thyroid, parathyroid, and many other compartments. Most insects also have a neck, but because the neck of fish is less noticeable, it can be fair to say that fish don't have a neck. In other words, at the morphological level, it can be said that fish lack a neck, for a neck requires a structure that is clearly detached from the head. The head of fish looks like an extension of the body.

Although a neck is usually perceived while dealing with an entire organism as a whole, I also realized that many body parts or components of living organisms have features which, according to their scale, are similar to a neck. For instance, when I consider the digestive system or respiratory system, I found features looking like a neck. To put it another way, the esophagus of the digestive system is like a neck connecting the mouth to the stomach. Likewise, the upper part of the respiratory tract is like a neck connecting the nose to the lungs. Many biological features show examples of necks. Therefore, on December 7th, 2021, in an attempt to come up with a comprehensive and inclusive definition that can also help me to better present my understanding of biological turbulence, I coined the term "turbulent neck" to designate features or a distance located between a primary body (or the initial position of a primary body) and its innermost secondary bodies. In other words, a turbulent neck refers to everything located between the distance separating the primary body, head, or the beginning of a system of bodies and the innermost body or the body resulting from the first ramification or branching of the precursor of the secondary bodies in that system. It can also be viewed as structures formed along the distance separating a primary body and its innermost secondary body.

On a tree for instance, the turbulence neck is everything located between the soil and the first branch on the plant. On the scale of a primary branch, the turbulent neck is the portion of the primary branch separating the innermost secondary branch and the node where the primary branch connects to the trunk.

Organisms have different kinds of necks. Some necks are very long, while others are very short. In general, the largest parts of organisms are found downstream, or downward, or after the neck, the position of the head being located upstream. Although I will talk about other types of turbulent necks and their significance later in this book, at this point, I will review some of the longest and shortest necks found in the animal kingdom. Beforehand, I need to emphasize that, because turbulence is not limited to living things only, but also to celestial bodies, turbulent neck also exists with nonliving things formed in the universe. For instance, after the formation of celestial bodies, the turbulent neck of their precursor is invisible for it disappeared as the precursors of the celestial bodies split-gathered, and in this case, the turbulent neck is located between the primary body and its innermost secondary body, approximately around the line connecting the center of the primary body to the center of the orbit of the innermost secondary body.

8.2. Animals with the longest neck

Before presenting the animals with the longest neck, I will first introduce the human with the longest neck record. Indeed, the longest human necks are found among the women of the Padaung (or Kayan) tribe in northwestern Thailand and southeastern Myanmar (Guinness World Records, 2018). The length of their neck (equivalent to the distance from the suprasternal notch to the lower jaw) is 19.7 cm (7.75 inch). The increase of the length of their neck is said to result from the impact of traditional rings made from coiled brass wire that those women had to wear as early as from the age of 5-9 years, and more are added as they grow. According to the Guinness World Records, these rituals not only increase the length of the neck but also affect other parts of the bodies such as dental development, ability to hold the head on the neck, etc., suggesting that those necks were stretched beyond what is normally natural.

At this point, I will address the animals with the longest neck. Indeed, the length

CHAPTER 8: TURBULENT NECK

of the neck of animals varies according to their family, class, and other classification criteria. Some necks are said to be long because, with respect to the size of the organisms bearing them, they are so. However, when animals with undeniably long necks are concerned, people seem to more unanimously name animals like:
- Giraffe
- Ostrich
- Gerenuk (gazelle)
- Eastern snake-necked turtles
- Swan
- Flamingo
- Dromedary camel
- Alpaca
- Llama
- Extinct animals like Tanystropheus (a giraffe-necked reptile) and sauropod dinosaurs

Before reviewing the neck of these abovementioned animals, I need to emphasize that these animals are just examples, not necessarily the record holders if all animals were carefully studied. Let's start with the giraffe. Indeed, while the giraffe (*Giraffa camelopardalis*) (Fig. 8) is usually thought of when long neck animals are mentioned, many other animals across the globe have a long neck. The length of some giraffe's neck is about 8 feet, meaning about 2.4 meter (Wildlife Informer 2021). The longest necked giraffes are found among the adult male giraffes. Although giraffes (the tallest living land animal) are naturally from Africa, many zoos all over the worlds have giraffes that they show to visitors, usually for lucrative gains. Wildlife Informer reported that newly born baby giraffes can measure up to 6 feet tall. The long neck of giraffes and their need for blood to be pumped faster to reach the brain which is 2 meters above the head could explain why the blood pressure of giraffes is the highest of any terrestrial animal (Hussain, 2021). When I considered that information in conjunction with the fact that some dinosaurs have a neck longer than that of the tallest giraffe, I felt like, it is possible that the blood pressure and/or the need to have a higher blood pressure to push blood through the long neck of dinosaurs could have contributed to extinguishing some of them. But the fossil record and the way some dinosaurs are preserved particularly in cold environments suggested to me that the extinction may be explained by other factors, which I elaborated on elsewhere. The neck of some dinosaurs is reported as 5 times as long as that of some giraffes. For instance, some dinosaurs (e.g. Mamenchisaurus dinosaurs) are believed to have had a neck over 10 meters long, which is about 5 times the length of contemporary giraffes. Some people may even wonder why dinosaurs were formed in the first place. As you will see in an incoming future chapter, when turbulence occurs in a system, beyond a certain intensity, both large and small bodies are usually formed. The size of the large bodies is affected by the intensity of the turbulence and some characteristics of the matters used as precursors of the bodies to be formed. The availability of the matters for the precursor of the body is not a limiting factor for the biological turbulence, for water and soil were hugely available when living things were being formed. The way the biological turbulence was calibrated or scaled to make the organisms aligned with the opportunities and constraints that their environment can impose on them (sooner or later) is one of the key factors which limits the maximum size of the organisms. Just as in the turbulence of the precursor of the celestial bodies, large bodies are always in smaller numbers than small bodies, the number of the largest organisms formed on Earth were smaller than that of the small organisms. Hence bacteria can be found everywhere on Earth, yet dinosaurs, elephants, giraffes (Fig. 8), and other big animals are not everywhere, and some of the very few gigantic animals which were formed did not last long before being extinguished and those which are still alive are threatened of disappearance.

Fig. 8: Giraffe (Photo credit © Nathanael-Israel Israel)

After the giraffe, the ostrich (Fig. 9) is another animal with a long neck. Indeed, with a neck capable of reaching 3.2 feet. (1 meter) long and contributing to about half to the body length, the ostrich (*Struthio camelus*), a flightless bird, holds the record of the longest neck recorded on birds. As the "largest living species of birds found in Africa savannas", the common ostriches can weigh about 320 lb (145 kg) (Wildlife Informer 2021). With its long and muscular legs, and ostrich can run up to 43 mph (70 km/h).

CHAPTER 8: TURBULENT NECK

Fig. 9: Ostrich (Photo credit © Nathanael-Israel Israel)

Another animal with a long neck is the gerenuk (*Litocranius walleri*), also identified as Waller's gazelle. It is a long neck gazelle (having a neck of 0.8 feet, meaning 24 cm long). Its name (gerenuk) was derived from a Somalian language and means "giraffe-necked." It is naturally found in open scrublands and lowlands along the Horn of Africa. Because of their long neck, these herbivores are able to graze on plants and trees up to 6-8 feet high, and even higher.

Next is the Eastern snake-necked turtles. With a neck capable of reaching 0.6 feet (18 centimeters, as reported by Encyclopedia Britannica (2021), the eastern snake-necked turtles (*Chelodina longicollis*) are among the long-necked animals. Unlike the other long-necked animals I listed so far, and which are herbivores, eastern snake-necked turtles are carnivorous mostly found in Australian freshwaters. As far as turtles are concerned, they hold the record of the longest necked turtles living today.

The next example of a long-necked animal is the swan. Also called whooper swans, swans (*Cygnus cygnus*) are a bird species looking like ducks and geese (but their neck is longer than that of geese). They are naturally found in cooler or warmer temperate regions such as Australia and New Zealand, but rarely in the tropics including the entire continent of Africa; and their neck measures up to 3 feet (91 cm) long while their height can reach 4 feet, meaning 1.2 meters (Wildlife Informer 2021).

Flamingo is also an animal with a long neck. With a neck length of approximately 2.6 feet (0.79 meter), flamingos (*Phoenicopterus roseus*) are long s-shaped neck birds capable of reaching 4.7 feet (1.45 meters) tall. They are found on almost all continents including Africa, America, Asia, and Europe.

Also known as Arabian camels, dromedary camels (*Camelus dromedaries*) (Fig. 10) are long neck animals domesticated in the Middle East, Afghanistan, and the Sahara Desert and the length of their neck can reach 6.6 feet (2 meters) and their shoulder height is up to 10 feet (3 meters) tall.

Fig. 10: Camel (Photo credit © Nathanael-Israel Israel)

With a neck length of 3.8 feet (1.17 m), alpacas (Vicugna pacos) are native of South America, looking like long-necked camels, but lacking humps or bumps. They also resemble llamas. They also have long and straight ears.

Another animal with a long neck is the Llama (Fig. 11). With a neck capable of reaching 4.3 feet (1.3 meters), Llamas (*Lama glama*) are herbivore native of and domesticated in South America. Their height can reach 5.8 feet, meaning almost 2 meters.

CHAPTER 8: TURBULENT NECK

Fig. 11: Llama (Photo credit © Nathanael-Israel Israel)

Many extinct animals also had long necks. This is the case of Tanystropheus, a giraffe-necked reptile which neck is said to be 3 times as long as its torso (Stephan et al., 2021). It took 150 years after the fossil of this reptile was found in Europe, before some of its characteristics could be defined, such as whether that animal lived in water or on land. After decades of investigation, the longest Tanystropheus (*Tanystropheus hydroides*) is said to have been an aquatic reptile. The neck of this extinct 6-meter-long (20 feet) reptile was 3 meters (9.8 feet) long, which is longer than its body and tail combined (Elbein, 2020).

The necks of some sauropod dinosaurs were 15 meters long, which is 6 times longer than that of the world's record holding giraffe with the longest neck (Taylor and Wedel, 2013). No other animal has been found with a neck as long as that of the sauropods, all of which are extinct animals.

8.3. Animals with the smallest neck

Many animals do not have a preeminent neck. Some even looks like they have no neck at all. Animals with a short neck include: panthers, zebras, and foxes. Penguins are some of the animals with the small neck and, in fact, some people even consider them as lacking a neck. They are aquatic flightless birds that live almost exclusively in the southern hemisphere, except one species (*Galapagos penguin*), which is found north of the Equator, as reported by Wikipedia (2022). They pass half of their life in water and the other half on the land. The largest living penguin is called the emperor penguin (*Aptenodytes forsteri*), and adults of this kind are about 1.1 meters (about 3.5 feet) tall and weigh 35 kg (77 lb) (DK, 2016). The little blue penguin (*Eudyptula minor*), also called the fairy penguin, and which is about 33 cm (13 inch) tall and weighs about 1 kg (2.2 lb) is the smallest penguin species (Grabski, 2009). Although they are flightless birds, Gentoo penguins are the fastest underwater birds in the world and capable of attaining a speed up to 36 km (~ 22 miles) per hour (Rafferty, 2021a). Furthermore, emperor penguins are the world's deepest-diving birds, capable of diving to depths of about 550 meters (1,800 feet) (Rafferty, 2021b). This implies that the diving ability of animals is not limited by their flying ability, else the bird which is not able to fly cannot be the bird which holds the diving record among all of the diving birds.

CHAPTER 9

TO KNOW HOW GREAT SCIENTISTS CHALLENGE CONVENTIONAL SCIENCE WHILE INVESTIGATING THE ANATOMY OF LIVING THINGS, FIRST DECRYPT THE SECRETS OF THE TURBULENT HEAD, BELLY, TAIL, SPINE, ARMS, AND LEGS

Body parts of living organisms hold secrets to the decryption of the origin of life. In this chapter, I will introduce you to some features that I will use later to describe how living things were formed. The more you understand these features, the better you will understand how life was formed. In this chapter, although I talk about the head, belly, tail, spine, leg, and arm, I want to advise you not to just think that I am talking about mere anatomical or morphological structures of living things, but I am using those terminologies to relate to a much stronger phenomenon, the biological turbulence that occurred during the formation of all organisms. Hence, I used the adjective "turbulent" to describe those body parts. For instance, although the term neck is usually used for animals, the term turbulent neck can be used even for plants and even for celestial bodies. As you will see shortly, plants also have a turbulent head, belly, spine, arms, and legs.

9.1. Turbulent head: the primary body of an organism

The head of all organisms is like the beginning of many systems of bodies. For instance, if I consider animals, the head host the brain which is the center or controlling system of the entire body. The hypothalamus (which is a central part of the endocrine system) is hosted in the head. Likewise, key parts of the nervous system (e.g. brain, spinal cord, nerves) are located and/or start in the head. The head hosts the nose, which is the entrance of the respiratory system. It also hosts the mouth, which is the beginning of the digestive system. The eye (for sight), ear (for hearing), and tongue (for taste) are all located on the head.

Even the central part of other key systems of bodies (e.g. integumentary system, lymphatic or immune system, musculoskeletal system, reproductive system, urinary or excretory system, and many other special systems of bodies) are located in or on the head and/or have significant or controlling components in the head. A careful look at the organization of most of the systems of bodies in most organisms sounds as if the precursor of the head must have been one of the very first precursors of body birthed during the formation of living things and even during the developmental stages of organisms as of today. In other words, the precursor of the head would not have been formed after that of the neck, belly, feet, and hands were formed. Considering what I said and showed about the split gathering of the precursor of the bodies, I felt like the head is the primary body in an organism, while the torso or the rest of the body downward of the neck is a cluster of secondary bodies of the head. Each of these systems of secondary bodies are organized into a system of primary bodies and secondary bodies, which at their turn also are organized into other primary and secondary bodies and so on and so forth until the smallest level of subdivision.

Considering plants and fungi, I felt like the turbulent head is in the root system. For without it, all other parts could not exist and from it emerged the shoot system (meaning the above portion of the plants where branches, stems, and leaves are found). The root system is like the socle or foundation on which the shoot system is built. For plants, roots serve for many purposes, including finding and uptaking the water needed for the life of the entire plants. Just as the vascular system connects the head with the other bodies, so also a vascular system connects the root to the rest of the bodies and allows the flow of key survival compounds or molecules.

9.2. Turbulent belly: largest part of an organism

Looking at the details of the body parts of animals, I realized that the largest or biggest parts are around the middle of the animal in the belly. For instance, some of the largest organs like the liver, stomach, and lungs are around the middle of the belly where the body size is bigger. Even if the organs inside the belly are not considered, the belly alone is usually the largest part of most animals. Usually, no human part is as large as the belly. For instance, no head, neck, hand, or foot is as large as the belly. Considering what I demonstrated about the turbulence of the precursor of celestial bodies, I felt like the portion of the belly having the largest organs belongs to turbulence Zone 3, the zone where the turbulence was the most developed. In other words, what I called turbulence Zone 3 is like the belly of the turbulence of the precursor of the secondary bodies of a system.

Considering plants, the belly is about the location of the shoot system where the largest branches are found; it is the largest portion of a plant. On the scale of each system of organ, individual organ, cell, tissue, and other compartments, a turbulent

CHAPTER 9: TURBULENT HEAD, BELLY, TAIL, SPINE, ARMS, AND LEGS

belly can be identified. For instance, the form or shape of a heart, liver, lung, head, or many other body parts, is not the same all around. A certain portion is larger than the rest. The largest portion can be considered as the turbulent Zone 3 of the precursor of that body part.

Some of the animals having a relatively large belly include: elephants, big-belly seahorse, pigs, hippopotamuses, etc. Elephants are found in nature in Africa and Asia. Also called the pot-bellied seahorse, the big-belly seahorse (*Hippocampus abdominalis*) is usually in the list of the animals with a big belly not because it has the biggest belly among all animals but, compared to its body size, its belly is relatively very large. In addition to its big belly, the sea horse has a long-snout with its head tilted forward and a long-coiled tail. Some species of pigs such as the pot belly pig which was bred originally from Vietnam has a very big belly, which sometimes even touches the ground as it walks. Some cows also have a huge belly.

Fig. 12: Pigs (Photo credit © Nathanael-Israel Israel)

Although not easily distinguishable, toward the end of the belly, some smaller organs are found. For instance, intestines are very small. Considering their nature, I think they belong to Zones 4 and 5, meaning the outer zones of turbulence.

9.3. Turbulent spine, turbulent vertebrate, and turbulent skeleton: the backbone or support system

All vertebrates have a spine, which is the backbone of the skeleton and through which many nerves run. Due to its crucial importance, any damage to the spinal cord highly affects the health or well-being of the entire body. Some back pain, shoulder pain, hip pain, neck pain, and other kinds of pain related to bones are examples of how damage to a bone or joint can hurt.

Although some animals do not have a skeleton made up of bones, some have a body made up of cartilage, while others have it made of materials like chitin, which cover the surface such as the exoskeleton of some insects. For organisms that lack a skeleton, it may be hard to define their turbulent spine, but it can be associated with whatever internal structure holds the organism together and serves as the main support of body organization. Some living organisms may not have a turbulent spine at all.

The turbulent spine of plants is the trunk, or main stalk. Together with the system of branches, the trunk of plants serves as the skeleton or the backbone holding the plants, and without which they can collapse or crumble. In other words, the trunk, or the main stem, or the main stalk is for plants what the spinal cord is for vertebrates. Just as some animals do not have bones, some plants also do not have a trunk, but have to rely on a soft stem sometimes laying or vining on the ground. The connection of the trunk and the branches denuded from leaves is for plants what the skeleton system (all bones and cartilages) is for some animals. A piece of stalk, stem, branch, or trunk is for plants what a piece of bone is for bony animals.

9.4. Turbulent arms and legs: the ramification of branches of a turbulent body

Humans and most mammals have 2 arms and 2 feet. Each of the arms and feet have digits. For instance, at the end of each hand or foot of a human being, 5 fingers and 5 toes are found respectively. While some animals have less than 5 fingers and 5 toes, others do not even have them at all. For instance, although some people think that some fins of fish are like limbs or members, most people would say that fish do not have arms, feet, or legs.

The size of the arms or feet of animals varies according to their species. Some are very long, while others are very short. Some animals are very large, yet they have very short legs. That is the case of the hippopotamus, which, despite its large body, has very small legs. In general, for most animals, the arms are smaller and shorter than the legs. While some animals go on all 4, meaning walking on their 2 arms and 2 feet, others like human beings go on just their 2 feet. Despite their gigantic size, dinosaurs are believed to have gone on their 2 feet, for their arms (as revealed by their fossils) are so short that if these animals would have walked on all their 2 hands and 2 feet, they may be unbalanced.

Whether the animals walk on 2 or 4 extremities, arms and feet look like ramifications from the precursor of the thorax and abdomen respectively. In other words, arms emerged somewhere from the precursor of the thorax, while legs emerged from the precursor of the abdomen. If I can use a turbulence terminology, I will say that arms and legs are secondary bodies that emerged as outgrowths, or like a "shoot", or "bud" from a developing thorax and/or abdomen. In other words, the split-gathering of the thorax and abdomen yielded the arms and the feet. The

CHAPTER 9: TURBULENT HEAD, BELLY, TAIL, SPINE, ARMS, AND LEGS

arm and leg of a four-legged animal are called limbs.

What I just said for animals can also be applied to plants. Primary branches of plants are like arms (each of the upper limbs of the body of an animal from the shoulder to the hand) and legs (limbs located beneath, below, or behind on which most animals walk and stand), for which secondary and higher orders of branches can serve as digits just as fingers (slender jointed parts attached to a hand) and toes (digits at the end of a foot) are ramifications of the hands and feet of animals. But because plants do not walk, this analogy may sound odd, yet they are real in the turbulence world.

To summarize, arms are ramifications or branches of the precursor of turbulence Zones 1 and 2, while legs are ramifications or branches from turbulence Zones 3 and 4 and maybe Zone 5, which I think are the lowest or outer or outward part of the belly. Hands (end part of some animal's arm beyond the wrist, including the palm, fingers, and thumb) and feet (the lower extremity of the leg below the ankle, on which most animals stand or walk) are like secondary bodies emerging from the thorax and abdomen and at their turn, fingers and toes are like tertiary parts emerging from the hands and feet.

9.5. Turbulent tail: the hindmost of outermost portion of a turbulent body

Unlike human beings, most animals have a tail, which is usually the outer part of their body. Human beings lack a tail, not because we lost it a long time ago as some people think, but because no human being ever had one. Although it can be very long, sometimes even longer than the other part of the body of some organisms, the tail is usually thinner and smaller than the other parts of the body. I am not aware of any animal which tail is bigger than its belly. Additionally, the size of the tail decreases as its position is farther and farther away. In other words, the end or the farthest part of a tail is smaller in diameter than its part that is directly connected to the body around the position of the anus. As you will see later in this book, this feature is a mark of turbulence.

Although birds do not have tails exactly like mammals, which is an extension of the body, what is called tail for birds is a collection of extra-long feathers. Found all over Africa, and with a tail said to be 20 inches long (which is 4 times the length of its body), the long-tailed widowbird (*Euplectes progne*) has the longest tail feathers (World Atlas, 2021). My wife wondered if it can fly fast or slow. Although I don't have the answer for that question, I think it will not be a fast runner, for the tail is not the organ mostly needed to fly faster!

The previous source also reported that, the longest tail (capable of reaching 20 feet, which is about 7 meters) among fish was found with the whiptail stingray, a marine fish. With respect to their body length, other animals that also have a long tail include some cats, monkeys (e.g. *Colobus angolensis*), dogs, squirrels, rats, long-eared jerboa (*Euchoreutes naso*, a rodent), African elephants, red kangaroos, giraffes, Asian grass lizards or long-tailed grass lizards (*Takydromus sexlineatus*, endemic to South East Asia with a tail longer than 3 times its total body length), ring-tailed lemur (*Lemur catta*), thresher shark (*Alopias superciliosus*), etc. One of the longest tails recorded on animals has been on the sauropod dinosaur called Diplodocus, which tail can reach 45 feet long, equivalent to about 15 meters. Despite the aforementioned record on tails, many animals have a small tail (e.g. Manx cat), while some have no tail at all.

Because some people may wonder why and where I am going with all of these biometric details on various animal's features, I would like to say for now that, I am not presenting the number about body parts of animals just for fun, but because, as you will see later in this book, the sizing of animals and their abilities are a highly encrypted code of the turbulence that prevailed during their formation. As you keep reading, you will understand the significance of this chapter.

Considering my discoveries about turbulence, I felt like the tail is for animals what the bodies in turbulence Zone 5 are for systems of celestial bodies. In other words, if I can talk about biological turbulence, the tail of animals can be placed in Zone 5. I initially thought that the legs were supposed to be the Zone 5, but I realized that legs and arms are branches from the stack of precursors that yield the torso. Just as some systems of celestial bodies lack bodies in turbulence Zone 5, some animals do not have a tail, not because it was lost during any so-called evolutionary process, but these animals never had one. At this point, I would like to report some tail records. In other words, looking at the organization of tails and knowing that they are a continuation of the spine, I felt like the tail is an organ belonging to turbulence Zone 5, the outermost zone. In other words, bodies in turbulence Zone 5 are like body components of a tail.

Therefore, I defined a turbulent tail as the collection of the outermost bodies of a turbulent system. For animals, the turbulent tail is the real tail, while for plants, it can be some apical meristems, which are the tip of most plants. However, knowing the intense activity occurring in the meristems, although the position and the small size of its constituents fit the definition of a tail, its endogenous activities are much stronger than what can be expected from the precursor of Zone 5. Nevertheless, root apical meristems and the shoot apical meristems can be classified as a turbulent tail belonging to the biological turbulence Zone 5.

CHAPTER 10

EXTRAORDINARY SCIENTIFIC ADVANCEMENT THAT CHANGED THE SECULAR EXPLANATION OF LIFE FOREVER USING THE TURBULENT GEOMETRY OF LEAVES AND LEAF-LIKE SHAPES OF ORGANISMS

During the writing of this book, I had several small but significant aha moments, and what you are going to read in this chapter will introduce you to a little-known secret that opened my mind to a different world as far as the formation of life is concerned.

10.1. Leaf of plants

If someone has told you that the morphology of leaves can help break the code of life, I don't think you would have believed him or her. But now, hear it from me. Without any further delay, let me share with you what I found.

Most leaves (Fig. 13 and Fig. 14) consist of the following parts:
1. Apex
2. Midvein (primary vein)
3. Secondary vein
4. Lamina
5. Leaf margin
6. Petiole
7. Bud
8. Stem

TURBULENT ORIGIN OF LIFE

Fig. 13: Simple leaf

Fig. 14: Examples of leaves
Source: Dr. Nathanael-Israel Israel / Science180, www.science180.com,
© 2020-present, all rights reserved.

CHAPTER 10: LEAF AND LEAF-LIKE SHAPES OF ORGANISMS

10.1.1. Leaf venation patterns

As I describe the veins found in leaves of plants, I will also be handling the turbulent meaning of the patterns of veins. For instance, on the scale of a leaf, the primary vein is like the primary body, the secondary veins are like the secondary bodies, the apex is like the outermost body on a leaf, everything inside the leaf or lamina except the veins are for a leaf what rings are for a system of satellites.

Veins are not arranged the same on the leaves of all plants. For many venation patterns (Fig. 15) exist and next, I will present some of them. Just as the precursor of a body can split into daughter bodies, even a vein can split and yield others. For instance, a vein can split into 2 and yield a dichotomous venation pattern. Some secondary veins do not split at all, but arc toward the apex of the leaf, therefore forming an arcuate venation pattern. In some leaves, the veins are aligned parallel to one another without intersecting. That is the case of the leaves of some grasses. The veins of some leaves branch repeatedly and form what is called a reticulate venation pattern. In rotate leaves, veins are arranged as emerging from the middle of the leaf and radiating toward the boundaries. For the pinnate venation pattern, secondary veins are born from a midrib.

The variation of the venation pattern of leaves (Fig. 15) results from how the precursors of the veins split-gathered during the formation of the leaves. For instance, for a dichotomous venation pattern, the precursor of each vein always splits into 2 new veins. For the pinnate venation pattern, the precursors of the secondary veins emerge or branch out along the path of the precursor of the midvein or as the precursor of the midvein was growing.

Fig. 15: Leaf venation

10.1.2. Leaf arrangement and shape

Leaves are arranged differently on plants (Fig. 16). In the case of the opposite leaf pattern, some leaves are opposite to one another along the stem that holds them. This arrangement derives from a split-gathering of the precursor of 2 leaves from the same point.

Leaves come in various shapes. Some leaves have a tripartite lobation, others are elliptic, palmate, etc.

CHAPTER 10: LEAF AND LEAF-LIKE SHAPES OF ORGANISMS

Fig. 16: Leaf shape and arrangement

10.1.4. Leaf edge

While the edge of some leaves (Fig. 17) has even or smooth margins without toothing, that of some leaves is covered with hair. In contrast, the boundaries of some leaves have teeth, which forms are various. While some teeth are rounded, others are not. Some teeth are small, while others are large. Some teeth have deep indentations, while for some leaves, the indentations do not reach the center of the leaves. Some edges are so sharp and stiff (e.g. thistles or thorns) that they can cut just like a blade, nail, or screw. According to the combination of the leaf characteristics I just presented, some edges of leaves are termed dentate, denticulate, serrate, serrulate, sinuate, lobate, undulate, spiny, etc. For instance, leaves that have an even and smooth margin without toothing are termed entire leaves. Other leaves having hairs are termed ciliate. Dentate leaves are toothed and/or have edges looking like teeth.

Fig. 17: Leaf edge

10.1.5. Leaf apex (tip)

The apex or tip of a leaf is its end (Fig. 18). While some apexes appear pointed, others are not. Among the pointed apexes, some are long-pointed, while others are narrowly pointed. Some endings are sharp, elongated, and rigid. Some ends look like as inversely heart-shaped. Unlike the pointed tips, some are rounded or blunt. Finally, some apexes end abruptly with a flat end. The nature of the leaf apex gave rise to terms such as acuminate, acute, cuspidate, emarginate, mucronate, obcordate,

CHAPTER 10: LEAF AND LEAF-LIKE SHAPES OF ORGANISMS

obtuse, and truncate commonly found in botanical descriptions of plants. All of these endings of various leaf apexes can be related to the nature of the turbulence that birthed them. The behavior of the fluid in the precursor of a leaf near the tip can explain why some endings are acute while others are flat.

Fig. 18: Leaf apex

10.2. Leaf-like shape of animals

When animals are dissected in half through the middle of their abdomen and thorax along the direction going from the head to the tail, they can form a leaf-like shape. For instance, if you take a rodent (e.g. rat) and dissect its abdomen using a knife, and then remove the viscera inside the abdomen, it will look like a leaf (Fig. 19). The same observation is true for many other animals. The shape of most fish looks like an elliptical leaf. On the scale of the leaf, the apex is the turbulent tail, while the petiole is the turbulent neck. If a dentate leaf can be folded in a certain way to mimic an animal, the larger part of the leaf would be like the abdomen, the main dents or indentations would be like the hands and legs, the apex of the leaf would be like the tail and the petiole would be like the neck connecting the leaf to the head located upward of the petiole (Fig. 19).

If most animals can lay on their back or belly, then open their hands and feet wide, their shape would be like that of a leaf which main vein is the torso, while the secondary veins would be the arms and legs, whereas the tail would be like the apex or tip of the leaf. On November 28, 2021, I realized that the similarities between the shapes of plants and animals is due to the fact that they are produced by a biological turbulence, which, at its turn, is related to the turbulence that birthed the celestial bodies.

TURBULENT ORIGIN OF LIFE

Fig. 19: Animal skin (Photo credit © Nathanael-Israel Israel)

CHAPTER 11

HOW TO QUICKLY DEFEAT THE BIGGEST LIES PEOPLE SPEWED ABOUT THE FORMATION OF LIFE BY KNOWING THAT DEVELOPMENTAL BIOLOGY POINTS TO A BIOLOGICAL TURBULENCE–DID DARWIN MISS THIS OR WHAT HAPPENED?

11.1. Importance of development biology in my search for the origin of life
Advancement in developmental biology suggests that the development of organisms is well programmed in such a way that some genes associated with it are known, and biotechnologists understand which ones to repress or activate as they wish. Starting with an egg born from the fecundation of an ovule and a spermatozoid, various types of cells are formed during the development of most organisms. Organisms use different mechanisms to multiply their cells and modify their constitution and functions. Some of those mechanisms involve mitosis, meiosis, epigenetic modifications, genetic interferences, post transcriptional modifications, and post translational modification, etc.

Epigenetic modifications are hereditary modifications of the DNA that can change gene expression without altering the genome. Some of these modifications are reversible. For instance, certain enzymes are able to modify the DNA by grafting other molecules through reactions such as methylation, phosphorylation, acetylation, ubiquitylation, and other DNA modifications or modifications of proteins (such as histones) interacting with the DNA. These modifications prepare the DNA for various reactions or interactions with other neighboring molecules. Hence, epigenetic modifications cause the same DNA sequence to produce different products or reactions according to its modification. Therefore, the destiny of cells can be affected by the modifications that their DNA can face, even after the expression of some genes.

Some morphological and anatomical modifications that living organisms go through during their development are controlled by biochemical mechanisms. For instance, during the formation of organisms, embryogenesis is a fundamental step during which an egg (produced by the fecundation of an ovule and a spermatozoid) divides to yield different types of cells, tissues, and organs. Meanwhile, other mechanisms allow the organisms to grow their cells, tissues, and organs according to their mission and position in the organism.

Plants are unable to move by themselves, but they are endowed with systems (e.g. meristems) that allow them to always be able to produce new cells that can be differentiated into others. In contrast, once animal stem cells are differentiated, they are unable to be converted back into other types of cells. For instance, animals cannot convert cells that make up the nose into cells that make up the eyes. The presence and potential of meristematic cells in plants allow them to also produce new compounds or macromolecules to adapt to certain environmental conditions that they cannot escape by moving away. In contrast, because they can move when faced with critical environmental conditions, animals migrate or hide themselves. The lifespan of plants and animals are different, but in both cases, cells are born and later die under the control and programs of genes and other regulatory systems. While animals are born with all the organs they will need in life and just have to grow, increase their size, and mature those organs, that is not the case for plants. In fact, throughout their life, plants always have a stock of embryonic tissues that they can convert into any type of tissue and organ as needed. In other words, early in its development, an animal embryo produces all of the body parts needed during its life, plants continually produce new tissues during their entire life from meristems situated at the tip of some organs (e.g. shoot, root) or between mature tissues.

I showed in my books titled *"Turbulent Origin of the Universe"* and *"Turbulent Origin of Chemical Particles"* that, just as DNA can be modified even after its formation, so also chemical particles can be modified and specialized for various purposes. Atoms, molecules, and various chemical compounds were differently shaped according to the modifications that their precursors had gone through. Just as the nose does not descend from the ear or the eye and vice versa, so atoms have been formed through a clear pathway based on the modifications of their precursors, all of which are connected to the original matter in the universe.

As of today, series of biological reactions and other processes encrypted in the morphogenesis, organogenesis, and histogenesis contain key information about how life originated. Today, those processes are some of the means used to form or reproduce living organisms, even from an egg. A simple microscope can allow the visualization of some of the fundamental stages involved in the development of organisms. However, because the formation of the first forms of life is remote and occurred many years ago, it is impossible for mere human beings to recreate or experience all of the processes involved in the origination or beginning of life. But

CHAPTER 11: DEVELOPMENTAL BIOLOGY POINTS TO A BIOLOGICAL TURBULENCE

some of the developmental processes I pointed out above, and which I will detail shortly, give a clue as to how life started. To put it another way, some biological processes are witnesses of some of the processes used to form the first organisms in the beginning. For instance, mitosis is an example of how matter was added or supernaturally multiplied during the formation of life. Reproduction contains codes about how living things are supposed to multiply throughout the ages. A careful observation of some behaviors and physiological changes in dying organisms can allow one to figure out crucial components or systems required to maintain life and retrospectively understand how life could have been formed and maintained. If a stem cell is able to become any type of cell, we cannot rule out that all types of organisms could have been made using similar processes or similar initial clusters of matter but framed or organized differently. In the following segments, I will revisit a few aspects of morphogenesis that I think will be relevant as I handle the origin and formation of living things later in this book.

11.2. Morphogenesis

Morphogenesis is the biological process that causes a cell, tissue, or organism to develop its shape. Morphogenesis is believed to be a mechanical process involving forces that generate mechanical stress, strain, and movement of cells (Bidhendi et al., 2019). Considering my discoveries on turbulence, I understood that the mechanical stress, strain, and movement of cells associated with morphogenesis can be linked to the turbulence occurring in the cell. For when things are moving in a turbulent fluid, stress, and strain can be created, and knowing that life is born and developed in a fluid, turbulence cannot be excluded.

Scientists have tried to understand the mechanisms involved in morphogenesis. Among many things, they found that, during morphogenesis, many fields are organized around "dominant" ones; the "dominant" field is thought to have a *"high concentration of some substance or activity, which falls off in a graded way throughout the rest of the field"* (Waddington, 2020). However, the Encyclopedia Britannica (Waddington, 2020) mentioned that the *"main deficiency of the hypothesis is that no one has yet succeeded in identifying satisfactorily the variables distributed in the gradients". Attempts to suppose that they are gradients of metabolic activity have, on investigation, always run into difficulties that can only be solved by defining metabolic activity in terms that reduce the hypothesis to a circular one in which metabolic activity is defined as that which is distributed in the gradient"*. When I read this remark, I felt like morphogenesis and growth follow a pattern of split-gathering of a mother body into a system of primary and secondary bodies that current biological equipment is unable to measure properly.

During morphogenesis, patterns are observed, but the origin and dynamics of these are still not well understood. Pattern formation is said to be controlled among other things by the ability of cells organized in a field to sense and respond to their position along a morphogen gradient and the ability of the cells to communicate over a short distance through what is termed as cell signaling, which is crucial for the refining of the initial pattern. The morphogen gradient mentioned above hides an encrypted code of the impact of fluid layers on the types of bodies to be formed from them according to their position.

11.3. Cell potency and differentiation

The ability of cells to differentiate into other cell types is called cell potency (Mahla, 2016; Schöler, 2007). Derived from the Latin word "totipotentia" which is the "ability for all [things]", totipotency is "the ability of a single cell to divide and produce all of the differentiated cells in an organism; spores and zygotes being examples of totipotent cells" (Mitalipov and Wolf, 2009). For instance, in humans, "when a sperm fertilizes an egg, the resulting fertilized egg creates a single totipotent cell, a zygote, which in the first hours after the fertilization divides into identical totipotent cells, which can later develop into any of the three germ layers of a human being (endoderm, mesoderm, or ectoderm) or develop into other cells such as the placenta (Asch et al., 1995). In the first hours after fertilization, the zygote is believed to divide into identical cells. In humans, about four days after fertilization, and after several cycles of cell division, these cells begin to specialize, forming a hollow sphere of cells, called a blastocyst, which is believed to have an outer layer of cells, and inside this hollow sphere are clusters of cells called the inner cell mass, which are believed to form virtually all of the tissues of the human body (Kumar, 2008).

Totipotent cells are also known as "omnipotent". Cells that can differentiate into all cell types of the adult organism are identified as pluripotent and they descend from totipotent cells. Examples of pluripotent cells are those that derive from any of the germ layers. These kinds of cells are called meristematic cells in higher plants (vascular plants which have most body parts) and embryonic stem cells in animals. Stem cells are found in both embryonic cells and in adult cells. Cells that can differentiate into multiple different cells but closely related cell types are called multipotent (Schöler, 2007). Adult stem cells are limited in their potency, for unlike embryonic stem cells, adult stems cells are unable to differentiate into cells from all 3 germ layers. Therefore, adult stem cells are considered multipotent. To some extent, this is almost similar to how children adapt to things in life more than older adults who are set in their ways. In contrast, according to the previous author, oligopotent cells are more restricted than multipotent, but can still differentiate into a few closely related cell types. Unipotent cells can differentiate into only one cell type (their own) but are capable of self-renewal.

11.4. Germ layers: Ectoderm, Mesoderm, and Endoderm

During the embryonic development of most living things, the zygote goes through mitotic division and cleavage to form a germ layer, a primary layer of cells. During the development of the zygote, cells layers are formed, and go through processes that change their shape, expand some layers or sheets over other cells (e.g. epiboly), migrate some layers to other places (e.g. ingression), fold a cell sheet to form the precursor of some organs such as mouth, anus (e.g. invagination), split or migrate one sheet into two sheets (e.g. delamination), turn some cell sheets over the surface

CHAPTER 11: DEVELOPMENTAL BIOLOGY POINTS TO A BIOLOGICAL TURBULENCE

of other layers (e.g. involution), etc. The number of the layers depends on the organisms. For instance, while in diploblastic animals (e.g. cnidarians, a type of invertebrate) only 2 germ layers (ectoderm and endoderm) are formed, in most animals 3 layers form (ectoderm, mesoderm, and endoderm) and these layers will give birth to all of the animal tissues and organs through a process called organogenesis. Some animals (e.g. sponges) form only a single germ layer which later differentiates into cells lacking true tissue coordination. While the cells of diploblastic animals are organized into recognizable tissues, triploblastic animals develop recognizable organs, suggesting that the number of germ layers affect the complexity of the development of the body that resulted from them. It can take days to weeks before these 3 cell layers are formed.

According to Langman's Medical Embryology, of the 3 primary germ layers formed in the very early embryo, the ectoderm is the most exterior or the outside layer. In contrast, the endoderm is the innermost layer, while the mesoderm is the middle layer between the ectoderm and endoderm. In other words, the order of layers starting from the outside to the inside are the ectoderm, mesoderm, and the endoderm.

The differentiation of the ectoderm leads to the formation of some "nervous tissues (e.g. spinal cord, peripheral nerves, and brain), teeth, skin, exoskeleton, mouth, anus, nostrils, sweat glands, eyes, hairs, and nails (Gilbert, 2010). All forms of exoskeletons such as chitin of arthropods (e.g. insects, spiders, ticks, shrimps, crabs, lobsters), shells of mollusks, brachiopods, and some tube-building worms, and silica of some diatoms are all derivatives of the epidermis, as reported by the Encyclopedia Britannica (Waddington, 2020).

The mesoderm differentiates to form the muscles, (endo)skeleton (cartilages and bones), kidneys, adrenal glands, part of the gonads, heart, blood vessels and blood cells (Ruppert et al., 2004; Scott, 2010).

The endoderm is believed to initially consist of layers of flattened cells, which becomes columnar and develops into 2 tubes or tracts: the digestive and respiratory tracts (Gilbert, 2013). The cell layers of the endoderm are the precursors of the following organs:

- Gastrointestinal tract (esophagus, stomach, small intestine, and colon) and the glands opening into the digestive tract (e.g. liver and pancreas)
- Respiratory tract (e.g. trachea, bronchi, and alveoli of the lungs)
- Endocrine system (e.g. thyroid, parathyroid, and pancreas)
- Auditory system (e.g. epithelium of the auditory or Eustachian tube and tympanic cavity or membrane)
- Urinary system (e.g. bladder, kidneys, ureters, and part of the urethra)

In other words, nervous tissues result from the ectoderm, while connective tissues come from the mesoderm, and the gut and internal organs derive from the endoderm. While some animals have bones, others do not. For instance, invertebrates have no bones, while vertebrates have bones. Examples of invertebrates include paramecium, octopus, lobster, and dragonfly. Together, invertebrates constitute about 95% of the animal species.

11.5. Cellular dedifferentiation

Why do cells even differentiate in the first place? For a long time, scientists have tried to understand the process called differentiation, in which a given cell "knows" the type of specific cell it must become, yet all cells have the same DNA. The processes that explain why the same precursor of bodies can split-gather into the precursor of many daughter bodies that can differentiate into "matured" bodies having specific proprieties can also explain why the same cell in living organisms can multiply and yield many more cells which differentiate into various types of cells having different characteristics. It all can be boiled down to the process of split-gathering and differentiation according to a specific timing and program. According to the timing or the stage of differentiation of cells, only certain types of processes (including gene expression) can occur. Sometimes, only cell multiplication can occur, another time only cell maturation or growth can occur, another time, only death must occur. And between the birth of an egg and the death of the organism developed out of it, many stages of development and differentiation occurred at the cellular level and at the entire organism level.

Although differentiated cells cannot always revert back to other cells, in some cases, it is possible. For instance, also called integration, dedifferentiation is a cellular process often seen for instance in *"worms and amphibians in which a partially or terminally differentiated cell reverts to an earlier developmental stage, usually as part of a regenerative process"* (Casimir et al., 1988; Giles, 1971).

Dedifferentiation is also said to occur in plants (Giles, 1971). Likewise, dedifferentiation is also believed to occur with cells in cell culture as they can lose properties they originally had (e.g. protein expression or change shape) (Schnabel et al., 2002). It was reported that a method was discovered to convert mature body cells back into stem cells called induced pluripotent stem cells (Ferreira, 2014).

11.6. Regeneration

Everything in nature seems endowed with mechanisms (e.g. regeneration and restoration) to respond to stimuli in its environment. Just as how living things and even ecosystems can renew, restore, grow themselves or part of their body in response to natural and artificial fluctuations, disturbance, or damage, so also, to some extent, nonliving things react to stimuli in their environment to yield new compounds, bodies, and derivatives when "disturbed". That is why when a living organism is wounded, it tries to regenerate itself back just as a destabilized matter tries to do the same by interacting with others near it. Although these mechanisms are expressed in many forms, they point to the same process of search of equilibrium in the midst of the turbulence surrounding them.

11.7. Apoptosis

CHAPTER 11: DEVELOPMENTAL BIOLOGY POINTS TO A BIOLOGICAL TURBULENCE

Apoptosis is a programmed cell death that occurs in multicellular organisms (Green, 2011). In Greek, apoptosis translates to the "falling off" of leaves from a tree (Alberts et al., 2015). This reminded me of what the Bible said as it compared the falling of planets and stars at the end of the world to fig fruits falling from a fig tree: *"The stars of heaven fell to the earth like a fig tree drops unripe figs when shaken by a great wind"* (Revelation 6:12-14, Tree of Life Version).

Some biochemical changes associated with apoptosis are cell shrinkage, nuclear fragmentation, chromatin condensation, chromosomal DNA fragmentation, and mRNA decay. The average adult human is believed to lose between 50 and 70 billion cells each day due to apoptosis (ScienceDirect, 2019). About 20–30 billion cells are believed to die per day for an average human child between the ages of 8 and 14 (Karam, 2009). Unlike necrosis—which is a traumatic cell death that results from acute cellular injury and can cause the loss of a human's extremities or in worse cases, loss of life—apoptosis is said to be a highly regulated, controlled process that confers advantages during an organism's life cycle. For instance, the separation of fingers and toes in a developing human embryo is believed to occur because of apoptosis of the cells between the digits. While an excessive apoptosis can cause atrophy (shrinking) as seen with cerebral atrophy, which can worsen over time, causing the brain not to fully function properly, an insufficient amount of apoptosis can lead to an uncontrolled cell proliferation, such as cancer. In other words, uncontrolled tissue growth causes cancer. Likewise, cancer can be caused from the disruption of normal morphogenesis.

11.8. Germination is a type of split gathering

For a seed to germinate, it dies first, and then, roots and shoots appear. Likewise, for the precursor of a system of bodies to birth its daughters, it first dies, and during that process, the precursor of the secondary bodies emerged like a shoot coming out of a root, which here is like a primary body. Just like the tissues that are the first to emerge in the aerial part of a plant grows and starts branching, and sometimes becomes a trunk surrounded by branches, so also the precursor of the secondary bodies grows, elongates, and branches out to give different branches which, at their turn, will go through a similar cycle (but on a smaller scale) to yield secondary branches and so on and so forth until the plant finishes its growth cycle and the entire completion of its lifespan during which other organs like flowers, seeds, and other parts, emerge and mature accordingly!

Even on the level of animals, the same things happen. A mother and a father come together to yield an egg, which develops in the womb of the mother where it is split-gathered into several organs sometimes organized along the spine (or a main secondary precursor) which is like the trunk or backbone of the entire system. After their birth, the children orbit around their parents just as secondary bodies do around their primary bodies. At their turn, those children will grow, find their own mate, and have their own children which, at their turn, will continue the cycle. In the end, animals are organized into families, which are a type of physical manifestation of the way they were split-gathered, raised, and interact with one another.

CHAPTER 12

CAN YOU SCIENTIFICALLY CONNECT THE SPATIAL DISTRIBUTION OF ALL FORMS OF LIFE TO A BIOLOGICAL INTERMITTENCE OF SIZE DURING THE FORMATION OF LIFE? CHECKOUT WHY THIS MATTERS A LOT!

In *"Turbulent Origin of the Universe"*, I expounded on a phenomenon called intermittence, which is basically the presence or the formation of very small bodies between large ones in such a pattern hard to explain using conventional mathematics. Technically speaking, I explained in that book that intermittence discloses how long period events of low magnitude usually separate rare events of large magnitude in a process or a system. For instance, the size intermittency is about the presence of smaller bodies, smaller clusters of matter, or smaller systems of bodies between bigger or larger ones. That is why in a system of celestial bodies, larger bodies are usually separated by smaller ones. For instance, large planets are separated by small asteroids. In the Solar System, there are a few planets, but millions of asteroids (smaller than planets). In each planetary system, there are a very few large or big satellites, but many smaller satellites. Similarly, small clusters of stars and even isolated stars are usually found between galaxies and clusters of galaxies. I explained that the size intermittence of celestial bodies in the universe could be due to how the turbulent prima materia and its daughters at each generation of split-gathering were broken up and gathered together into daughter bodies or clusters of bodies. Intermittence could have been also caused by the fact that the "cleavage" lines, "cleavage" points, interface of separation, or the interface of breakup of the precursors of the bodies and their clusters was not neat or sharp, therefore leading to the formation of smaller bodies or systems of bodies between bigger or larger ones.

I managed to group the turbulent intermittence of nonliving things found in nature into 9 generations accordingly to the order of the split-gathering of their precursors:

- 1st generation of turbulent intermittence, leading to the presence of smaller clusters of galaxies between larger ones and smaller clusters of globular clusters between larger ones
- 2nd generation of turbulent intermittence: intermittence of stellar systems
- 3rd generation of turbulent intermittence: intermittence of stars
- 4th generation of turbulent intermittence: intermittence of planetary systems and asteroid systems
- 5th generation of turbulent intermittence: intermittence of satellites
- 6th generation of intermittence: intermittence of minerals, mineraloids, and rocks
- 7th generation of turbulent intermittence: intermittence of atoms, molecules, and chemical compounds
- 8th generation of turbulent intermittence: intermittence of subatomic particles
- 9th generation of turbulent intermittence, leading to the precursors of invisible bodies (which will never be scientifically discovered) between and within larger ones

As I was investigating living organisms, I realized that some of the traits I discovered with the intermittence of celestial bodies are also found with all forms of life. I found that, intermittence is also very important for the explanation of the origin of organisms, for it also occurred during the formation of living things under the influence of a biological turbulence that shaped their precursors, meaning the matter they are made of.

For instance, as I was carefully looking at organisms, I realized the nearly constant presence of countless small living things between large ones. Even when I considered the plants and animals separately, I noticed that it is rare to find 100% pure species or individuals not mixed at all with others than to find the latter. I also realized that individuals of each of the 6 forms of life known as of 2025 (i.e. plants, animals, fungi, protists, archaea, and bacteria) are usually found mixed with those of other forms of life. For instance, although it is possible, it is rare to find a huge population of one form of life completely and purely separated from other forms of life. To abound in the same direction, it is rare to find animals naturally living in a place where there is no plant. In nature, it is also rare to find vast populations of plants where there is no animal, microorganisms, fungi, or bacteria nearby. It is rare to find large animals in a natural environment where there is no small animal. It is even impossible to find giant carnivores in a natural, not highly degraded environment where there is no small animal, particularly herbivores they can prey on.

CHAPTER 12: BIOLOGICAL INTERMITTENCE OF SIZE AND SPATIAL DISTRIBUTION OF THE FORMS OF LIFE

In other words, I have come to realize when organisms were being formed, several small ones were formed between larger ones, and different forms of life were formed between others. Hence, except for some rare exceptions of pure populations of certain forms of life, most plants, animals, fungi, protists, archaea, and bacteria are usually found in association with others, meaning living together with others even if it is a small population of mixed individuals. Where fungi are found in a natural environment, plants are usually near. Where there are bacteria, other forms of life (i.e. plants, animals, fungi, protists, and archaea) are not very far and if they are far, at least more than one type of bacteria is found in that place. In a place where protists (e.g. algae) are naturally present, other forms of life (i.e. plants, animals, fungi, archaea, and bacteria) are not very far away.

In a forest, savannah, wetlands, and other types of ecosystems, there are various types of plants mixed, and although sometimes some plants or trees are found together without being mixed with others, in most cases, plants of various species and families are frequently mixed with others. Where the population of plants is or can be said to be 100% pure without the mixing of other types of plants, other forms of life different than plants (e.g. animals, fungi, protists, archaea, and bacteria) will be found. For it can be inconceivable to find a huge area covered by plants without finding at least a single bacterium, insect, or other types of organism between them, on them, in the air, or underground. Even in the closest bush to your home, plants of one kind are usually mixed with those of other kinds and between some of them, organisms of various kinds (e.g. worms, insects, rats, snakes, etc.) can be found. Between big trees, small trees are found and between larger animals, smaller animals are usually found. Those who have a chance to grow a garden can testify of how the desired crops or vegetables are usually not the only thing to grow in the land they are planted. Other plants sometimes termed weeds usually appear. As farmers or gardeners try to handle the weeds, insects also appear, laying their eggs on vegetables or on the leaves of the plants and it does not take too long before some gardens become the home of undesirable weeds and insects, and even rats, snakes, raccoons, deer, birds, and squirrels, which also want to claim a piece of the planted crops or vegetables. My point here is not the undesirability of some plants or organisms in cultivated lands, but the fact that living things are usually mixed with others in a way that between large ones, there are small ones as well. And interestingly, there are usually more smaller ones than larger ones. In other words, between a few large plants or trees, several smaller plants or trees are usually found. Although possible, it is rare to find ecosystems where the number of larger organisms is higher than that of smaller ones. Hence it is much easier to find a natural land having a few big trees, but countless smaller trees or grasses between them, rather than the reverse. Although human beings usually try to weed their farms, gardens, and yards by cutting, digging, hoeing, or even spraying chemicals to try to kill plants and trees they do not desire between those they want to grow, some remnants of those being fought always come back in some form or as nature manages to land some pollen or seeds back to repopulate the fields, gardens, or yards that human beings think they can clean from so-called weeds. For example, because some homeowners want their yard to look nice and green without any dandelions, they treat it to get rid of those weeds, but because neighboring houses do not treat their yard to get rid of their dandelions, all it takes is a wind or an adventurous little kid blowing the puffy white matured dandelion and the seeds go flying to land wherever they may, including into the treated yard and the untreated yard.

Likewise, in the bush where animals can be found, although some animals of the same kind always live together, in general, they are usually mixed with others. Herbivores may try to flee carnivores, but the later always manage to find themselves near their prey. Unless a system is highly degraded and near its extinction, it can be impossible to find carnivores living in a natural land where there is no herbivore nearby for them to eat. Hence, various kinds of mammals can be found living in the same bush or forest with various kinds of reptiles, birds, insects, etc. Even in a small sample of soil, countless species of insects and other invertebrates can be found. The microbiologists who wanted to sequence the genome of microorganisms in soil samples must have been surprised at the diversity of genomes that can be found in some soils, not because they were contaminated, but because organisms were meant to live in mixture not only to maximize the use of natural resources, but also to align their existence with the law of intermittence which took place during their formation, etc. To make a long story short, in natural ecosystems, large or big organisms are usually separated by many small ones. Ecological niches are usually more intertwined than what some people may think. To some extent, even among the rich, there are poor and even if the rich try to live far from the poor who some rich don't usually help as they should, they cannot accomplish their goal without involving the poor.

Before I close this segment on the intermittence and mixing of forms of life, I would like to say a few things about similar characteristics found even with chemical particles which are some of the foundational or the raw materials used to form living organisms. Indeed, even among the chemical particles, the larger ones are few in number, while the smaller ones are countless. Furthermore, chemicals are mixed in such a way that it is very difficult to find pure clusters of one chemical element not mixed with others. Those working in ores and mines can testify that even minerals having so-called 100% purity are very rare, if not impossible to find. It is not by chance that chemical particles tend to form systems of bodies termed as atoms (which are a mixture of various subatomic particles), molecules and other forms of chemical compounds (a mixture of different atoms not usually of the same kind), etc. For chemicals were formed by being mixed with one another and they have no choice than to "bind" with those near them in clusters which encode some of the processes used to form them.

To summarize, natural biological systems were not formed or meant to be segregated into unseparated pure clusters of forms of life. For such organization is against the laws which formed things in nature. Although some organisms are capable of helping disseminate seeds to some ecosystems, the distribution of organisms is a process that occurred since the formation of the first organisms. The

CHAPTER 12: BIOLOGICAL INTERMITTENCE OF SIZE AND SPATIAL DISTRIBUTION OF THE FORMS OF LIFE

mixing of small organisms with large ones is not something that the organisms just chose to do, but it is an organization which resulted from how organisms were distributed since the day they were formed. Migration of animals and the ecological niche of some animals also later played a role in clustering some animals according to their environments of course, but still, those ecological realities are not sufficient to explain the underlying rules forcing animals to live in mixture with one another, even if it is a "pocket" of population of one form of life mixed with a "pocket" of the population of another.

The examples I just gave concerning intermittence of geographical distribution is also found with other parameters such as speed. For instance, while some organisms can run very fast, some do not and others do not even move at all. The study of the size intermittence in conjunction with the distribution of speed and many other variables helped me to unearth some of the codes encrypted in the origin of the celestial bodies. Because I sensed some of the codes with animals as well, I will also first review a few records of the largest, smallest, fastest, and slowest animals before I plainly start diving into how all forms of life originated.

CHAPTER 13

IF YOU THINK THAT SIZE DOESN'T MATTER OR THAT ITS IMPARTATION ON LIVING THINGS WAS BY CHANCE, THEN PAY ATTENTION TO THIS INTERESTING DETAIL ABOUT THE LARGEST AND SMALLEST ORGANISMS

Content available at www.Science180.com/LargestSmallestOrganisms
The size of organisms is not a propriety evenly distributed throughout nature. Just by focusing on size alone, some people may rightly say that nature is not "fair" in the appropriation of resources to bodies in the universe. From the nonliving things (e.g. celestial bodies and chemical particles) to the living things or beings, size varies. Hence, some celestial bodies are larger than others. For example, the size of the Sun is much bigger than that of Jupiter, which, at its turn, is much larger than that of Earth, which, at its turn, is much bigger than that of the Moon. Likewise, on the scale of chemical particles, some elements like Uranium and Plutonium are much larger than Hydrogen (usually considered as the smallest chemical element, yet the most abundant among them all). Still talking about chemical particles, some are not big enough or well organized enough to be called chemical elements. Hence some are called subatomic particles, meaning smaller than atoms.

    Likewise, when it comes to organisms, some are very large, while others are very tinny to such a point that an advanced microscope (e.g. electronic microscope) may be needed before fully seeing it or fully unearthing some of their characteristics. In this chapter, I will present a few organisms that are considered to be holding some of the largest and smallest records for size. Later in this book, I will explain the significance of such records. To see the rest of this chapter, go to: www.Science180.com/LargestSmallestOrganisms

TURBULENT ORIGIN OF LIFE

CHAPTER 14: SPEED OF ORGANISMS

CHAPTER 14

CAN YOU JUDGE THE ORIGIN OF AN ORGANISM BY ITS SPEED?

Content available at www.Science180.com/SpeedOfOrganisms
Speed is one of the variables I used to decode the processes involved in the formation of the celestial bodies. It has nothing to do with the size of the bodies, but with the position of the bodies with respect to their primary body. For the celestial bodies, orbital speed (which the speed celestial bodies use to orbit or move around their primary body) is one of the types of speed I studied. I discovered and reported in *"Turbulent Origin of the Universe"* that, in each system of celestial bodies, the orbital speed of secondary bodies (i.e. a body orbiting a primary body) decreases as the distance away from the primary body increases. For instance, the primary body of the planets in the Solar System is the Sun, while the primary body of the satellites is their primary planet. More specifically, when I take the satellites in a planetary system for instance, as their distance from their primary planet increases, their speed decreases. Likewise, as the distance separating planets from the Sun increases, the speed of the planets decreases. I also demonstrated in *"Turbulent Origin of the Universe"* that the fastest celestial bodies are not always the largest ones in their system nor the smallest ones.

While some animals walk (e.g. humans, goats, cows, etc.), some hop (e.g. frogs, kangaroos, and rabbits), others fly (e.g. birds), others swim (e.g. fish), and others just crawl (e.g. snakes). Therefore, when it comes to animals, speed can be about how fast they can run, hop, fly, swim, or crawl according to their class. Some of the fastest speeds have been recorded when animals are chasing their prey or racing for some reason. Animals chasing a prey run faster when the prey is in front of them is also cruising at its fastest speed to escape. When animals are in fear because of a danger or an unusual opportunity in front of them, they usually run fast. Their adrenaline kicks in to help them pick up speed. As you will see below, size and speed of animals are not perfectly associated. Hence, some large animals go slower than some small ones. However, the fastest animals are neither the largest nor the smallest, but somewhere in the between. For instance, elephants and giraffes are some of the largest living animals today, but they are not the fastest animals on Earth.

The rest of the content of this chapter can be found in the electronic version of this book at: Available at www.Science180.com/SpeedOfOrganisms

CHAPTER 15

CAN YOU REALLY DECODE THE ORIGIN OF LIFE WHILE NEGLECTING ITS SPIRITUAL COMPONENT OR WHAT MUST CHANGE IN THE BELIEVERS-NONBELIEVERS RELATIONSHIP BEFORE RATIONALISTS AND FREETHINKERS TAKE THE SPIRITUAL DIFFERENTLY?

Life cannot be apprehended if any of its major components is ignored. Life is connected to spiritual things in such a way that presenting life while ignoring to address its spiritual aspects is partial and incomplete. In this chapter, I will introduce you to key spiritual facts that can help to better understand and decode the origin of life.

15.1. Life has a spiritual component that some people have chosen to ignore
Up until this point, I have not properly discussed nowhere in this book the spiritual aspects of life yet, but in this chapter, I will do so. However, in most secular books, life is not properly viewed through the glance of the spiritual world. Yet, life could have never existed without the involvement of spiritual things and spiritual beings. Therefore, in my efforts to explain how life originated, I chose not to ignore the spiritual realm.

Although many scientists do not believe in it, human beings are not just made of the dust they return to the Earth after the death of their physical body that some people think is the death of everything once contained in that physical body. The physical body contains the spirit, which does not die when the physical body ceases to live, but which enters another realm, which cannot be properly described without using spiritual terms (e.g. spirit and soul) that some people do not want to hear about. In other words, the definition of life (particularly that of human beings) is incomplete if spiritual components are not properly addressed. For instance, some people believe in the existence of demons and angels, all of which are not physical beings whose forms of life are different from ours. This implies that, besides the 6 forms of life I presented early in this book, other forms of life exist, including spirits, which science will struggle to comprehend, for they are made up of unfamiliar materials that science ignores. Although they know very well that life has a spiritual component, which most religions try to address one way or the other, the scientific community has "managed" to "evacuate" debates or discussions pertaining to spirituality from their theories of life. With that in mind, several questions arose to my attention:

- Can the world expect any comprehensive explanation of the origin of life by the scientists if the latter continue to ignore the spiritual aspects, which all tribes and cultures in the world have massively acknowledged as crucial?
- How can most scientists believe in a form of a god or higher power, yet, they fight to remove any notion of God or gods from their theories of life?
- How can countless scientists believe in demons and angels, and some are even initiated into demonology, witchcraft, exorcism, magic, and other kinds of spiritualism, and yet most of them (some of whom control scientific boards of decision) chose to ignore the spiritual aspects of life in their theory about life?
- How can some scientists be adept of the religion of their choice, yet, they refuse or are afraid to talk about God or their gods in their theories of life?
- Isn't the presence of churches, mosques, temples, and other kinds of worship centers everywhere across the globe an encrypted message that, regardless of their religions, worldviews, and doctrinal mistakes, most human beings are aware of a kind of supreme being or spiritual master that something in them eagerly wants to connect with?
- Can the diversity of religious views and other philosophical ideologies be enough to rule out that there is no Creator?

15.2. Extraterrestrial life

While they fail to understand their own life, some people are very busy trying to unearth mysteries about the life of beings beyond earthly realms. Because of their ignorance, some people interact with strange forms of life every day, but they just don't know what they are dealing with. Many people have even entertained angels but they were unaware of doing so. Most scientists believe that life is only found on Earth, yet others postulate that extraterrestrial life exists. Therefore, billions of US dollars have been poured into research to investigate and discover extraterrestrial forms of life on other celestial bodies. Because they refuse to believe in the existence of God, angels, and demons, all of which belong to a world or realm "different" than ours, some people reject the existence of a living realm beyond that which they are familiar with and which they can only physically touch. It is as if some people invested billions of dollars into finding things they do not want to believe exist, and

CHAPTER 15: NEGLECTED SPIRITUAL COMPONENT OF LIFE

when they do not find them the way they want, they keep investigating differently instead of changing their mind and letting the data lead their wrong perception toward the unchangeable reality. Although I do not embrace the theory elaborated by the proponents of the extraterrestrial life painted in some secular books seeking to explain life that they refuse originated from God, I do believe that other beings such as angels exist beyond the limits of Earth and on Earth, and that scientific equipment and methods may not be able to detect them. Furthermore, like I explained in other books (*e.g. "Origin of the Spiritual World"*, and *"Science180 Accurate Scientific Proof of God"*), some UFO (Unidentified Flying Object) sightings that have been increasingly talked about in the news are examples of extraterrestrial beings, such as angels whom scientists are failing to properly interact with.

15.3. Life is a spirit

Life cannot be viewed as a physical thing or physical being only, but as a spiritual entity hosted or incarnated in a physical matter. This also implies that life could also exist without the physical matter that usually contains it and which most people erroneously perceive as life itself, while the real life is behind the scene. The spirituality of life is also why dead people are not really dead but sleeping or translated into another dimension that can be reversed or restored via supernatural processes involving resurrection, which some people do not want to hear about. This also explains why just before and after people die, their body is still the same, but because their spirit has departed, their life known on Earth is extinct, yet its spiritual component can appear somewhere else. People working in the medical field have testified that they have seen patients peacefully transitioning from this world with some telling them that they were seeing angels, and talking to some supernatural being before transitioning from this life. In contrast, other patients were frightful, screaming, and swinging their arms in the air as if they had the strength to fight off any being and cursing at that being (which they alone were seeing as they were agonizing) while begging people standing in the room saying "Don't let it get me" in their last days on this Earth.

Similar experiences of some dying people have been recounted by many across the globe and even in various cultures, suggesting that, life after death not only is real, but also everybody does not go to the same place after death. For, if some people were smiling as they were dying, while others were screaming out of fear while they were dying, it is more likely that they were not going to the same place after life on Earth. During the last moments of some dying people, many have confessed things that they never would have mentioned during their lifetime. Some even confessed doing very evil things in such a way that it seemed like they thought that those confessions may erase the consequence they may face after death and send them into a better place. Because they fear the consequence of some confessions made by their dying relative, some people have to do whatever it takes to stop the agonizing person from talking too much. For instance, in some countries where witchcraft is practiced, people have testified that, on their death bed, witches have told stories about evil things they have done in their lifetime as if someone is forcing them to reveal some few things before dying. Some have even revealed experiences about people (including their own children) they have killed, cursed, and sickened, etc. In some cultures, people have testified that cloths or other things were stuck into the mouth so their dying relative would stop talking too much, because in their dying state they could not hear or respond to any instruction begging them to stop revealing negative things that can damage their legacy or complicate life for their family who may face even legal battles from the victims related to those confessions.

What I have been saying about the spirituality of life also means that, just as all organisms are not contained in the same form or same morphological and physiological boundaries, some forms of life (which are "purely spiritual") are not contained in any physical matter findable on Earth. That is the case of angelic beings, which were formed under conditions different from those on Earth. And if I can go deeper, I would say that angels were formed even before the formation of the Earth was completed and before the formation of any physical being was present on it. Many religions across the globe believe in spirits of living organisms, yet spirits are not made of, or clothed with, the same material as that of most organisms on Earth. Nevertheless, some spirits can transfigure themselves or take different forms as they wish, but still within the limits set for them; for creatures are not empowered to do everything they want limitlessly.

15.4. Death is not the cessation of life

I have come to realize that, just as some people ignore how life originated, they also ignore what happens when someone's life on Earth "ends". Therefore, some people think that every living thing starts from the time it breathes for the first time on Earth and ends on this Earth as they take their last breath. Although some people think that death is the cessation of life, death as known to human beings is just a door opening some life components into another form of life. For death is not the cessation of existence of everything that once lived in a living thing. Although some scientists think that death ends or dismisses all biological processes of organisms, meaning that for them, death is the end of life, it is important to notice that some people came back to life after being pronounced dead medically or clinically. History has recounted stories according to which some people have been dead for days but suddenly came back to life after some supernatural encounters. This also implies that science does not fully know its limits as far as life and death are concerned, for after all, life and death are not in the power of men although, unfortunately, some people do not hesitate to take their own life and/or the life of others. Else, some wealthy and influential people could have afforded to buy or pay anything required for them to live longer and even eternally if they could. But there is a limit after which every human being must exit this world.

After death, the remains of organisms enter a biogeochemical cycle which breaks them down into all of the chemical elements that those organisms may have consumed, ingested, breathed in, or inserted into their body during their lifetime.

CHAPTER 15: NEGLECTED SPIRITUAL COMPONENT OF LIFE

While the physical body is going through decomposition after its death, the spiritual body is in another state that I will elaborate on later. Although they don't have the same explanation of the spiritual world, most religions in the world believe that human beings have a spiritual component. It is only in science that some people have managed to remove the spiritual from the domain of definition and research of things. Yet, most scientists do some intellectual activities beyond what wild animals can do, pointing at a certain level of spirituality of human beings even if they do not openly and publicly confess their belief.

Likewise, when people try to describe life without involving some spiritual aspects, they end up explaining apparent connections between some physical things, while the deep underlining causes are neglected. For it is the spirit that gives life. Therefore, no one can fully explain life without addressing certain spiritual things. In other words, although I will be trying to explain life using chemical, biological, and physical terms or phenomena, the real engine behind life is none of those human efforts to interpret nature and the things and beings it contains. For, no matter how many billions of years people think chemicals could have taken to organize themselves into the biological systems necessary for life, life could have never existed without invisible forces that the naked eye cannot see. I will talk more about this toward the end of this book!

15.5. Nonliving things can perform some functions found with the living only
By the time I reached 20 years of age, I understood that spiritual entities can even influence, curse, and perturbate the physiology of other beings, even that of plants. When spirits enter a living organism, they can do what they want if their host does not restrain them to some extent. By my teenage years, I learned that spirits can enter and manipulate anything they enter, regardless if it is a living or nonliving thing. For instance, in the years 1970-1980, the government of an African country which name I willingly chose to omit here, seized thousands of acres of land from farmers in order to plant palm trees to be exported. Because they were unhappy about the confiscation of their lands, the aforementioned farmers cursed the palm trees. For decades, those palm trees never produced fruits, which are the products for which they were planted. I remember driving by those trees many times but on one day, I was shocked when, for the first time, I was given an explanation of why none of the thousands upon thousands of trees never produced a single fruit over the decades. Most of those trees were more than 10 meters high and could be seen just with leaves, but no fruit. That is the power of a curse, which can also fall onto some human beings, which in the end, cannot produce anything in their life. After failed attempts to make those trees to reproduce, the government ordered them to be cut down so the land could be at least used to build homes for the population, which was increasing in number, and who had witnessed that evil demonstration of power over the palm trees. Just as the curses pronounced over those trees prevented them from bearing any fruit, so also when spirits enter organisms, they manipulate the will of their victims to act and bring to pass evil agenda. Some spirits and curses can stay quiet in their hosts until the appointed time for them to act. Because they

are under the influence of spirits, some nonliving things can perform some functions executed by the living. Likewise, some wild animals can perform certain tasks originally thought to be doable by men only. For instance, human beings may think that they are the only beings who go to stores to shop food and other goods for their children. But, in fact, although they don't have a store like humans do, wild animals know how to seek food and properly take care of their children until they are grown enough to be independent. Some animals are so social that, even if they are adults, they stick with their herd or family. People like to use the example of mother eagles which will go around looking for food for their baby eagles, then bringing the food back and putting it in the mouth of their baby repeatedly until a day comes for the mother eagle to remove the babies from the nest, and "throw" them in the air as if the mother does not know what she is doing, just to see the mother eagles catch the babies before they hit the ground, and take them back up high in the air again, repeating that same experience over and over until the baby eagles know how to fly. Animals may not think as humans do but they know and have programs (beyond instinctive actions or reflexes) in them which allow them to seek the best for their children. Human beings also need to be reminded that wild animals are not the only ones that act using reflexes, for many acts or behaviors crucial for the life of humans are based on reflexes. I can go on and on to give you more examples but I have to stop here. Considering all I said in this chapter, I concluded that life is more than chemical materials packed together.

CHAPTER 16

HOW TO SCIENTIFICALLY TALK ABOUT THE FORMATION OF LIFE AND HAVE EVERYBODY BOW TO THE UNIVERSAL POOL OF QUALITIES AND THE UNIVERSAL HOLISTIC BEING (THAT HAS NOTHING TO DO WITH RELIGION OR FAITH)

16.1. Universal pool of qualities of organisms in the universe
On September 25, 2021, I learned that, under the biological turbulence and the turbulent program of life that accompanied it, abilities were given to organisms differently. I coined the terms "universal pool of qualities", "universe of qualities", "universal pool of abilities", or "universe of abilities" to express all the qualities found in all organisms. I understand that what is a quality for one person may not be so for another one. But my definition of qualities is based on the fact that it is possible to label or define potentials of things and beings without maximizing biases toward their intrinsic values according to a universal standard of perfection that no human being can claim to fully master.

The functions, attributes, traits, or abilities achievable by physical and spiritual beings (most of which are not gathered into a single organism known to most human beings on Earth) are diverse and include the ability to:
- breath
- swim
- walk on 4 feet
- walk on 2 (legs) and have an erect limb posture (and be up like human beings)
- jump
- crawl on 4 feet (e.g. lizards), or on more than 4 legs (as in the case of some

TURBULENT ORIGIN OF LIFE

mollusks) or even without legs at all (as in the case of snakes)
- fly
- run
- burrow
- glide
- climb
- think
- feel
- see
- foresee
- see beyond the visible spectrum
- hear
- taste
- speak
- sing
- live underground
- live in fresh water only
- live in marine water only
- live in both fresh and marine water
- live on earth or on land only
- live beneath the earth or underground only
- live in water, on the earth, in the earth and in the air (e.g. amphibians such as frogs)
- produce own food through photosynthesis (e.g. plants)
- get energy through food that one cannot produce
- eat grass (e.g. herbivores)
- eat meat (e.g. carnivores)
- eat both grass and meat (e.g. omnivores)
- communicate vocally
- hibernate
- transfigure (like angels)
- echolocate (like bats)
- chew the cud (like ruminants)
- respond to environment changes
- grow
- reproduce sexually
- reproduce asexually
- maintain stability of internal organs
- adapt to the environment by changing behaviors, abilities, or structures

CHAPTER 16: UNIVERSAL POOL OF QUALITIES AND THE UNIVERSAL HOLISTIC BEING

- repair one self
- sensitive to environment
- respond to things within their internal and external environment
- discharge or remove waste products into or from the environment
- perform various other physiological functions, etc.
- invent, discover, and research (like human beings)

etc.

The selected functions I mentioned above could not have been performed without the organisms having a set of some of the following organs or features I mentioned below according to their morphology. Some organisms have some of the following features while others do not:

- amyloplast
- centriole
- centrosome
- chloroplast
- cytosol
- endoplasmic reticulum
- Golgi apparatus
- lysosome
- mitochondria
- nucleus
- nucleole
- peroxisome
- plasma membrane
- ribosome
- vacuole
- bones (e.g. bony skeleton)
- cartilages (e.g. cartilaginous skeleton)
- fins (like fish)
- scales (like fish)
- stomach having many compartments or sacs (e.g. polygastric like cows)
- stomach with a single sac (e.g. monogastric like pigs)
- organism made of only one cell (i.e. single cell organism)
- organisms consisting of many cells (i.e. multicellular organisms)
- cells are compartmentalized
- cells protected by a cell membrane
- cell wall made of cellulose only (e.g. plants)
- cell wall containing chitin (e.g. fungi)
- sweat glands

- mammary glands (e.g. breasts to produce milk)
- DNA wrapped inside a nuclear membrane (e.g. eukaryotes)
- DNA not wrapped inside a nuclear membrane (prokaryotes)
- DNA loose (e.g. prokaryotes)
- DNA compacted and wrapped into chromosomes
- many chromosomes (some have many chromosomes)
- very few chromosomes
- chromosomes compacted a certain way (chromosomes are not compacted the same way)
- long chromosomes (some animals have long chromosomes)
- very short chromosomes
- presence of hoof
- hoof separated and completely divided
- etc.

While some people think that the presence or absence of some of the above listed features or abilities is a bad thing, in reality, while some organisms may have some abilities, others may have just its opposite; and in the end, most organisms do not have an ability and its opposite at the same time. For instance, an organism like a fish, which lives and swims in water, is not empowered to fly in the air like birds. In general, most organisms just have a set of abilities, morphological features, physiological features, meaning an arrangement of qualities and features instead of their combination. This strategy also causes living organisms to live without contradicting themselves. Part of this reality is that features and abilities were formed according to specific pathways, and when some pathways are on, others are shut down, meaning off.

Even among human beings, although they have the same general morphology and physiology, some are taller than others; some can cope with cold more than others. Some can live easily in very hot environments than others. Recently, based on years of investigation, it has been shown that the strength of human beings can be summarized into 34 groups or themes of talent, but nobody is strong in all of them. The strength of one person can be the weakness of another. The arrangement of the 34 strengths or talents in different sets defines most of the behavioral variations among human beings. Other psychological and biological factors are behind these variations of course, but my point is that human beings are different on many scales. Because people are different, the theme sequences of the strengths are not the same. The order of the themes in someone's report is a footprint of the strengths behind who he or she is. Below, I listed the 34 themes as portrayed by the Clifton StrengthsFinder, which is recognized as the culmination of the more than half century pioneering work of Donald O. Clifton (1924-2003), known as the "father of the strengths-based psychology and creator of the Clifton StrengthsFinder" (Buckingham and Clifton, 2001):

1. Achiever: Ability to "have a great deal of stamina, work hard, and be satisfied

CHAPTER 16: UNIVERSAL POOL OF QUALITIES AND THE UNIVERSAL HOLISTIC BEING

from being busy and productive".

2. Activator: Ability to "make things happen by turning thoughts into action" even if that means to be "often impatient".
3. Adaptability: Ability to "prefer to go with the flow, be "now" people who take things as they come, and discover the future one day at a time".
4. Analytical: Ability to "search for reasons and causes, and think about all the factors that might affect a situation".
5. Arranger: Ability to flexibly "organize and figure out how all the parts and resources can be arranged for maximum productivity".
6. Belief: Ability to "have unchangeable central values out of which emerge a defined purpose for one's life".
7. Command: Ability "to have presence, take control of a situation and make decisions".
8. Communication: Ability to usually "find it easy to put one's thoughts into words and be good conversationalists and presenters".
9. Competition: Ability to "measure one's progress against the performance of others and to strive to win first place and revel in contests".
10. Connectedness: Ability to "have faith in the links between all things and to believe there are few coincidences and that almost every event has a reason".
11. Consistency: Ability to be strongly "aware of the need to consistently treat people the same, set up and adhere to clear rules".
12. Context: Ability to "enjoy thinking about the past, and understand the present by researching its history".
13. Deliberative: Ability to be very "careful while making decisions or choosing", and to anticipate obstacles".
14. Developer: Ability to "recognize and cultivate the potential in others, to spot signs of each small improvement and derive satisfaction from these improvements".
15. Discipline: Ability to enjoy routine and structure and to create order in their world.
16. Empathy: Ability to "sense the feelings of other people by imagining themselves in others' lives or others' situations".
17. Focus: Ability to "take a direction, follow through, and make necessary corrections to stay on track and to prioritize" before acting.
18. Futuristic: Ability to be "inspired by the future and what could be and to inspire others with one's visions of the future".
19. Harmony: Ability to "look for consensus. not enjoy conflict but to seek areas of agreement".
20. Ideation: Ability to be "fascinated by ideas and find connections between seemingly disparate phenomena".
21. Includer: Ability to "accept others, show awareness of those who feel left out, and make an effort to include them".
22. Individualization: Ability to be fascinated with the "unique qualities of each

person and to have a gift for figuring out how people who are different can work together productively".
23. Input: Ability to crave to know more, to enjoy collecting and archive all kinds of information".
24. Intellection: Ability for intellectual activity, introspection, and appreciation of intellectual discussions.
25. Learner: Ability to have a "great desire to learn and want to continuously improve".
26. Maximizer: Ability to "focus on strengths as a way to stimulate personal and group excellence and to seek to transform something strong into something superb".
27. Positivity: Ability to "have an enthusiasm that is contagious, to be upbeat and get others excited about what they are going to do".
28. Relator: Ability to "enjoy close relationships with others and to find deep satisfaction in working hard with friends to achieve a goal".
29. Responsibility: Commitment to stable values such as "honesty and loyalty and ability to take psychological ownership of what you say you will do".
30. Restorative: Ability to skillfully deal with problems and be good at figuring out and resolving what is wrong.
31. Self-Assurance: Ability to "feel confident in one's ability to manage one's own life, to have an inner compass that gives one confidence that one's decisions are right".
32. Significance: Ability to work independently and "desire to be very important in the eyes of others and be recognized".
33. Strategic: Ability to "create alternative ways to proceed, and when faced with any given scenario, quickly spot the relevant patterns and issues".
34. Woo: Ability to "love the challenge of meeting new people and winning them over; and derive satisfaction from breaking the ice and making a connection with another person".

Before I continue, I must say that my quoting of the aforementioned 34 themes is not an endorsement nor a recommendation for anyone to feel obliged to take the test for these, but is just an effort to show how human strengths can be and have been clustered into meaningful themes. Unfortunately, most people do not know their strengths or how to use them to maximize their life. As I carefully reviewed each of these features and abilities that organisms can have, I realized that the frontier of the classification of life is not solely based on what organisms can do or not. For instance, human beings are not the only intelligent beings. Some wild animals display certain traits that some human beings fail to do. In other words, wild animals have their own universe of strengths, some of which some human beings may envy. What can I say about the endurance and the fearlessness of the horse, the courage of a lion, the speed of the cheetah (some human runners would like to run as fast as some wild animals), the beauty of some flowers and landscape, the smell of some flowers, the docility of some domestic animals, the ability of some

CHAPTER 16: UNIVERSAL POOL OF QUALITIES AND THE UNIVERSAL HOLISTIC BEING

organisms (e.g. plants) to produce their own food and not depend on any organisms to live, etc. Furthermore, animals are not the only organisms that move, nor plants the only ones that photosynthesize. For example, Encyclopedia Britannica reported (Sagan, 2021) that "many bacteria both swim (like animals) and photosynthesize (like plants), yet they are considered neither animals nor plants. Many algae also swim and photosynthesize simultaneously", yet they are not considered as plants.

Considering some spiritual realities, I understood that spiritual beings (e.g. angels) are also diverse in their morphology, functions, missions, and other attributes associated with them. As I explained in other books (e.g. *"Turbulent Origin of the Universe", "Reconciling Science and Creation Accurately",* and *"Origin of the Spiritual World"*), even angels are not the same. Some have wings, while others do not. Some are made of fire and still carry fire, others do not. Some have more power than others. Some are corrupted (e.g. demons) and others are pure (e.g. holy angels). I know some people do not like talking about angelic beings, but it is important not to forget them when living things and beings are concerned; for they are real and, whether we are aware of them of not, they still influence living things.

16.2. Universal holistic being

The single organism that possesses all of the universal pools of qualities, abilities, traits, functions, features, and characteristics found with living beings is what I called the "universal holistic being". To put it another way, I coined the term "universal holistic being" to define the being that has all of the characteristics found in all created (known and unknown) organisms. I do not mean that such an organism necessarily exists or has really existed or will exist (or do I consider it otherwise? – but I just felt on October 20, 2021 that, to be exhaustive and precise in my definition of life, I needed to coin a term for a being that can fully represent everything ("good" or "bad") thinkable about the organisms in the universe. By this definition, I excluded God (or do you prefer I don't), for although some people do not believe He exists, or that He is good, I do not think He contains anything bad in Him or can be defined using terms applicable to the things and beings I will endeavor to explain in this book. Many books have already been written about God, and those who are interested in knowing more can consult them. Furthermore, because my goal in this book is not to demonstrate the origin of God or engage in some religious debates, I felt like I better focus on the formation of the living things found on Earth. In this book, I chose not to dwell too much on living things that may exist on other celestial bodies beyond the Earth, for the Earth is not the only place hosting life.

Because some of these features are sometimes contradictory, to fully meet the requirements of its, or his, or her potential, the universal holistic being must be able to change forms and functions as needed. For instance, while some organisms are very small, others are very large. Because the universal holistic being cannot be small and large at the same time, it has the ability to change into small and large organisms as it wishes or as the conditions in which its precursor found itself imposed or

TURBULENT ORIGIN OF LIFE

demanded. In other words, the universal holistic being is capable of transforming, transfiguring, or becoming anything possible in the realm of spectra of the living beings. As of today, although some people may not believe in the ability of certain things and organisms to be converted into anything if the right instruction is given to them by and with the right power, all at the right time. For instance, those who are well versed in spiritualism including witchcraft know that it is easy for a human being to take the form of an owl or that of other birds. Some prophetic people and some animists know very well that everything in nature has a form of spirit or spiritual component, which, to some extent, can be manipulated to change them or cause changes to other things and beings in the universe. I have witnessed human beings do supernatural things, causing things to appear and disappear, fixing things supernaturally, defying natural laws, to the point that, as for me, I know that meristems of plants and stem cells of animals are not the only things that have the ability to become anything. Even adult human beings can be transformed into anything if the right instruction is given to them. Likewise, to a much higher level, in the beginning, there was a program that had the ability to produce anything out of any matter it could act on. As I explained in the first chapter of this book, conventional scholars have limited the definition of life to 6 groups of organisms (plants, animals, fungi, protists, archaea, and bacteria) indeed, but here, I expanded it to all forms of beings including spiritual beings. The rest of this book is basically about explaining and demonstrating how the turbulent program of life acted on matters to form the various types of living things present in the universe today and also those that are extinct.

CHAPTER 17: TURBULENT PROGRAM OF LIFE AND MY DEFINITION OF LIFE

CHAPTER 17

CAN YOU ACCURATELY DEFINE AND DECODE LIFE BY IGNORING THE TURBULENT PROGRAM OF LIFE THAT IS MORE ADVANCED THAN ANY COMPUTER PROGRAM?

17.1. Turbulent program of life

Just as life is run by a program, in the beginning, life was also formed by a program I termed the "turbulent program of life". Indeed, the turbulent program of life is a program that formed the matrix or combination of all possible functions, features, and characteristics that living organisms (known and unknown to human beings) can take to express themselves. The turbulent program of life acted by means of a biological split-gathering of the matters of the precursors of organisms into organisms.

As the term indicates, the biological split-gathering is a set of processes that explain how the precursors of organisms were split and collected together as they were being formed. Just as a series of split-gatherings of fluid layers of the precursors of celestial bodies led to their organization into various organized systems of bodies (galaxies, stellar systems, planetary systems, asteroid systems, chemical particles, etc.), so the way the original biological clusters of matters were split-gathered (as can be seen during developmental processes today) explains how each organism reached its current body plan and characteristics, which, as of today, is reproducible via procreation and others means of germline "perpetuation". Here, by "original biological cluster of matters" I meant the matters on which the program of life acted on to produce organisms.

Any cluster of matter on which the life program acted became alive. I defined the turbulent program of life as a mysterious biological program which, at the beginning of the formation of life, ran matters to cluster them and impart onto them different shapes, abilities, and qualities I explained above in such a way that, the matter it acted upon became alive, capable of performing biological functions according to the extent of the impact it received. The variation of the impact of the turbulent program of life explains the diversity of the forms of life that resulted from its actions. The similarities and dissemblance of the formed living things explain why some people ended up classifying organisms into different groups to which various names were given.

Because anything upon which the "life program" acted gained life regardless of its shape, some clusters of matter that can hardly be identified as neither plant nor animal were also empowered to become living things. Examples of these are some algae, sponges, and mosses, which scientists have been struggling to properly classify and identify. This phenomenon is more abundant in aquatic environments, where various forms of life are found. Some marine organisms behave or look like both plants and animals, and others as neither plant nor animal. As the forms of life were being shaped on the outside, internal changes and arrangements were made to the same matter. In the end, both changes and rearrangements occurred on both the outside and inside of the organisms.

Just as inside galaxies there are stellar systems inside of which asteroid systems and planetary systems are found without forgetting chemical elements constituting them, so also as the precursors of the forms of life were being shaped by the biological turbulent program of life, various types of clusters of biological systems were formed inside of them in many inclusive fashions. For instance, as organisms were being shaped, different systems of bodies or apparatuses were organized at the same moment that the individual organs constituting them and the different components and macromolecules forming these organs were being molded. In other words, the split-gathering of the turbulent program of life assembled body parts and abilities inside the matters used to form organisms. Putting this another way is to say that different kinds of ability and power were given or granted to everything in nature from the nonliving to the living, according to the type of turbulence that their precursors went through. This was the act of an original supernatural, spiritual, and energetic force or power behind the existence of every natural thing or being.

The turbulent program of life affected two aspects of the formation of life: morphological aspect and physiological aspects. Hence, I refer to the morphological turbulence as the turbulence which defined the morphology of organisms and the physiological turbulence as the turbulence which defined the physiology of living things.

17.2. My definition of life

Based on everything I said since the beginning of this book and much more you will see in the rest of this book, I felt on October 20, 2021 that here would be a good place and time for me to lay out my definition of life before I continue addressing how living organisms were formed. A living organism, living thing, or a living being

CHAPTER 17: TURBULENT PROGRAM OF LIFE AND MY DEFINITION OF LIFE

is an entity (physical, spiritual, and of unknown nature) made of an arrangement or a set of the characteristics of the universal holistic being under the influence of the turbulent program of life. In other words, a living thing, or an organism (spiritual or physical) is an expression of the turbulent program of life. Just as a DNA can be expressed into mRNA (and other types of RNA), which can be translated into proteins, which can also take different forms, structures, or configurations, so also the turbulent program of life was expressed to produce different organisms. Another way to illustrate this is that, just as the same initial matter in the universe was split-gathered to produce various celestial bodies under the influence of the mother of all turbulences, so also various substrates (e.g. water and soil) were split-gathered under the influence of the biological turbulence and the turbulent program of life to produce the plethora of organisms in the universe. The objective of the rest of this book is to detail how life originated. Toward the end of this book, I will revisit this topic and handle the forms of life before birth and after death.

CHAPTER 18

WHY AREN'T SCIENTISTS ACROSS THE GLOBE THINKING ABOUT THE SPLIT-GATHERING OF THE UNIVERSAL POOL OF ABILITIES IN LIVING AND NONLIVING THINGS BEFORE WASTING BILLIONS OF DOLLARS ON LIFE-ORIGIN RESEARCHES WE DON'T NEED?

As I was reviewing the distribution of the qualities or attributes of living and nonliving things in the universe, it appeared to me that they could have been split-gathered over the things and beings in the universe, however, none of these creatures have all of these attributes. For instance, some wild animals are able to sense things that most human beings cannot. For example, long before they reached land, tsunamis have been detected by some animals, which, for some unknown reasons, knew a danger was coming and began seeking higher ground before the water came crashing into the land, while human beings had no idea danger was heading their way and in fact, when some saw the ocean receding, they ran toward the ocean out of curiosity and unfortunately lost their lives when the water came rushing in. Some animals can see very far (e.g. eagles), while others can smell things located very far from them (e.g. some big carnivores). To illustrate, a herd of bison often smell the wind to detect predators that may be hiding ahead of where they are headed. Some animals can perceive or detect remote things including radiation. Others can even see in the darkness (snakes, raccoons, and opossums), others use echolocation to measure distance (e.g. bats). Some wild animals work very hard (e.g. ants), others are very wise. Some animals are clean (e.g. goats, sheep, and cows), while others are very strong (crocodiles and lions).

Some organisms can live in water only (aquatic organisms), but cannot live in the air, and vice versa. Others can live on land only (land organisms), but not in the air

CHAPTER 18: SPLIT-GATHERING OF THE UNIVERSAL POOL OF ABILITIES IN LIVING AND NONLIVING THINGS

or water. Some can live in both water and land (e.g. amphibians) but they cannot fly, but jump. Some can fly and walk, others can only swim. While some animals are benign and contain compounds that can benefit others, some are very dangerous and deadly, even having poisonous venom (e.g. snakes, scorpions, and some spiders) that can kill others immediately, yet they have the poison within themselves, but do not die from it.

Even when I looked at the physiology of plants, animals, and other forms of life, I understood that they (even the poisonous ones) have unique abilities. Even snakes which are generally poisonous, have some talents like wisdom that can be envied. In fact, despite their mischievous traits, snakes are very wise to the point that even in the Bible, Jesus advised His disciples to be wise as serpents and innocent (harmless) as doves and sheep (Bible's Book of Matthew 16:10). If you did not know that snakes are very wise, you may want to do some research on how they behave very carefully, thoughtfully, and according to their abilities. Some plants are so beautiful that human beings had to move them from the bush and domesticate them so humans can enjoy seeing, smelling, or using them for various purposes. Likewise, some animals are so beautiful and endowed for instance with a fantastic voice that human beings enjoy listening to them. What can I say about some songbirds? In fact, it has been known in some parts of the world that some people use birds in competitions to hear them sing.

As far as the nonliving things are concerned, each one has unique capabilities. Either they are beautiful, or they are useful for something specific. Although some things and beings are toxic of course, everything in nature has a reason to exist and most things have something to be praised for (unless their original state has been significantly denaturalized). We may not know why they exist, but everything in the universe deserves a place and nothing should be considered useless. What can't I say about the usefulness of water, gold, diamonds, carbon, oxygen, nitrogen, phosphorous, sulfur, copper, and many other precious materials to living things? What about the beauty of some landscapes and natural habitats that some people have to pay a high price just to visit?

Before I close this segment on the abilities of wild organisms, I must say that, just as nothing is equally distributed in the universe, as much as some wild animals have some outstanding qualities, on the other end of the scale, others lack them. Hence, some animals are foolish meaning they lack wisdom (ostrich), and others are lazy. Some still are very dirty (e.g. pigs, cats, and dogs although most people consider them as the best pets or even "man's best friend", while in other cultures, people find those animals very delicious yet others detest them). Some cannot walk, fly, or swim (like fish) but just crawl (e.g. snakes). While some have bones, others do not, hence, some cannot properly stand as others can. In other words, even among the organisms, some are limited by the qualities that others have, yet they all are called living organisms. Likewise, body parts and functions are not equality distributed. Some organisms are deficient in certain abilities and parts that others have. To some extent and/or beyond some levels or limits, even the most precious things in nature can become toxic and harm others.

As far as human beings are concerned, it has been proven that they are born with various talents or abilities on top of which some can educate themselves to be able to perform certain tasks according to the stage of the civilization they live in. Just like I explained for wild organisms and nonliving things, while some human beings are evil, or wicked, others are very gentle and filled with gracious abilities. Efforts were made to discover and classify human strengths and talents into groups and it was showed that, no single human being is able to perform every task at the top level of perfection. In other words, some people have more of some specific talents than others, and vice versa. Some people can perceive certain things in the future accurately (prophets or seers), others can minister to their peers with different gifts. In fact, people are given different combinations of gifts or talents as they are born. Although some people can switch more easily from one talent to another, no one can be the best at all talents or gifts known to humankind. Anyone who will say that he has done everything to the uttermost level all the time is just boasting. Even when I consider some religious people who believe that they can do anything through their God who strengthens them, I have come to realize that God has not empowered any human being to do everything. Only God is able to do everything more than any human being can think.

Considering the qualities that most human beings seek in one another and even in plants, animals, and in other forms of life, I also realized that no single organism contains them all. For instance, some organisms are more resistant to certain diseases and environmental stresses than others. Some are more productive than others. Some are resistant to drought, while others are resistant to wet climates. Some are resistant to heat, others are resistant to cold. Some human beings are more beautiful, handsome, gentle, loving, caring, peaceful, smarter, selfless, kind, honest, sympathetic, empathetic, responsible, courageous, and patient than others.

To shorten a long observation I have made, I felt like the program of life split-gathered different good qualities into the created things in such a way that if all good qualities available in all living and nonliving things can be gathered into a single being, that being will have an ability next to that of God. This can be why, in nature, it is hard to find an organism that has all attributes. In contrast, what on organism lacks is compensate by something else. Organisms are atrophied in one thing but then hypertrophied in another thing. In other words, the hypertrophy of one thing seems to "compensates" the atrophy of another thing. For instance, some people think that ostrich is a bird better fit for running at the expense of flying. Put another way, people think that that the ostrich's ability to run very fast is a compensation of its inability to fly. People also use this fact to explain why penguins have fins that allowed them to swim, but they cannot fly. To some extent, the morphology is a response to the local environment. But the initial morphology of living things was defined by the biological turbulence that formed the first specimens of their kinds.

Although I understand that some people do not believe in God, and some will never do so despite the evidences that could be provided to them, I felt obliged to

CHAPTER 18: SPLIT-GATHERING OF THE UNIVERSAL POOL OF ABILITIES IN LIVING AND NONLIVING THINGS

say a few things about God here. Despite the divergence of their belief, most people who believe in God say that He is strong, smart, powerful, etc. Indeed, God is smart enough not to put all of His abilities into a single thing or being. Hence, He distributed part of it into His creatures so that He could fill the Earth with His power and have different creatures representing part of who He is. In the end, the universe declares the glory of God. Yet, human beings, even those who have studied nature, failed to realize that, behind everything in the universe, there is a fingerprint of God, and we should seek and honor God more than honoring and praising the things He has created. So no one can come before Him on the merit of what he or she had, God, in His majesty, power, honor, glory, and many more attributes, has deprived every single human being and everything He has created or formed from something (regardless of how gifted that person or thing may be). Therefore, no human being or any other living thing (including the holy angels) have all the skills or talents to qualify before God without His grace. If you want to learn more about my viewpoint on God, check out my book called "*Science180 Accurate Scientific Proof of God*" in which I presented a scientific argument for the existence of God.

On September 22, 2021, I figured out that, no single organism has the "universal pool of qualities", but the "universal pool of qualities" was split-gathered differently in the organisms according to various factors including their spatial location and size of their precursor. Putting this fact another way is to say that the program of life (which gave different forms of life to the organisms in the universe also) conferred on them different sets of abilities taken from the "universal pool of qualities" during the split-gathering of the precursors of these organisms. In other terms, as the precursors of organisms were being shaped into their daughter bodies, the "universal pool of qualities" was distributed and limited differently according to some characteristics of the precursors of life.

Even if I combine all of the qualities found in the organisms of the universe, I found out that they are not the total or the sum of the attributes of God (as revealed in the Bible). In the end, except for the Supreme Being which some call God, who is omnipotent, no organism or even human being has all of those attributes. From human beings to the smallest form of life, something or an attribute is always missing. No created thing is 100% perfect! Even animals called clean do not have all of the aforementioned attributes found among the living things. For each organism has just a share or portion of the universe of attributes or overall abilities or functions findable with the living things in this world. Hence, I believe that, during the formation of organisms, attributes were also split-gathered differently among living things. Even inside each organism, all of the cells, organs, and apparatuses are not endowed to perform the same function.

During the formation of organisms, some body parts may not have appeared instantaneously at the same time as if every organism was fully formed in the blink of an eye (although I cannot exclude that this could have been the case for some organisms). Just as an egg can go through developmental stages to become a fully mature organism ready to be borne, so also, although extremely fast, components

TURBULENT ORIGIN OF LIFE

of organisms were split-gathered on various scales (e.g. cells, tissues, organs, organelles, apparatuses, whole body levels, etc.), under the influence of many codes including the biological turbulence code and the mission codes, which I will explain shortly. When some bodies were being formed, the turbulent program of life could not have been expressed beyond certain limits because the biological turbulence had to stop for reasons including the lack of room to develop the turbulence further and other limits imposed by the characteristics (e.g. size) of the matter being used. In other cases, the biological turbulence continued its course and let more attributes of the program of life to be expressed or imparted into the bodies being formed. Because some of the attributes or characteristics of the program of life excluded others, all organisms were not able to have everything or perform everything. Yet, some were endowed to act in ways that may seem contradictory under some circumstances. That is the case of some organisms like amphibians which can live in both water and land. The same thing can be said about bats which can fly in the air like birds, yet they are mammals having other traits that most birds do not have. Those who do not understand these mysteries try to break down and classify the anatomical and physiological functions of organisms as if they were acquired during a very long process that caused the so-called evolution to occur. Yet, organisms were formed very fast and mostly independently from one another just as I explained that the origin of the celestial bodies in the universe cannot be understood by thinking that any of the current ones is the source or sink of others.

CHAPTER 18: SPLIT-GATHERING OF THE UNIVERSAL POOL OF ABILITIES IN LIVING AND NONLIVING THINGS

'Science180 Academy' Success Strategy
SCIENCE180 PUBLISHING: AUTHORS WANTED

Science180 Publishing, the American publishing company that published the groundbreaking discovery about the origin of the universe, of life, and of chemicals spearheaded by Dr. Nathanael-Israel Israel, really wants to publish your book(s) regardless of your field of expertise. This is a unique opportunity for:
- established authors
- people aspiring to become authors
- people who have written a book or are wanting to write one and need help with anything regarding publishing
- people who are not well known, inexperienced
- people whose books are viewed as nonconformist, controversial, or unconventional
- people who do not have enough resources or knowledge to navigate the publishing process
- people who are struggling to find an affordable, experienced, and high-quality publisher

Although Science180 Publishing is based in the USA, it can publish your books within your budget regardless of your geographical location. Science180 Publishing is highly interested in your document and possibly helping you publish it. Please visit Science180Publishing.com to explore how we may assist you. No matter the content of your book, as far as it is original, not promoting anything illegal, not duplicating anyone else's idea, Science180 Publishing can help you publish it in the USA. Please contact us asap and see how we can help.

To start your journey of publishing your book with Science180 Publishing, please visit Science180Publishing.com today.

CHAPTER 19

WHAT RAPID PARADIGM SHIFT CAN WE EXPECT FROM THE DEMONSTRATION OF THE TURBULENT LAW OF MISSION AND CALIBRATION OF LIFE? DON'T SAY I DID NOT WARN YOU!

19.1. Turbulent law of mission of life
On November 29, 2021, it appeared to me that every life has a purpose defined and limited by the level of expression of the "turbulent program of life" or the processes that formed it. On that day, I also realized that the biological turbulence (which acted on the matter used to form organisms under the influence of the turbulent program of life) did not split the matter used to form the body anyhow, but according to a law that considered the mission of the beings to be formed.

Indeed, as I carefully looked at and investigated everything in nature, I realized that, from the nonliving things to the living ones, everything has a specific function. Even things and beings that some people may think are useless exist for a reason. While some ecologists may argue this differently, some religious people may also want to handle the mission of living things differently. Therefore, here, I will use a language that can strategically relay all viewpoints. For instance, as I extensively showed in my books on the origin of celestial bodies (e.g. *"Turbulent Origin of the Universe"* and *"Origin of the Spiritual World"*), if the size and other characteristics of the Sun and the Earth were different from what they are now, life (as known on Earth today) could may have been impossible, or it must have been different. In other words, every submicroscopic, microscopic, and macroscopic matter in our environment is made for a reason according to laws that shaped nature. Similarly, as I was looking at the diversity and characteristics of the known organisms, I felt like their organization and functioning must have been scaled according to laws that respect the nature of the chemicals and spiritual entities they are made of. To put it

CHAPTER 19: TURBULENT LAW OF MISSION AND CALIBRATION OF LIFE

another way, if organisms were made to disobey natural laws (which bind the so-called nonliving world as under or by a yoke), they could not have been sustained in their environment. Ecologists would have a lot to say about the relationships between organisms and their environment. Putting these complex issues another way is to say that the biological laws (that have made and still allow organisms to function) hide encrypted messages of the processes that formed the matter and everything else in the universe. Hence, organisms are shaped and scaled differently according to the characteristics of the physical and spiritual entities involved in the molding of their precursors. Here, by precursors of organisms, I do NOT mean any living organism that has ever lived and evolved into any currently known or extinct organism, but I mean the original matters that were quickly mixed with spiritual entities to shape all organisms on Earth and beyond (for life is not limited to Earth only). Some of the processes that took place during the reproduction of organisms today could have taken place quickly or even instantly for the first time that the original or first organisms were formed. In other words, the reproduction of organisms in the universe today contains footprints or fingerprints of some processes that could have taken place during the formation of the first-born or first-made individuals of the organisms. However, because no organism was formed at the beginning yet, nothing like mating or copulation (as done by some forms of reproductions today) did not happen at the beginning before things were formed. It was after organisms were formed that reproduction was enacted as part of the strategies to perpetuate the existence or progeny of the organisms until the time when most of them will be extinct. For, because no organism of Earth can live forever without being affected by the environment, which also participates in their aging, reproduction has been a key that allows some organisms to have descendants until today. And because the environment has also been changing, adaptive abilities allowed some organisms to survive some challenging conditions that could have exterminated them.

When I looked at the organization of the living and nonliving things, I felt like their attitude, characteristics, abilities, and all other things that can be said about them are for a purpose. Those who studied animal ecology can testify about how the hierarchy and interactions between organisms and their environment are highly aligned with nature to such a point that it would be unthoughtful to deny that nature and its inhabitants have a purpose. Although some people have no clue about what exists beyond the Earth, or even beyond their city or country, we can all agree that without a certain number of plants, animals, and other forms of life on Earth, human beings could not have achieved what they did.

Did you ever think about what could have happened to human beings if the plants that we feed on could be able to run away from us every time we want to harvest them, or if they could scare us as some big carnivores ready to devour our life? Did you ever imagine if the seasons can refuse to occur in their due time but make us pay some fees before they give us some heat, cold, rain, or allow the plants to grow so we can consume them? Did you even wonder what could have happened to us if the Sun refused to give us its light but make us pay a high bill before it shines on us? Did you ever wonder how we could be living if gravity does not even allow us to walk by either making our legs so heavy like a ton of bricks or by making our weight so light that we could be flying in the air like leaves or feathers blown by a mighty wind, or floating in the air like what astronauts orbiting the Earth in the space station experience for months (sometimes without knowing the long-term impact on their health)?

Unfortunately, some people have gotten so accustomed to the privileges they have on Earth and the free things offered to them by what they wrongly call "mother nature" that they neglect to reflect on their own mission on Earth just as they fail to acknowledge how the so-called lowest things and beings (e.g. wild animals and plants) around them are fulfilling their mission to us usually for free so much so that we take their service for granted, as a right or something not worth pondering on. Sometimes, because they fail to understand the mission of organisms around them, some people think that some organisms are not good enough and that some must have evolved from others (to become better), and organisms they think are less complex or less advanced are wrongly viewed as inferior, or as things still evolving to attain their ultimate maturity. What a mistake, display of ignorance, and ungratefulness from the part of those who think like this! This way of thinking is like that of leaders who think that they are more important than their followers or the people who helped them get that power and without which they could not stay in their so-called powerful position. What power does a political leader have which is not in the hands of the people? Which kind of people are smart enough to know the outrageous flaws in their leaders and still stay obedient and faithful to them?

Because the things and beings that human beings call inferior do not usually respond to the wrong label and profiling by which they are being erroneously called, some people think that the so-called inferior beings do not hear what we are saying while others think that those entities are accountable to us and must obey our rules even the contradictory ones. Why are we and will we be surprised when many organisms become extinct and that our ecosystems are being degraded to an extent that the planet is increasingly becoming inhabitable without forgetting the climate that is changing and to which some people pay attention without properly thinking about the cause of all that, which I put under the umbrella of human ignorance and misunderstanding!

Most of those who seem to understand something about life embrace wrong missions, which are not usually aligned with the holistic order among all things! Hence, some people scale their actions, intentions, understanding, lifestyle, professions, and other activities beyond their measure. For most people fail to know that the coordinated diversity and order in nature is hidden behind an encrypted, proper scaling of everything according to a measure that is not always equal, but very fair, so that everything formed in nature can exist in alignment with its environment, missions, opportunities, and constraints of its environment, which sometimes host spiritual entities that most people usually neglect to properly access,

CHAPTER 19: TURBULENT LAW OF MISSION AND CALIBRATION OF LIFE

assess, and appreciate. To shorten a complex observation, without seeing life through the glance of a system of properly-scaled missions of various entities, and processes including turbulence (biological, chemical, spiritual, physical, etc.), it will be impossible to fully understand the origin and fate of life and of the whole universe. My investigation on the origin of the celestial bodies helped me a lot to better appreciate the importance of the scaling and environment of formation of things on their diversity, forms, functions, and other characteristics. For instance, because of a little variation of the intensity of the turbulence and other characteristics of the precursors that shaped some celestial bodies, significant differences can be seen in their characteristics including their appearance. Likewise, I felt like differences (even small variations) in the biological turbulence and the size of the matter used to formed organisms could have had a significant impact on their anatomy and physiology.

I also came to understand that, the laws that shaped and arranged organisms were not under the power of those organisms. In other words, the laws that formed the organisms were what defined them. And because these laws must respect their foundation, which was partially related to the laws that formed the universe (when the organisms were not formed yet), it was needed that all organisms must be properly scaled so that their size, missions, functions, and other characteristics could holistically fit into the global or universal perspective. Else, the world would have been denuded of meaningful forms. For although some people think that nature is very chaotic, which I came to understand is an expression of the ignorance of natural laws by some human beings, it is very important to notice that nature is highly organized to a level that the problem of its understanding by some human beings is what is chaotic. Ignoring these facts, some people try to fit the relations, similarities, and differences between organisms into a limited logical umbrella according to which everything must pass through a certain stage of so-called development, survival, or maturity, before becoming what some human understanding wants them to be. Yet, the diversity of the forms of life in nature points at the fact that anything is possible and anything can start living if a certain "dose" or "amount" of the program of life is endowed to it accordingly. That is why some very tiny organisms (which could not even be seen with the naked eye) have life just as some gigantic organisms (e.g. whales, elephants, and sequoia trees), which also have life, although their biological functions and characteristics are different.

All organisms do not have the same number of organs or body parts. Even those that do, they do not have identical proprieties. This is because the biological turbulence that took place during their formation also restrained the expression of the turbulent program of life, which in the end was also affected by the mission of each type of organism. Even among individuals of the same species, the expression of life can be different. All human beings do not act the same. For, we have different missions, were formed or grown under different conditions, and we adopt different lifestyles, etc. Likewise, while some organisms (e.g. most animals) have a head, others such as plants do not. Even among those which have a head, its composition is not the same. The brain in the head is not the same from one species to another and sometimes, not even from one individual of the same species to another. Among the species, the organs of the brain are designed and wired differently. Hence, a variation of some brain-controlled behaviors of some animals. Furthermore, some people choose to live their life the way they want, even if their choices can hurt them.

Some animals have bones, others do not. Even among those who do, the size of their body parts is different and are calibrated accordingly. A mosquito cannot have a head as big as that of a human being, and vice versa. From one kind of organism to the other, significant variations exist even when the same organ is considered. Sometimes, because of the similarities of the calibration or scaling of the processes involved in the formation of some organisms, resemblances exist between some organisms. In other words, organisms found in different environments and which have never been connected can look similar to some extent if the processes that formed them were also similar. To put it another way, anatomical and physiological similarities between organisms is much more a matter of the similarities of the processes, ingredients, environment, or circumstances of their formation than a matter of descendance. What I illustrated above about the anatomy and physiology of the head can also be applied to various organs among the organism.

Before the publication of my work on the origin of the universe and everything in it, including living things, most scientific efforts to explain life have turned around genetic codes. However, there are other codes that work together with the genetic codes. Spearheaded in this book, these less-known codes include the turbulent code and the mission code, which together have defined the expression of the turbulent program of life in all kinds of organisms and other entities in the universe: plants, animals, spiritual beings, celestial bodies, etc. Because I addressed the turbulence that formed the celestial bodies in other books (see *"Turbulent Origin of the Universe"* and also *"From Science to Bible's Conclusions"*), I will here focus on the turbulent code and mission code as they pertain to only organisms known to most human beings, leaving the details of the spiritual beings (e.g. angels) for another time. The mission code implies also the limits imposed on things according to their nature. Also, it is important to understand that some organisms have added things to their original mission in such a way that some organisms that were supposed to be only herbivorous became omnivorous, eating all kinds of things, including meat of organisms of their own kinds, while other organisms downplayed their mission by acting below or beneath the standards endowed to their kinds such as some human beings who belittle their potential and live lifestyles beneath their power, while others put themselves above others as if human beings have been given a power to crush others and promote only people looking, acting, thinking, believing in the same things like them!

For instance, although some organisms share sequences of genetic codes, their genes are not always expressed or used the same way, meaning forming the same size of living matter performing the exact functions all the times. For beyond,

CHAPTER 19: TURBULENT LAW OF MISSION AND CALIBRATION OF LIFE

beneath, and even before the genetic, epigenetic, and all other known biological codes, there was (and there is still) a mysterious turbulent code and mission code, which did not and cannot allow the same program of life to be enacted and reproduced genetically the exact same way from one organism to the other without a calibration according to size and other initial conditions. What I am saying is that the turbulent code and mission code were and are still above the genetic code, which some people think is sufficient to linearly or branchingly explain life. These turbulent and mission codes also explain the wide diversity of characteristics of the celestial bodies across the universe, yet, these bodies perceived as inert (although they may be living according to other standards that human beings ignore) do not seem as under the influence of a traditional genetic code, even though they have a way to maintain and adjust their characteristics to changes trying to affect them. In *"Turbulent Origin of Chemical Particles"*, my book on the origin of chemical particles, I explained how chemical particles (subatomic, atomic, molecular, and compound) hide a mysterious code, which I termed as an abiotic "hereditary" code. All of these turbulent codes and mission codes have not been understood or discovered before because the scientific community had not properly comprehended turbulence neither on the astronomical scales (where massive amounts of data have been collected on celestial bodies) nor on the microscopic and macroscopic scales (despite the tons of data collected on chemical and biological entities). Therefore, although turbulence is present in organisms all around us just as in the "nonliving" systems of bodies in the universe, nobody has clearly seen it until now. Now that the turbulent codes and mission codes are revealed, I expect a lot of satellite discoveries to be made very soon and a significant paradigm shift in many scientific disciplines.

Under the umbrella of molecular biology, so-called biotechnological "advancements" may have well made people believe that it is just ok to extract a gene from bacteria and introduce it into a human being, and vice versa, because these genes may look alike. But what most people do not know is that, the many complications in life, human health, reduction of life expectancy in some so-called developed nations despite so-called scientific advancements, and the appearances of various "new" forms of deadly viruses (even in this age in which some people think we know a lot) can be linked to how little we know about the scaling of life and biological systems. Even common sense tells us that the same engine part of a motorcycle cannot work in a vehicle. Even among vehicles, the same body part cannot work well (or even at all) from one brand of vehicle to the other. Yet, biotechnologists have tested things on mice and then implemented them on human beings as if a test that works on a Nissan Infinity must also work on every other kind of vehicle, or as if a light bulb that works in a torch must also work for a human eye, or as if a monkey's brain transplanted into a human head can produce the same intellectual result! By the way, early 2022, a team of American surgeons transplanted a pig's heart into a human's body for what they called "the first successful pig to human heart transplant" in the USA", but about 2 months after that surgery, the patient died. Afterwards, more calls were made to do more of such similar experiments. What an ignorance or a misunderstanding!

To summarize, organisms (e.g. plants, animals) are what they are (morphologically, physiologically, spiritually, etc.) because, under the influence of the biological turbulence, which shaped them, the turbulent program of life was aligned and scaled according to the characteristics of their mission and environment, which, at its turn, was formed according to laws dictated by the mother of all turbulences, which prevailed during the formation of the universe. The missions of organisms have limited and empowered them according to several factors. The missions assigned to each organism have been causing them to respond differently to the abilities and other characteristics of the program of life. Hence, some organisms have some organs that are not functional such as some animals' eyes which cannot see. Others have ears but cannot hear. Some birds have wings but cannot fly. Even some human beings have some morphological organs that they cannot use to the ultimate degree. Blind people have eyes but cannot see. Deaf people have ears but cannot hear. Some crippled or disabled people have legs but cannot walk. The challenge I addressed in the book is to illustrate how the turbulent codes and mission codes have been expressed to form the various organisms present in nature in a way "similar" to how the mother of all turbulence explains the origin of the universe.

The Bible for instance says that one of the missions of green plants is to be eaten by human beings, who initially were not supposed to eat meat, but only vegetables and nuts. We now know that without plants, the oxygen supply that most organisms depend on may run dry. The Sun and Moon are meant to illuminate and to dominate the Earth. Without the Sun, plants cannot photosynthesize, for the solar energy is needed to power the biological reactions needed for the formation of organic compounds. And without plants, most animals would die. Yet, human beings walk on plants, cut them without wondering if they feel any pain, and act as if they are superior to plants in every area, not knowing the power that plants have to feed the beings that eat them. I am not saying that human beings should not eat plants, but they are unaware of the mission of these autotrophs. Likewise, some carnivores would do everything to eat the herbivores in their lands until one day, those herbivores are finished, and consequently, those carnivores may have to move or die, unless they can go back to eating grass, which was supposed to be the primary food for most animals. Looking at their abilities and capability to use their brain to invent and innovate, human beings were supposed to dominate everything on Earth. Yet, some humans are under the influence of nature and are lacking to properly express or manifest themselves. In other words, although human beings were supposed to dominate (hence they have more abilities), most of them lack strength, endurance, patience, self-control, love, kindness, perseverance, and other crucial virtues, which should empower and allow their inner being to express better phenotypes on the outside. In the end, some wild animals (e.g. carnivores) tend to dominate some ecosystems that some human beings are even afraid to visit. To

CHAPTER 19: TURBULENT LAW OF MISSION AND CALIBRATION OF LIFE

make a long story short, each organism has a specific mission that many usually fail to discover. Inside each organism, cells, tissues, organs, organelles, and apparatus also have their own missions, which performance is sometimes impaired by some so-called scientific advancements trying to engineer or "improve" life that they ignore. If I can talk about spiritual beings, I could say that some of them also fail in their mission, transfigure themselves, taking wrong assignments, which, together with some human errors, are working to paralyze or destroy other forms of life in the universe. Later in this book, I will revisit these realities, which cannot be properly understood by only sticking with physics and other branches of modern science. Before closing this segment on the turbulent codes and mission codes, I would like to humbly ask you if you know your mission and assignment in this life. For nobody is born by chance, and everybody has something specific to do during his or her lifespan for the benefit of all. Whether it is becoming a president one day or the doctor that may find the cure for cancer or whether it is the stay-at-home parent that gives a good education and instills great morals into their children, every human being has a mission in life.

19.2. Turbulent law of calibration

When I looked at the morphological and physiological aspects of organisms, I felt like the living components of organisms were calibrated according to their missions and the environment in which they were placed to fulfill or execute their mission. In other words, the characteristics (including the mission) of organisms were calibrated or attuned to the environment where they would live their life. Although some people think that the environment defined the form of life, I realized that the forms of life were made according to the missions they can fulfill in their environment. In other words, the environment did not make organisms, but organisms were made to meet the requirements of their environment so that the constraints of the environment do not impair their abilities.

The processes involved in aligning the lifestyle of organisms with their environments is part of what I termed the "turbulent law of calibration". In other words, body parts were calibrated according to their missions in the environment they were made. By calibration of living components, I meant a set of laws that defined the size, morphology, physiology, and other properties of organisms and the components of their body parts according to their missions and the conditions (including environmental ones) in which they were formed. The previously mentioned calibration is not just about setting the size, surface, volume, energy, abilities of the organisms, but about setting everything pertaining to the organisms.

In *"Turbulent Origin of Chemical Particles"*, my book on the origin of chemical particles, I explained how the chemical elements that constitute living organisms were not selected by chance. Indeed, most organisms are dominated by carbon, hydrogen, oxygen, nitrogen, phosphorus, and many other elements. In my book on the origin of chemical particles I detailed the abundance of chemical elements in human beings and on many celestial bodies in the universe. For instance, oxygen is the most abundant chemical element in human beings (61%). More than 97.9% of the mass of human beings is made of nonmetals. About 94% of the mass of human beings is made of 3 nonmetals: oxygen, carbon, and hydrogen. By decreasing order of their abundance, the 15 most abundant chemical elements in human beings are:

1. Oxygen: 61%
2. Carbon: 23%
3. Hydrogen: 10%
4. Nitrogen: 2.6%
5. Calcium: 1.5%
6. Phosphorus: 1.1%
7. Sulfur: 0.2%
8. Potassium: 0.2%
9. Sodium: 0.14%
10. Chlorine: 0.12%
11. Magnesium: 0.027%
12. Silicon: 0.026%
13. Iron: 0.006%
14. Fluorine: 0.0037%
15. Zinc: 0.0033%

When I studied the characteristics of these chemical elements, I noticed that their choice ensured that life could exist without too much hindrance. To put this another way, each chemical element plays a specific role in human beings and its abundance was fine-tuned for a mission.

Just as the planets in the Solar System are not the same, yet formed after similar processes under different initial conditions, so also the forms of life in nature are different yet formed under the influence of the biological turbulence, which caused the turbulent program of life to produce different beings according to several parameters. Just as the distance between the celestial bodies is not the same, so also the distance between body parts (at the cellular, tissue, organ, organelle, and apparatus levels) is not the same. For those distances were calibrated by the processes that took place as the components of each system were being formed, each according to its turn and position in the precursor of the precursor of each system of bodies. Even on the spiritual level, human beings are not the same. Some have more spiritual gifts than others, but in the end, what matters the most is how each of us uses his or her gift to improve his or her life and that of others around us and many other things associated with life.

The law of calibration caused the size of the organisms to depend on the turbulence of their precursors. In other words, the size of the precursors of the organisms affected the expression of the turbulent program of life. Likewise, the characteristics of the components of the living organisms were affected by the size of the precursors of the organism. For example, the size of the organs of animals depends on the type of turbulence that the matter used to build them went through. The head of an elephant is thousands of times bigger than thousands of heads of

CHAPTER 19: TURBULENT LAW OF MISSION AND CALIBRATION OF LIFE

crickets. The size of the head of an elephant is correlated with the size of its body. The size of an elephant does not and cannot fit the body of a cricket. The eyes of an owl (night bird) for instance are thousands of times larger than that of an ant. All of this is because the turbulence and the size of the matter used to form the organisms affected the size of the organs that they ended up being made of.

CHAPTER 20

GIGANTIC ERRORS MADE ABOUT LIFE-ORIGIN AND HOW THIS SCIENTIFIC FORMULA ACCURATELY PROVES THE GENERIC TURBULENT PROCESS OF THE FORMATION OF ALL FORMS OF LIFE ... (IS THERE ANY NOBELIST NOT DEEPLY SHOCKED YET?)

My goal in this chapter is to break down the processes that I came to realize took place during the formation of all forms of life. I will be giving specifics about the biological split-gathering of the matter used to form living things. I will also shed more light into extraterrestrial beings or living things, including angels, which some people have encountered.

20.1. Turbulent programming of life

As I carefully investigated the universe, I understood on December 8, 2020, that the turbulent nature of life is encrypted in the turbulent code, which, to some extent, is found not only with the living things but also with the so-called nonliving things. Because I deeply investigated the turbulence of the precursor of the celestial bodies, I learned a lot of things that I applied to decode the biological turbulence. For instance, I realized that, just as an abiotic program of turbulence was run through the turbulent prima materia (the original matter in the universe) to produce the precursors of bodies, which in the end were split-gathered into various celestial bodies and chemical particles, so also on the scale of living things, a biological program of turbulence was run throughout the waters in oceans, seas, and rivers (precursor of aquatic beings) and the soil or dust of the Earth (precursor of terrestrial living things) to split-gather them into the precursors of different living organisms according to their types. Just as for the celestial bodies, intermittence

CHAPTER 20: GENERIC TURBULENT PROCESS OF THE FORMATION OF ORGANISMS

existed and caused the formation of small bodies between bigger ones, so also for the living organisms, small organisms were formed between large ones. Hence in the same field, giant organisms like elephants were formed, while organisms smaller than ants and the currently known microorganisms were formed as well. In the waters, in the air, in the Earth, and on the Earth, various organisms were formed.

Just as a computer program can be written and after launching it, it runs by itself and executes all of its codes, the biological systems and beings in the universe are running today under the influence of turbulent programs coded during their formation. Some of those programs have been maintained and/or functioning under programs of reflexes, while others are influenced by the will of humans and spirits, but everything is under the control of a primary plan. In other words, the whole universe was formed under the control of a main program, which can never be found just by using scientific means only, for the mystery of life is above modern science. For the things "known" through scientific investigations are just a scratch on the surface of the deep reality inside the substance of everything. In other words, physics on which most human beings tend to focus is a limited aspect of the massive iceberg of spiritual things. Unfortunately, most human beings think that the spiritual world excludes the physics, whereas in fact, the physical world is just a tiny perception of the surface of some of the visible aspects of the spiritual realm.

Just as cells are believed to carry hereditary information in their DNA, so also, they carry turbulent information in the way the fluid layers of their precursors are and/or will be stratified during the biological split-gathering leading to their birth. In the fluid layers of the precursors of celestial bodies, structures including vortices and other structures of various forms and shapes existed. But when it comes to the fluid layers of the precursors of organisms, instead of just vortices, structures like precursors of cells and precursors of different body parts were formed inside of them, vortices and other forms of turbulent structures could also be found as well. Before my work on the mother of all turbulences, very little was known about turbulence. Throughout this book, I explained some of the programs involved in the formation of living things. Coined by me, turbulent developmental biology aims at describing the processes involved in the turbulent origin of diverse living things in the universe. The turbulent architecture defines the turbulence basis for the differences between the living organisms in the universe.

20.2. Split-gathering of the precursor of organisms into precursors of primary bodies (largest bodies) and precursors of secondary bodies (smaller bodies)

On the morning of Saturday, September 25, 2021, a lot of ideas raced through my mind concerning how, in the precursor of some animals, the program of life was not executed to its utmost level, therefore making some animals unclean and others clean. During my time on bed that morning, my mind turned toward how the Sun is very big, yet less dense. Likewise, the giant planets (e.g. Jupiter and Saturn) are very big but less dense than the terrestrial planets. Although there are many terrestrial planets (e.g. Mercury, Venus, Earth, and Mars), only one (the Earth) is habitable. In other words, among all of the celestial bodies in the Solar System, only one is "clean" and perfect to host human beings and the other forms of life found on Earth. Here, when I say that the Earth is "clean", I mean that it is the only one fit for the kind of life it hosts, for no other celestial bodies in the Solar System can accommodate all of the forms of lives found on Earth. The Earth is not the largest celestial body, yet it is the most suitable for us. The other celestial bodies in the Solar System are not fit for life as known on Earth, yet they play specific roles for the maintaining of such life. For instance, human beings cannot live on the Sun, yet they could not live on the Earth without the Sun. Likewise, many living things are unclean, but they play some roles that benefit others. Ecologists can say a lot about how no living organism is as useless as some people ignoring their value may think.

On the scale of living things, I found that the clean animals (see chapter 28 for more details) are not always the biggest of the others in their family or kind. The clean ones are neither the smallest. For instance, sheep, goats, cows, and deer, which are example of clean animals, are not the biggest or the smallest mammals. Some of the largest mammals are hippos and elephants, yet they are not clean. In fact, the largest mammals that ever lived on Earth are dinosaurs, and none of them is still alive today, implying that their bodies did not survive some of the conditions that the current living organisms (or their ancestors) have survived. In other words, size is not synonymous of cleanliness or of longevity. Else some giant tortoises could have been the cleanest animals. Among the fish, the largest ones are not those that have fins and scales. Else, whales and sharks could have been clean, but they are not. Some of the largest fish do not even have bones. Carp and tilapia, all of which are clean animals (according to the standards in the 11th chapter of the Bible's Book of Leviticus) are not the biggest fish. Some of the unclean fish such as catfish are bottom eaters and are very dirty although some people love eating them so much. Carp, tilapia, and many other clean fish live in the water, on the top part of the water, and do not spend most of their time in the mud or eating dead animals in the water. What can't I say about some animals like lobsters, shrimp, crabs, all of which the Bible classified as unclean, yet, some human beings across the globe prey on them to the extent that some restaurant brands are even named after some of these animals? When it comes to birds, chickens are examples of clean ones, yet they are not the largest birds, nor the smallest ones. Some eagles, ducks, and ostriches are bigger, stronger, and faster than chickens, yet for the reasons I explained above, they are not among the clean.

I came to realize that the largest animals in each kind (fish, birds, mammals, and insects) are like the primary bodies and the largest secondary bodies formed by the processes or program of life that split-gathered portions of matter to form different kinds of animals. For instance, just as all stars are not the same, so also the largest animals in each kind are not the same. For instance, if I consider the mammals, there are many kinds and the largest of them vary. Furthermore, just as the size of the largest secondary bodies in a system of bodies varies, so also the size of the largest secondary living things according to their kinds varies. Hence, among the mammals

CHAPTER 20: GENERIC TURBULENT PROCESS OF THE FORMATION OF ORGANISMS

for instance, there are many kinds of large mammals. Examples of the largest mammals include: bears, elephants, rhinoceros, buffalos, giraffes, whales, etc. Concerning the fish, sharks are some of the largest ones. When it comes to the birds, ostriches, eagles, hawks, geese, and ducks can be named as some of the largest ones. These largest animals according to their kinds are like primary bodies and their largest secondary bodies. This does not mean that the largest ones are the ones from which the smallest ones emerged, but that as the clusters of matter used to form the animals were being split-gathered, the precursor of the largest animals may have been the leftover of the primary body according to each cluster. Just like we have many stellar systems with several planets and asteroids around, many large animals and the several smaller animals (and other smaller forms of life) around them were also formed according to their kinds. Just as the Sun is a star and its nature is different from planets and asteroids, yet they were made from the same precursor of the Solar System, which was split-gathered into the precursors of several celestial bodies, so also when living things were formed, the matter they contain were once just soil and/or water (according to the environment) that was split-gathered into different clusters that became precursors of various organisms having similarities and differences. This also implies that organisms that physically look completely different may still have been formed from a same cluster of matter. What I just explained for animals can also be said for all the other forms of life. For instance, when I refer to plants, some trees (like the sequoia and the baobab) are larger than others.

In *"Turbulent Origin of the Universe"*, I showed that the precursor of the celestial bodies was the "turbulent prima materia" (the original or initial matter in the universe). But in the case of living things, the precursor of matter used to form them was not the "turbulent prima materia" but the matters already present in the environment they were formed. Hence, the precursor of most land organisms (animals, plants, and other forms of life found on the Earth) was the soil taken from the Earth, but the precursor of aquatic animals (organisms found in the water) was the water. Considering the similarities between fish and birds, the precursor of the birds was water. Just as the precursor of the Solar System was split-gathered to form various celestial bodies with different characteristics, so also a portion of soil and of water was split-gathered under the influence of the program of life to yield various forms of living things in the water, on the Earth, and in the air.

At the time the Earth was about to put forth or bring forth living organisms, a biological turbulence took place under the influence of the program of life, which imparted living abilities to the matter being gathered together according to their position and how their precursor was split. Creatures of various kinds were formed, having different abilities and other characteristics according to their location and/or position in the cluster of matter used to form them.

Just as on the scale of the celestial bodies, stars are surrounded by planets and asteroids, some of which have their own satellites, so also on the scale of living things, some very large ones were formed and surrounded by relatively fewer large ones and many smaller ones. Just as the number of the large celestial bodies is smaller than the smaller ones, the larger organisms (according to the forms of life) are fewer than the smaller ones. For instance, there are less bigger animals than smaller animals and less tall or giant plants than smaller plants. In most natural ecosystems, you will come across many small plants before finding big or large plants. The same thing goes with animals. In a forest or savannah, smaller animals usually outnumber large ones. Even with human efforts to select or nurture which plants and animals they want on their farms or in their forests, savannahs and other ecosystems, smaller organisms usually outnumber larger ones. This is the consequence of the turbulence that shaped the universe since its beginning and which imprint was also strong during the formation of life. As it was 9:50 AM on the morning of that September 25, 2021, I needed to stop writing what was flowing in my mind, for it was my day of rest, a day I usually rested from scientific work, but I had to write these things down for they were racing in my brain, and I could not restrain them. This inspiration contributed to helping me to see the formation of life through the glance of a biological turbulence which fingerprint bears similarities with the turbulence which shaped the celestial bodies and chemical particles.

20.3. Turbulence zones in biological turbulence

The turbulent program of life followed a turbulence law according to which different organisms were created according to the intensity of the turbulence that animated the precursor of their body. In other words, in some turbulence zones, large amounts of soil and/or water were split-gathered together by the turbulent program of life to form living things. In the end, some living things were formed very big, while others were shaped very small. This is true not only for animals, but also for plants and other types of beings. It appeared to me on September 17, 2021 that the largest animals (e.g. dinosaurs, elephants, whales) were formed in the most developed biological turbulence zones of the clusters of soil or water that were used to form them and other animals. For the turbulence of the precursor of the celestial bodies, the most developed turbulence was found in what I termed turbulence Zone 3. However, for living organisms, the characteristics of the biological turbulence were different, but I will not deepen them here. However, considering the turbulence of the precursor of the universe, and the similarities and differences I discovered between the turbulence in living and nonliving things, I divided the biological turbulence zones into 5 zones:

- Zone 1 where smaller organisms were formed
- Zone 2, a transition zone leading to the developed turbulence in Zone 3
- Zone 3, where the turbulence is the most developed, and which birthed the largest organisms between which smaller ones were also formed because of the law of intermittence
- Zone 4, a transition zone out of turbulence Zone 3 and
- Zone 5, a zone of minimum turbulence and located outer of all other zones.

CHAPTER 20: GENERIC TURBULENT PROCESS OF THE FORMATION OF ORGANISMS

While I expect the largest organisms to be formed in the biological turbulence Zone 3, I also expect smaller living ones (including microorganisms) to be formed in less turbulent zones or between larger bodies. It is not by chance that microorganisms are prolific, for some of them are for biological systems what hydrogen is for celestial bodies and chemical elements. In other words, microorganisms were formed at many places just as the smallest chemical element (hydrogen) is present almost everywhere in nature.

All birds, mammals, reptiles, fish, and insects do not have the same size. Some birds (e.g. eagles, ostrich) are very huge, while others (e.g. hummingbirds) are very small. Some reptiles (e.g. boas or pythons) are very huge, while others (e.g. worms) are very small. Some insects are gigantic, while others are tiny. Some fish (e.g. great white sharks) are huge while others (e.g. trout, minnows, or guppies) are small. It is all about the intensity of the turbulence used to form them according to the size and position of/in the cluster of matter upon which the turbulent program of life acted. In other words, the size of the clusters of matter affected the outcome of the organisms born from them; and the characteristics of the organisms was affected by the position of their precursors in the clusters of matter that born them in association with others. To some extent, this is similar to how the position of the fluid layers of the precursors of the celestial bodies affected the characteristics of their daughter bodies.

Moreover, some plants are tall, yet their leaves are tough. Others are down to the ground, yet their leaves are coriaceous. That is the case of some Crassulaceae also called stonecrops. Plants that have large leaves are not always those with the softest leaves. For instance, the leaves of the giant cola (*Cola gigantea*) and the teak (*Tectona grandis*) are large, yet they are dense and lignified. Rotproof woods (woods which are hardly attacked by insects) usually hard. An example of such plants is *Prosopis africana*. Although it belongs to the family of palmaceae, which is not considered as having a typical wood, the African fan palm (*Borassus aethiopium*) has a very good wood used for woodwork and capable of lasting hundreds of years without problems. In other words, the relationship between the size of plants, leaves, and their density is not absolute, but relative. Yet, this relativity hides the code of how life was formed.

At the level of an organism, the head is like the primary body. The neck is like the portion of space separating the primary body and its innermost secondary body. The outer part of the neck can also be viewed as part of Zone 1. Just as some systems of celestial bodies do not have bodies in all turbulence zones, so at the individual levels, all organisms do not have body parts in all turbulence zones. For instance, in most fish, the neck, which represents the junction between the head (primary body) and the rest of the body (considered as secondary bodies), is less pronounced or even nonexistent. The shape of a fish does not even make it easy to know where the neck is. Likewise, although most wild animals have a tail, some animals (e.g. human beings) do not have a tail, although the position of the tail can be said to be around the same place it is found on monkeys. To shorten a long story, just as all turbulence zones are not present in all systems of celestial bodies, likewise, all biological turbulence zones are not present in all organisms. Hence, some organisms look symmetric and do not have body parts in all turbulence zones. The shape of algae for instance looks like a baguette, not like an organism having a turbulent belly. In the end, on the scale of a single organism, the size and the nature of body parts formed in the biological zones defined the shape of the entire body.

Moreover, as I was deeply reflecting on the biological turbulence zones, it seemed to me that even the lifespan of human beings is organized like a series of turbulent events connected to what, on August 5th, 2021, I called "familial turbulence". For instance, to some extent, children in a family are like products of fluids layers born according to their rank. To explain what I mean, I would like to say that, the order of the birth of children is related to the order in/of the split-gathering of the reproductive fluids away from their parents' reproductive organs, the main difference is that they are usually born about 9 months after their precursor leaves the genitals (testicles and ovary) of their parents. If they are produced through artificial insemination, the length can be longer. After their birth, they also live like other types of lives during which phases are passed as stacks of fluid layers being split according to their position. From the baby stage until death, passing by the adolescence stage, adulthood stage, and the end-of-life stage, many behavioral changes can be noticed. I showed that, the trouble of puberty and middle age crisis are activities of turbulence Zone 3 of a human being's lifespan. When I see humans being more turbulent in their puberty years, a period that gives a lot of headaches to parents, than in their late years or young years, I felt like this behavior respects the fact that turbulence is more developed in Zone 3, which is neither the innermost or outermost zone just as the middle age is neither the youngest age nor the oldest one. In other words, the trouble that young adults have in life is like the behavior seen in turbulence Zone 3 leading to turbulence Zone 4. Life between 50-80 years old is like turbulence Zone 4, while beyond 80 years old is like turbulence Zone 5. In other words, old people are usually slower because their speed has decreased just as that of the bodies in turbulence Zone 5 does. Today, those who reach 120 years of age or older are like the outermost bodies on the current lifespan scale. In this dimension, death is like another split-gathering during which the flesh is separated from the spiritual being, which layers will also later be reorganized according to their destiny. The cries of babies at birth and those associated with pain as people die are marks of split-gathering into and out of this world as layers of life are being separated and gathered differently. Most separations in life are painful! The icons below (Fig. 23) are a reminder of some of the aforementioned stages of turbulence during the lifespan of human beings ranging from reaching puberty, becoming a teenager, being a single adult seeking a mate, getting married, getting pregnant or impregnating someone, becoming a parent, and having babies to care for, to reaching old age, while the family grows.

CHAPTER 20: GENERIC TURBULENT PROCESS OF THE FORMATION OF ORGANISMS

Fig. 23: Selected milestones during the lifespam of most human beings

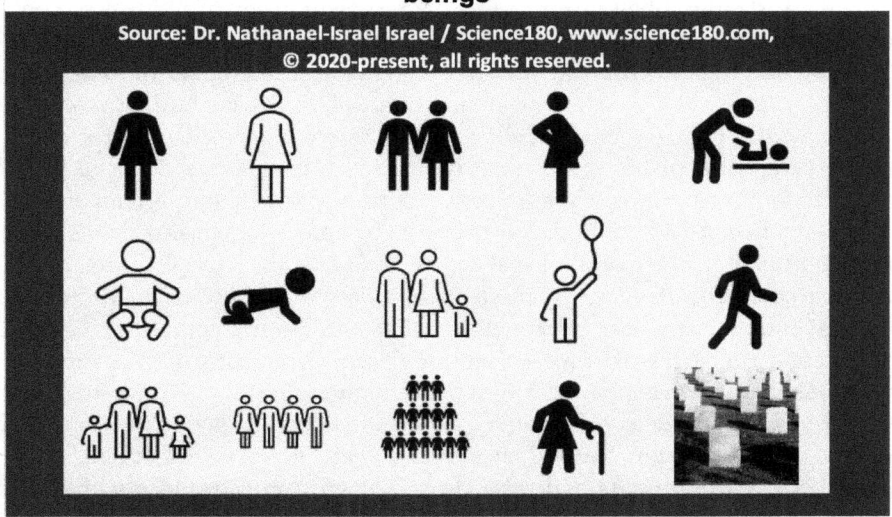

20.4. Provisioning of chemicals during the formation of the architecture of organisms

Life cannot be properly understood without decoding the turbulent architecture of every living component (from the subcellular levels to the organism level, passing by the organelle, cell, tissue, and organ levels). Epigenetics has revealed that genes are not the only means through which heritable traits can be transmitted from one generation to the other. Likewise, I also found out that the turbulence code allows the transmission or "inheritance" of physical and biological characteristics by organisms and even so-called nonliving things without the mediation of genes. While the matter used to form the celestial bodies was carried from one position to the other by the processes that split-gathered the original matter in the universe, for living things, I discovered that the matter used to form them was carried by processes that ejected matter as precursors of secondary bodies were escaping the precursor of their primary body. The expression of genes and the chemical particles deriving from alimentation (i.e. provision of nourishment such as food and drink, and other necessities of life) and breathing, all together, help living organisms to get the raw material they need to build their bodies today. Today, some organisms are able to eat because they already have a system to digest their food. But before they were fully formed, organisms had no way to eat, meaning that alimentation was not the main means by which chemical compounds were furnished to the precursor of the bodies as they were being shaped. I felt like, in addition to turbulent means including shear flow and other movements of fluids under the influence of a biological turbulence, supernatural mechanisms were involved in the movement and provisioning of matter to the precursor of organisms as they were being formed.

TURBULENT ORIGIN OF LIFE

Just as of today, a seed is able to germinate, withdraw the chemical elements it needs from its environment to grow and become an adult producing offspring, so also during the formation of all forms of life, the turbulent program of life acted on the matter in the environment like a seed of life, which was able to selectively uptake or withdraw from the environment what was needed for each organism to be formed according to the size of the cluster of matter that will form each body and according to other factors I already introduced. This was true for plants as well for animals. Because each form of life has distinct chemical composition, I deduced that the biological program of life was very selective in the types of chemicals it incorporated into the architecture of the bodies formed. Even among organisms formed in the same environment, the chemical makeup of their bodies is usually different. The uptake of chemicals elements by the turbulent program of life to build organisms must have been very fast so much so that the "germination", growth, and reproduction processes of some organisms (e.g. plants) that can take months to occur could have even happened within a few minutes if needed. Today, the growth rate of plants for instance is defined by factors including the genotype of the plant, the atmospheric pression, the soil characteristics including water content, nutrient content, etc. Some bacteria and other forms of life can reproduce within a few minutes. These facts suggest to me that if some key factors can be changed, plants can take very less time to complete their cycle, therefore implying that a certain tuning of the turbulent program of life and the environmental conditions during the formation of the first specimens of life could have sufficed to form life very quick, even within days as other sources suggested.

20.5. Why a turbulent neck, belly, arms, legs, and tail?

In a previous chapter, I talked about turbulent neck, belly, tail, arms, legs, etc. Here, I will explain why organisms have a turbulent neck. I initially labelled this segment as "why do some animals have a long neck?", but then, I realized that even nonliving systems of bodies like stellar systems and planetary systems also have their own turbulent neck. Therefore, to be inclusive, I changed the title to reflect every type of turbulent neck.

As I woke up from my bed on Saturday October 16, 2021 around 7 AM, a day I was supposed to rest as has been my habit for years since I became a Messianic believer observing many Jewish traditions (e.g. the feasts of the Lord) that some uninformed people (including gentile Christians) think are obsolete, I got a flow or rush of ideas in my mind as I was still laying on my bed, trying to massage my hand and upper legs which, for some reason, had hurt me a lot in those days. Because some people may wonder what the feasts of the Lord mean or are, I would like to list them here:

- Feast of Passover (Pesach) celebrated on Nissan 14,
- Feast of Unleavened Bread celebrated on Nissan 15-21,
- Feast of First Fruits (corresponding to the celebration of the Resurrection of Messiah) celebrated in Nisan,

CHAPTER 20: GENERIC TURBULENT PROCESS OF THE FORMATION OF ORGANISMS

- Feast of Shavuot (Pentecost) which is 50 days after the Feast of First Fruits,
- Feast of Trumpets on Tishri 1,
- Yom Kippur (Day of Atonement) on Tishri 10, and
- Feast of Sukkot (Tabernacles), a week-long feast which begins on Tishri 15.

For more details, please consult "Reconciling Science and Creation Accurately" and

"Turbulent Origin of the Universe" in which I scientifically demonstrated how God created the universe in 6 days. Anyway, on the morning of October 16, 2021, it appeared to me, and this was not the first time either, that it is not by chance that a specific distance separates the Sun from its innermost celestial body orbiting it, which, as of 2022, is said to be Mercury, a planet. A distance also separates the planets from their innermost satellite. On the scale of living things, a certain distance usually separates most primary bodies from their closest or innermost secondary body. For instance, on a tree, a distance separates the base of the trunk or main stem from the first branch. With animals, a neck usually separates the head from the rest of the body. Most animals lacking a neck are reptiles (e.g. snake) and they do not have a much larger body part downward of the position that is supposed to be the neck. Later in this book, I will explain why some animals lack a neck, while others have a long neck, and how the length of the neck relates to the genesis of these animals.

In general, the head is like the primary body of animals and the neck relates to the distance travelled by the fluid layers of the precursor of the secondary bodies (located below or downward of the head) before it started splitting into larger secondary body parts. This does not mean that the head was completely formed before the other parts, or that the neck contains no body part, but that, just as in the case of the Solar System for instance, some planets were formed before the formation of the Sun was completed, so also the precursor of some upstream or upward organs could have been split and formed before that of the head finished its formation. As any comprehensive developmental biology would reveal, all organs and even their precursors were not formed at the same time. As the embryo of living things grows today, some organs are born, while others continue to grow. Take for instance a baby in the mother's womb. Once the baby is conceived, the baby does not automatically have all of its body parts on day 1 of conception, but generally over 9 months, the baby fully develops what it will need to survive outside the womb. Like I already explained in the chapter on lessons I learned from developmental biology, early during the formation of life today, layers of cells are formed at a certain stage, and the way they multiply and split-gather into body parts defines the final morphology of the resulting organisms. During the process, the cell layers that would later birth the head are different from that which would birth the belly, etc.

For plants, the root system is like the primary body and the distance separating the upper part of the root or the base of the main stem or trunk from the lowest branch is associated with the distance travelled by the precursor of the shoot system (aerial part of the plant) before starting to split-gather into different parts. Long after the formation of the branches, due to growth and development, the distance separating them increases of course, implying that the distance separating parts of a plant is not always that travelled before these parts were formed. Just as how celestial bodies are expanding and the whole universe is expanding, so also the distance I mentioned above for living things are expanding and are not exactly the same as they were when the first innermost secondary body split.

As I was thinking about this on Saturday October 16, 2021, my mind was turned toward the shape of a mushroom. It immediately appeared to me that the vertical hyphae looking like the main "stem" of a mushroom usually appears relatively long because the precursor of the secondary bodies (the upper part above the "root" took longer before producing the mushroom head, which at its turn was not even able to split and give many layers of daughter bodies separated by increments of distance. Instead, as the mushroom head was about to split, it just expands and gives layers looking like a dome but which are unable to stack one on top of the other so branches can be found like in other plants. Hence, mushrooms do not branch out like most plants do. Below are examples of turbulent necks (Fig. 24). Snakes seem to have no neck although its spine may show something like a neck.

Fig. 24: Examples of turbulent necks
Source: Dr. Nathanael-Israel Israel / Science180, www.science180.com, © 2020-present, all rights reserved.

CHAPTER 20: GENERIC TURBULENT PROCESS OF THE FORMATION OF ORGANISMS

I also noticed that, most animals with a long neck are among the largest in their families. Among mammals, giraffes and dinosaurs which have long necks are among the largest or tallest. Most animals that have a small neck are short. Some animals that have no well-defined neck (like snakes) are long and do not have a large belly, but a body which size is almost the same all throughout from the head to the tail. For instance, snakes have no neck and look like bodies which precursors were formed in a turbulence Zone 5, where small bodies are formed and things appears tinny. The longest snake (30 feet long) in the world is said to be the reticulated python (*Python reticulatus*), which is native to southeastern Asia and the East Indies. Yet, that snake does not have a big belly like some mammals. The belly of some snakes is noticeable only after they swallow a big animal. Some snakes are able to swallow an adult human being, yet their body cannot by itself appear big anywhere, but just looks long without the appearance of an extrusion, bulge, or bump anywhere. For the precursor of serpents did not go through a very strong biological turbulence to the point to yield a bump in their body as a witness or manifestation of turbulence Zone 3 on the scale of the entire body. This also explains why they don't have legs and arms as most other animals.

As I kept reflecting on why some animals have a long neck, it appeared to me on November 26, 2021 that a certain distance was needed to be travelled before the precursor of the living organisms could be amalgamated to a point that allowed a developed turbulence capable of forming the thorax and the abdomen which, among other things, comprise the belly, which is usually the largest part of most animals' body. In some previous segments, I explained how the neck is like a portion of body separating the primary bodies (e.g. head) from the innermost secondary body like the chest and the arms. Its length can be associated with the processes that took place before the innermost secondary body was formed. In other words, the length of animals' neck was defined by the processes and time it took for the cells that were differentiated into the precursor of the body parts downward of the head to elongate before being split-gathered and differentiated into bodies which size is higher than that of the neck. Put another way, the longer the cell division continued and cells must have piled up before the moment of their branching occurred, the longer the neck of the corresponding animals could be. If the precursor of the neck escaped the precursor of the head very fast, it could have traveled a longer distance before its turbulence could allow the formation of larger body parts downward of the neck. Likewise, if the speed of cell division and elongation of the precursor of the neck is fast, the neck can end up being long as well. The intensity of the turbulence of the precursors, which can also be correlated with the size of the animals, could have affected the length of the neck. In other words, the intensity of the turbulence that was required to form some large bodies could have defined the length of the neck of the resulting bodies, for it took some time before the major biological turbulence could have developed downward of the neck, just as the turbulence in Zone 3 usually needs some time before fully developing. The neck of giraffes and dinosaurs can be long for such a reason. However, just as in the case of some celestial bodies, bodies were formed in Zone 3 without any body being formed in Zones 1 and 2 (such as the case for the Plutonian satellites), so also, for some organisms, the neck is very small, short, and barely connecting the head to the torso or the larger parts (e.g. belly) of the organisms. That could have been the case of some fish and mammals which lack a pronounced neck. Very soon, I will explain how this phenomenon also applies to plants that branch out at the base of their main stem.

Instead of realizing that animals with a long neck were formed like that since the Day 1 of their formation, some people have theorized that long necks were acquired throughout a so-called evolution and adaptation as a way for those animals to cope with grazing needs for reaching the tops of trees, even that of tall trees. For instance, although giraffes are herbivores and their long neck gives them an advantage over other animals, that gain should not be perceived as a product of evolution. Based on the evidences I presented in this book and others I wrote on the origin of the universe, the long neck of some animals was produced by evolution or an adaptation for survival. Unless people understand how life originated, they will never understand why some animals have a long neck, while others do not. The same logic explains why some plants have to grow a tall main stem before having their first branch, while others produce their first branch almost near the soil. Some plants produce their leaves right on the surface of the soil or the base of their main stem as if they do not need to grow their stem high above the ground before branching out. In the chapter on the formation of plants, I will provide more details.

Because the largest bodies are usually found in the most developed turbulence zone, which is usually Zone 3, I felt like the belly can be placed in Zone 3. However, because the belly of animals like snakes is not bulged, I felt like these reptiles may not have been formed under the influence of a significant or developed turbulence. Hence, I felt like snakes and other reptiles that have no leg and crawl like snakes would have been formed under turbulence circumstances different than that of Zone 3. Furthermore, I noticed that some reptiles that go on 4 legs have a more pronounced belly than snakes. Hence, I felt like reptiles which go on 4 legs have been formed under conditions similar to or almost that of Zone 3 for they have a kind of belly which prevents their body from looking like a cylindrical or linear structure. Maybe those reptiles are born from the intermittence of precursors of bodies formed in Zone 3. The small intensity of the turbulence of the precursor of snakes and all other animals lacking legs or arms can explain why their belly is so small and their body usually relatively long. Likewise, the relatively small intensity of the turbulence of the precursors of reptiles that go on all 4 legs explains why their belly is also small compared to that of some mammals. As compared to the length of their bodies or belly, snakes may have a relatively longer esophagus than that of many other animals. It may not be surprising that cattle, which are considered as clean by the Bible, have a large and more compartmentalized belly, implying a product of a more developed biological turbulence. Considering the outermost parts

CHAPTER 20: GENERIC TURBULENT PROCESS OF THE FORMATION OF ORGANISMS

of the bodies of most animals, I felt like the tail was formed in turbulence Zone 5.

Many other animal features can aid in determining whether their precursor went through a developed biological turbulence or not. For instance, like I will explain later, the ability to have fins and scales on the body of some fish is a proof of the advancement of the process of the formation of some fish like tilapia and carp. Likewise, the ability to have divided and completely separated hooves, and the ability to chew the cud is a sign of the mature formation of some mammals like cattle. The ability to jump on 4 legs like crickets is a sign of the developed formation of some insects like crickets and grasshoppers. No reptile is labeled as clean because none of them reached the advanced level of formation attained by some other kinds of animals.

At this point, I would like to explain why the arms of animals are smaller than the feet. Indeed, using the logic behind the turbulence zones, the arms of animals are usually (if not always) larger than the feet because they were formed in turbulence Zones 1 or 2, while feet were formed near turbulence Zones 3. In other words, the precursor of feet inherited, or is influenced by or have a memory of the developed turbulence going on in the belly or the abdomen, hence it formed larger limbs than arms, which were influenced by the lower turbulence in the precursor of the thorax. Hence during embryogenesis, the precursor of the feet emerged near the precursor of the abdomen and were formed after the precursor of some parts of the abdomen (e.g. precursor of the belly) were in place.

The tenderness of some animals' meat and the hardness of some plants can inform us about the type of split-gathering that their precursors went through, for how the matters of living things were packed can affect their density and hardness. For instance, all meats do not have the same tenderness. Some are softer, easier to cook, cut, and chew than others. Some of the parameters involved in the tenderness of meat are the way the fibrous tissues of the meat were connected and intertwined. In other words, the way the matter in meats was gathered together as they were being formed affected how dense or hard the meat can be. In general, some poultry and fish are among the tenderest meat. Some of the fastest animals have a very hard meat (high density). For instance, the lion is among the strongest and fastest animals, but it is not the biggest animal. Likewise, lion meat is very hard or tough. This kind of information can have something to do with the way their tissues were organized during their formation.

20.6. Generic processes of the formation of the forms of life

Considering all of the things I have been explaining so far, here is what I think was the generic processes by which living things were formed on Earth. After the Earth was formed, the precursors of many celestial bodies in the universe were still going through the processes that would birth them. At that moment, although every celestial body was not fully formed yet, life on Earth was formed as follows. Before the formation of life on Earth, water and soil were already present on the surface of the Earth. Unlike the original matter at the beginning of the universe, which was the precursor of the celestial bodies in the universe, the water and soil on Earth were the matter used to form most of the organisms on Earth. I said most of the organisms on Earth because some organisms live on Earth today but were not originally formed on Earth. That is the case of angels, which were not formed on Earth, but which interact with things and beings on Earth today. This implies that life is not limited to Earth only. I will illustrate this shortly. Furthermore, forms of life that were on other celestial bodies or other realms were formed using the matter already present in those bodies. As for the Earth, not all of its soil and water were used to form life on Earth, but part of the them. The portion of the soil and water which was used to form life is a component of the precursor of life on Earth. In other words, when I talk about the precursor of organisms, the very first version of it was the soil and the water used to form them. However, it is important to understand that some beings are not made up of material things only, but also of spiritual things such as spirit, which cannot be defined using conventional chemical or physical means. Before I continue, let me remind you of the chemical composition of oceanic water and the Earth's soil.

Indeed, in my book on the origin of the chemical particles, I explained that oxygen and hydrogen dominate the chemical particles in the oceans. The abundance by mass of the 15 most abundant chemical elements in the oceans (Dayah, 1997) is:

1. Oxygen: 86%
2. Hydrogen: 11%
3. Chlorine: 2%
4. Sodium: 1.1%
5. Magnesium: 0.13%
6. Sulfur: 0.093%
7. Potassium: 0.042%
8. Bromine: 6.7E-03%
9. Carbon: 2.8E-03%
10. Strontium: 8.1E-04%
11. Boron: 4.4E-04%
12. Calcium: 4.2E-04%
13. Fluorine: 1.3E-04%
14. Silicon: 1E-04%
15. Nitrogen: 5.0E-05%

The abundance of the 15 most abundant chemical elements in the Earth's crust is as follows (Dayah, 1997):

1. Oxygen: 46%
2. Silicon: 27%
3. Aluminum: 8.1%
4. Iron: 6.3%
5. Calcium: 5%
6. Magnesium: 2.9%

CHAPTER 20: GENERIC TURBULENT PROCESS OF THE FORMATION OF ORGANISMS

7. Sodium: 2.3%
8. Potassium: 1.5%
9. Titanium: 0.66%
10. Carbon: 0.18%
11. Hydrogen: 0.15%
12. Manganese: 0.11%
13. Phosphorus: 0.099%
14. Fluorine: 0.054%
15. Sulfur: 0.042%

As a reminder, by a decreasing order of abundance, the 15 most abundant chemical elements in human beings are:
1. Oxygen: 61%
2. Carbon: 23%
3. Hydrogen: 10%
4. Nitrogen: 2.6%
5. Calcium: 1.5%
6. Phosphorus: 1.1%
7. Sulfur: 0.2%
8. Potassium: 0.2%
9. Sodium: 0.14%
10. Chlorine: 0.12%
11. Magnesium: 0.027%
12. Silicon: 0.026%
13. Iron: 0.006%
14. Fluorine: 0.0037%
15. Zinc: 0.0033%

Soil and water contain all of the chemical elements found in human beings and in other organisms. However, although carbon is not the most abundant chemical element in soil or water, it is the second most abundant chemical in human beings. We know that, as of today, carbon or organic products are not directly fabricated by human beings and most other forms of life except for plants. This means that, during the formation of the organisms, either a special mechanism was in place so the little carbon in the soil and water could be extracted and inserted wherever needed in the bodies of the organisms or alternatively the biological turbulence which formed life may have been able to form or add new chemical elements as needed, but this time under conditions limited to the matter being formed. Carbon and all other chemical elements needed to form the organisms were pulled from the matter used to shape the precursor of these organisms.

Based on details mentioned in some religious book (e.g. Bible), I felt like some forms of life (e.g. angelic lives) appeared on other worlds before the Earth. For instance, angels were formed elsewhere before life appeared on Earth. Here, I will not dwell on those spiritual aspects, but I felt like it is important to make a note of it so those who are interested in the spiritual beings may know when they appeared. Later in this book, I will provide more details on angelic lives. At this point, let me focus on the formation of life on Earth only.

At the beginning of the formation of life on Earth, a biological turbulence broke out in some water and soil on Earth. The turbulent program of life started acting upon the turbulent structures being formed due to the turbulence occurring in the water and the soil. Quickly, the impact of the turbulent program of life on the entire Earth birthed the precursor of the biosphere which, after going through changes and impartation would become the biosphere, which is the total sum of all of the organisms on Earth. Because of the variation of the environmental conditions on Earth at that time, the turbulent program of life did not act on all of the precursors of the biosphere the same way. Therefore, according to the variations of the changes it was going through, the precursor of the biosphere was split-gathered into different clusters of soon-to-be living things, which became the precursors of ecosystems. Due to their size and the diverse impact of the biological turbulence, the precursors of the ecosystems were split-gathered into the precursors of communities, which were split-gathered into the precursors of populations, which were split-gathered into precursors of organisms and all of these events happened quickly. In other words, by the time of the formation of first specimens of living things on Earth, a portion of the soil and waters (e.g. the oceans waters, the water of rivers, lakes, and all other types of waters on Earth) were turbulently moved so that the precursors of organisms and systems or clusters of organisms could form. Because of their position in the global ecological niche, plants were the first forms of life to be fully formed. This was crucial so that by the time animals and other forms of life that would depend on plants were formed, they could have food to eat. The clustering of some organisms in ecosystems today is a footprint of how some of them were staggered or organized when they were first formed. Later in this book, I will detail how most of the biological components of plants were formed.

As the turbulence was going on in the water, the precursor of aquatic waters was formed. Among other things, the close resemblance of fish and birds made me think that waters could have also been the precursor of birds. Other forms of life besides fish also originated from the waters. But most land animals originated from the soil. As the precursors of organisms were being split-gathered, the largest precursors were those of plants and animals. This is because most of them were formed where the soil and water were most moved by the biological turbulence. I kept using the word "biological turbulence" instead of simply saying "turbulence" to describe the formation of organisms because, unlike what I explained for the formation of celestial bodies, the formation of life requires the additional act of the turbulent program of life, which imparted specific biological abilities upon the precursor of the matters used to form life. In other words, without the role of the turbulent program of life, a turbulence in the waters and at the surface of the soil can yield structures that may even look like the carcasses of some living things, but these structures will have no life. For instance, when animals die, their remains have no

CHAPTER 20: GENERIC TURBULENT PROCESS OF THE FORMATION OF ORGANISMS

life and are like an arrangement of matter denuded of life. Likewise, as of today, scientists can use the exact matter found in organisms to build things, which may look exactly like organisms, but they are unable to give them life. For example, together with artists, scientists can make sculptures of animals and nice artistic plants, which, from afar, may look identical to what those animals and plants look like in the real world, but unfortunately human beings cannot give them life. For life is not just a matter of chemical compounds put together nicely, but its formation and maintenance require a program of life, which was initially imparted through the biological turbulence, which did not just form physical bodies, but also left in them an expression of the turbulent program of life. These processes occurred on the precursors of the organisms in the precursors of all populations, in the precursors of all communities, in the precursors of all ecosystems in the precursor of the entire biosphere of Earth and beyond. Because the characteristics of the clusters of precursors formed by the biological turbulence were not the same from one environment to the other, and because the forms of life which ensued or followed from these precursors were different, all organisms were not borne at the same time. According to the similarities in the processes birthing them, organisms were born at different moments, yet the process was rapid or quick.

To recapitulate, I would say that the hierarchy of life components according to various scales resulted from how the biological turbulence was run through the environment where life was formed. The inclusion and hierarchy of living things such as organelles forming a cell, while many cells can be organized into a tissue, and diverse tissues can act together to form an organ; and many organs are collected together to form an organ system or an apparatus; furthermore, many apparatuses or systems of bodies are combined to form an entire organism, while many organisms can form a population, while many populations can form a community, and many communities can form an ecosystem and all the ecosystems of Earth are termed the biosphere, which is just a way to view the product of the biological split-gathering of the "turbulent program of life" run over the soil and water on Earth at the beginning of the formation of life. In the spiritual world, a similar program was run to form many spiritual beings, but because the scope of this book does not allow me to delve into the spiritual world much, I will not say any more about spiritual realities.

Just as how celestial bodies can be studied on different scales (individual body, planetary system level, stellar system level, galactical level, clusters of galaxy level, etc.), so also, at the ecological level, living organisms can be studied on various scales, including the individual level, population level, community level, ecosystem level, biome level, and the biosphere level. While ecologists define the biosphere (limited to certain areas on Earth) as the largest scale of organization of organisms, or the total sum of all ecosystems on Earth, I know that life is not limited to Earth only, but some organisms unknown to modern science can be living on other celestial bodies and even beyond the visible universe known to scientists. Some scientists including astrobiologists may have interacted with such beings, but they were unable to detect them because these beings are different from those present on Earth. Just as at each of the aforementioned levels or scales of study for nonliving matter, the existence of similarities cannot be sufficient to conclude that one celestial body or system of celestial bodies descended from the others, so also at the aforementioned levels of living organisms, the existence of similarities between living organisms is not sufficient to conclude that one descended from the others. For instance, not because two planets or planetary systems in two different stellar systems (and even in the same stellar system) look alike or have the same internal characteristics that one can conclude that they descended from the same parent or from one another. In the next chapters of this book, I will review the processes which formed most of the parts of living organisms, from the cell, to the entire organisms, passing through the tissues, organs, organelles, systems of organs and all other forms of internal and external organizations of living things. Likewise, I will later revisit this issue as I deal with evolutionism.

20.7. Pregnancy and birth of nonliving things

Before I close this chapter, I felt like I needed to put into perspective how some of the turbulent processes of the birth of the celestial bodies are related to those of the birth of living things. Indeed, just as the age of human beings and other animals is counted from the time they were born, meaning separated from the womb of their mother, so also the age of celestial bodies is counted from the time they split from their mother and were fully formed. This implies that, as long as a fluid layer is in a stack of fluid layers of its mother, it cannot be considered as a celestial body yet. It has to be separated from the stack and take its own path toward its orbit and go through other processes that will complete its formation or developmental processes. Precursors of celestial bodies which were being gathered together were like "embryos" in their mother's womb going through changes, yet, they were not fully formed yet and could not be labelled as a fully mature or complete being yet, until their developmental processes were completed and they were born. In the case of human beings, pregnancy last about 9 months. But in the case of the celestial bodies, the length of the pregnancy of their mother is like the entire time the precursor of the celestial bodies were in a fluid layer not separated or gathered together yet. Even after the layers separated from their mother (as a child being born and separated from the mother's womb or placenta), the precursors spent some time to complete their gathering together before being called fully-formed celestial bodies. It is like an infant who is born with a baby's organs but who needs to go through morphological and physiological developmental processes before reaching his or her own full potential. Just as the stay of babies in the womb of their mother conferred to them intrinsic abilities and programming, which they carry throughout their life, so also the time that fluid layers spent in the stack of fluid of their mother precursors also programmed them to move in specific ways and to have specific characteristics matching their history and the impact of their environment on them. In other words, pregnancy is not found only with living things. For even before they

CHAPTER 20: GENERIC TURBULENT PROCESS OF THE FORMATION OF ORGANISMS

were born, nonliving things had precursors which went through processes similar to those of pregnant living things. But because celestial bodies and chemical particles were formed long ago when no human being was formed yet, and because since the beginning most human beings have failed to understand the origin and formation of the universe and everything in it, they have failed to know that the world went through a period when precursors of everything looked like pregnant bodies programmed to deliver different daughter bodies according to the rules I extensively explained in my books. Nonliving things also gave birth to "offspring", which, though denuded of life as perceived by human beings, have proprieties "inherited" from their mother precursor! When I considered this fact and some creation myths concerning pregnancy, I felt like some ancient people may have had supernatural encounters in their days which may have told them a partial story about the formation of the universe.

Just as living things do not birth all of their children at the same moment, but progressively through their reproductive period, so also, in the beginning of the world, the precursors of nonliving things did not birth all of their daughter bodies at the same time. In my books on the origin of the universe, I showed that it took a few minutes for some celestial bodies to be born, while others took days and even weeks according to their position in the chain of birth of their mother. Just as how all of the children in a family do not have the same age, size, and characteristics, so also the celestial bodies born from the same mother have various properties (some of which are similar, while others are different) which cause the daughter bodies to be diversified. Innermost planets lacking satellites are like some children that are sterile. Because most innermost planets lack satellites, I wondered if the sterility rate of the first children in most families may be higher than that of the children born in the older age of their parents. In other words, the way that the first children were formed could have predisposed them not to reproduce themselves or certain things as their siblings that were born later.

Just as there was a time in the past when no living thing was formed yet, but then came a time when mother precursors were impregnated, then went through their pregnancy, then gave birth to children, which went through stages of development and growth, then matured, senesced (aged) and died, so also there was a time in the past (before time could be counted as of today) when nobody (except those in the spiritual realm that some human beings cannot apprehend in their mere current form) was formed in the universe yet, then came a time when the universe was impregnated with a substance I called "turbulent prima materia". Many precursors of bodies (similar to mothers) were pregnant then and they carried their "pregnancy" for some time before progressively birthing their daughter bodies (which split-gathered) according to their order or position in the lineage of their mothers. Then, those daughter bodies went through developmental and growth processes until they reached their maximum level of maturity. Just as living things age and then die, the universe has already matured and is nearing its death or end. Some death pains for the Earth for instance are equivalent to the increasing wars,

TURBULENT ORIGIN OF LIFE

climate changes (that some people deny), migratory problems (that some people ignore), hunger (that is increasingly happening even in most the most developed countries), earthquakes (which are occurring at an increasing rate), hurricanes, tornados, floods, bad weather, lawlessness, diseases, and other calamities occurring at a growing rate higher than ever seen before. These disasters are like the illnesses some people go through before dying. Now, I will specifically explain how the constituents of most organisms were formed and how each form of life was made for the first time.

CHAPTER 21

'I WILL NOT ANSWER THAT RECKLESS MISTAKE PEOPLE MADE ABOUT THE FORMATION OF MACROMOLECULES': THE 180SCIENTIST HAS NO TIME FOR THIS IRRATIONAL AND NONSENSE ARGUMENT

Living organisms are made of key macromolecules (i.e. biological compounds or molecules having a very large number of atoms); and because the origin of life cannot be addressed without properly tackling the source of these molecules, I wrote this chapter to help you understand how these indispensable molecules were formed and how they fit into the big picture of the beginning of life.

Indeed, in the early days of my efforts to explain the formation of life, the first ideas that came to my mind were about the formation of macromolecules. For instance, more than 95% of the content of this chapter was written in 2013 at the beginning of my exploration of the universe. Examples of macromolecules include:
- nucleic acids (polymers of nucleotides) such as DNA and RNA
- proteins or polypeptides (polymers of amino acids)
- carbohydrates or sugars or polysaccharides (polymers of sugars)
- lipids (sometimes, lipids are not classified as macromolecules)

These macromolecules are crucial for the functioning of organisms. In this chapter, I will relay my understanding of how they could have been formed.

21.1. Carbohydrate formation

Also called sugars or saccharides, carbohydrates fundamentally consist of carbon (C), hydrogen (H), and oxygen (O). As my biochemist teacher used to say it, carbohydrates are like hydrated carbons, which formula can be generalized as

$(C.H2O)_n$ or $C_m(H2O)_n$. They can serve as support, energy reserves, and other crucial functions for organisms.

Because in the beginning the machinery to metabolize carbohydrates were not formed yet, the first carbohydrates in organisms were not metabolized, but were directly formed by bringing together the required atoms constituting them. It was after the formation of the first organisms that the mechanisms to multiply and build body parts or body components were put in place so life can be renewed and built from the template initially formed. Since then, organisms can metabolize carbohydrates and incorporate them into their bodies according to their needs and stages of development. Some organisms that cannot do so have to uptake these carbohydrates in forms of food or drink, and other means of alimentation.

To my understanding, under the influence of the biological turbulence, some atoms of C, H, and O were associated, rearranged inside the precursor of organisms to form carbohydrates. According to the environment they were born and the missions they were being made for, some carbohydrates were tightly combined, sometimes forming compounds like cellulose (highly present in support organs). In contrast, some carbohydrates (e.g. glucose) were loosely combined and easily assimilated by organisms. Some sugars were made into longer polymers than others according to the number of monomers that were brought together to form them. Some carbohydrates such as deoxyribose (found in DNA) and ribose (found in RNA) were also implicated into the formation of nucleic acids. In other words, ribose and deoxyribose are pentoses (meaning carbohydrates made of 5 atoms of carbon) and are involved in the formation of the RNA and DNA respectively. Although the DNA is the fundamental molecule encoding most of the genetic information of organisms, carbohydrates were also formed with them, else, sugars could not be also found in the structure of nucleic acids. It is possible that some carbohydrates could have been formed before some DNA sequences, hence they are found inside DNA. Unlike ribose, the deoxyribose does not have a hydroxyl group (-OH) in its pentose. It is likely that the hydroxyl group was added after the ribose was formed. The chemical characteristics of the atoms involved in carbohydrates allow them to hold and conserve energy while also being able to release it when needed.

21.2. Nucleic acids (DNA and RNA) formation

It was on November 19, 2013, when the things you are about to read in this segment came to my understanding while I did not plan to intentionally think about them. The 2 main types of nucleic acids are DNA (deoxyribose nucleic acid) and RNA (ribose nucleic acid). Although the name nucleic acids sound as if they are found in the nucleus only, they are in fact in other organelles such as the mitochondria and chloroplasts. In other words, mitochondrial and chloroplast DNA and RNA are found in the mitochondria and chloroplast respectively. This also supports the idea that DNA was not made in one organelle and then transported to the others. For all cells contain their own DNA and some DNA are specifically formed in certain organelles only. For instance, when the precursors of prokaryotes were being

CHAPTER 21: FORMATION OF MACROMOLECULES

shaped, they were not able to form the structures that the DNA of some eukaryotes have. For instance, prokaryotes lack a nucleus, hence their DNA is spread throughout their cytoplasm. In other words, the ability to have a nucleus even surrounded by a membrane requires additional processes, which did not take place with the precursors of prokaryotes. Even among the eukaryotes (where a nucleus is found), the structures, size, and compaction or compression of the DNA is not the same from one organism to the other. I will detail the reasons of these trends very soon.

Nucleic acids consist of carbon, hydrogen, nitrogen, and oxygen. They were formed by the polymerization of monomers called nucleotides connected to one another by phosphodiesters, which could have been formed by the reaction between the hydroxyl group. Nucleotides consist of a nitrogen base (also called nucleobase of nitrogenous base), a carbohydrate (or sugar), and a least a phosphate group. Here, the sugar is a pentose (made of 5 atoms of carbon).

DNA and RNA differ by the constitution of their sugar and the nitrogen bases. In DNA, the sugar (deoxyribose) is like a ribose lacking an oxygen atom, while in RNA that ribose has an oxygen atom. Although it can be easy to think that the ribose resulted from the deoxyribose to which an oxygen atom was added, that idea may be wrong. Likewise, people thinking that the deoxyribose came from the ribose by removing or cleaving an oxygen atom may also be equally wrong, for the cleavage of an atom requires some sophisticated enzymes, which activities necessitate the preexistence of molecules even more complex than DNA and RNA. Likewise, the addition of oxygen to a deoxyribose requires the presence of some enzymes that could not have been present if DNA and RNA were not previously formed. In other words, trying to explain the formation of ribose and deoxyribose by the addition or removal of an oxygen atom may cause someone to turn around and fall into a snare like trying to explain which came first between the chicken or the egg. Hence, I think that the first ribose and first deoxyribose used in the DNA of the first organisms were directly formed by the biological turbulence acting on the turbulent program of life on the scale of the precursor of these sugars.

Two types of nitrogen bases are found in nucleic acids: purines and pyrimidines. Purines (e.g. adenine and guanine) have two rings, while pyrimidines (e.g. uracil, thymine, and cytosine) found in nucleic acid have one ring. In other words, purines seem more complex than pyrimidines, and they form stronger and more resistant base pairs. Therefore, it is possible that purines are more advanced, but that does not necessarily mean that they were formed after the pyrimidines.

In RNA, thymine (which is found in DNA) was replaced by uracil. Thymine is like a methylated uracil, meaning that thymine is like uracil to which a methyl group (CH_3) was added. However, this may not necessarily imply that the formation of thymine was after that of uracil. Thinking that uracil was formed before thymine is one of the assumptions causing some people to think that RNA was formed before DNA.

Unlike DNA which has 2 strands of nucleotides, RNA has one strand. In DNA, the purines are connected to the pyrimidines via hydrogen bonds to form base pairs.

TURBULENT ORIGIN OF LIFE

I felt like the chemical elements constituting the nucleotides could have been assembled and more nucleotides joined together to form the monomers, which in DNA were connected to one another to form a double strand. Phosphorous was the connecting element between the nucleotides. Phosphorous could have been the "last" chemical element incorporated into the formation of nucleic acids. Complementary strands of DNA were brought closer so the formation of the double strands could be established. The affinity that phosphorous has for nitrogen (a key element of nucleotide) may also partially explain its incorporation into the nucleic acids.

After the DNA strands were formed, they were folded according to the biological turbulence in their environment. The wrapping of the DNA also helps protect it from damage, from unneeded expressions, and also from compacting the immense amount of information in them in a tinny volume. As nucleic acids were being formed, other chemical elements were also being combined to form other macromolecules (e.g. proteins) and biomolecules indispensable for the formation of the organisms. When the DNA was being folded or coiled to get its final form, some of the protein in its vicinity got incorporated into it. Later, some of those proteins would be involved in the expression of the genes contained by the DNA.

The DNAs of the organisms vary because they were formed under different initial composition and for different roles. For the things that each type of organism will face in its lifespan depend on their environment and what they are. Hence, the DNA which contains a crucial code of life was calibrated accordingly. The first nucleic acids were formed separately before the processes or command to reproduce life was completed. Likewise, the replication of DNA and its transcription into RNA and the translation of some RNA (e.g. mRNA) into proteins and all of the post translational modifications of proteins occurred after life was formed. Nevertheless, all of those processes were formed very fast.

Many types of RNA were formed. Some were messenger RNA (mRNA), others were ribosomal RNA (rRNA) and some are transfer RNA (tRNA), and each of them have specific functions. For instance, mRNAs are translatable into protein, rRNA intervene during translation, and tRNAs help transport some amino acids to be incorporated into chains of protein being produced. Other types of RNAs are required to regulate the expression and functions of some genes. Considering what is needed for a mRNA to be translated into a protein, I don't think the first proteins were made through the currently known translation mechanisms, but they were formed by the biological turbulence under the influence of the turbulent program of life; for some proteins are required to make the translational machinery to work. Hence, I think that the first proteins were directly made by the program that formed life out of the available precursor.

To prevent the chains of mRNA from being attacked and damaged by some enzymes (e.g. RNAase, which are enzymes that can break down RNA), a long sequence of adenine was put at the end of them. Even until today, before mRNA are translated into proteins, a long chain of adenine is always added to their sequences after transcription, knowing that adenine is a very strong base, it appears

CHAPTER 21: FORMATION OF MACROMOLECULES

to me that the system that formed the machinery to be used to reproduce life is well thought.

In summary, to form nucleic acids, atoms of carbon, hydrogen, oxygen, and nitrogen (CHON) could have been assembled to form nitrogen bases. While that was happening, 5-carbon sugars could have been assembled and integrated into the precursor of the nucleic acids being formed. By then, nucleotides were formed. Finally, phosphate groups were formed or brought into the precursor of the nucleic acids to join the nucleotides. The amount of sugar incorporated depended on the type of nucleic acid. In the case of the DNA, the polymers of nucleotides and complementary strands were brought together to form the base pairs, while in the case of RNA, the strands of nucleotides were not paired. Other forms of nucleotides were formed in other biological complexes involved in other reactions. In other words, the nucleotides were not formed in DNA only. Some of the molecules found in DNA were also present elsewhere in the organisms.

The diversity of the conditions which reigned the action of the biological turbulence which took place when the precursor of the DNA was being formed can explain the variation of the characteristics of the DNA of organisms. Furthermore, the law of mission caused some organisms to be endowed with specific abilities according to their environment. Hence organisms living in the poles have ways to cope with the cold winter weather, while organisms living at the equator have their own ways to adapt to very hot conditions.

Why is the DNA of prokaryotes less compacted than that of the eukaryotes? According to the size of some precursors and the biological turbulence that shaped them, some organisms ended up having DNA not wrapped and/or compacted inside a nucleus but have it spread across the cytoplasm. That is the case of all prokaryotes. For instance, bacteria do not have nucleoproteins and a nuclear membrane, which are key components of a chromosome. In contrast, eukaryotes have many chromosomes. In other words, the DNA of prokaryotes is not as compacted or concentrated as that of eukaryotes. This fact implies that the process that squeezed and compacted nucleic acids such as DNA was not as strong in prokaryotes as it was in eukaryotes. To put it in other words, the gathering together of the DNA of prokaryotes was loose while that of eukaryotes was dense. Even along the genome of eukaryotes, the DNA is not compacted the same. Some sequences are more compacted than others, underlining the fact that everything was not gathered together the same. Consequently, because they do not have the chromosomic package like eukaryotes, prokaryotes do not divide mitotically.

When I was studying the celestial bodies, I noticed similar trends but on different scales. Some bodies or even parts of some bodies are more compacted than others. For instance, some planets in the Solar System are denser than others. Within each planet or celestial body, all the components are not compacted the same. On Earth for instance, some places are more compacted than others, some places are rockier than others. It all depended on the factors that squeezed the precursors of these bodies. Likewise, the density of the compaction of biological components (including nucleic acids) can be related to the forces that were at play when their precursors were being gathered together. The more a component (including precursor of DNA and RNA) was squeezed, intertwined, or winded into complex networks, the denser or more compacted the daughter bodies can be. Hence, before the duplication and replication of some DNA, other proteins are needed to unwind the compacted DNA. To some extent, the compaction of a DNA protects it from things that can break or damage it.

Because the DNA of prokaryotes is "less" complex or "less" compacted than that of eukaryotes, some people mistakenly think that prokaryotes have evolved into eukaryotes. Advocates of the so-called bacterial evolutionary theory think that some types of bacteria could have merged symbiotically after entering the bodies of each other and trapping genes. However, considering the turbulence evidences that I uncovered, although some bacteria may have lived in symbiosis with others, I do not think that eukaryotes were formed by the merging of prokaryotes. The DNA is one of the cellular components that was not equally distributed and squeezed or compacted across all kinds of organisms. Some organisms have longer DNA than others, and some have more supercoiled DNA than others as well. The density and the intensity of the compaction, concentration, or coiling of DNA from one organism to the other cannot be sufficient to deduce their origin or descendance. If I can use the example of the celestial bodies, not because stars are less compacted than most planets that one can say that planets derived from stars. Like I explained in other books, including my book called *"Turbulent Origin of the Universe"*, stars and planets in the same stellar system came from a common precursor (which is the precursor of their stellar system) of course, but the precursor of the star and its planets had to go through different stages of split-gathering before yielding the star and the planets. The precursor of the Sun is different from the precursor of a planet. Likewise, although according to their kinds, prokaryotes and eukaryotes can be traced back to a certain precursor, which cannot be expected to be the same at all stages of their formation, the precursor of eukaryotes is different from that of prokaryotes. In fact, during the formation of organisms, there came a point of split-gathering when countless precursors of prokaryotes and eukaryotes were present in the environment where they would be formed. According to their size and localization, each of those precursors were differently split-gathered, yielding body components that were not always the same. Hence, all prokaryotes are not the same nor are all eukaryotes the same. Although the underlying rules that shaped all organisms have some similarities, the way these rules were applied or impacted the matter which was used to form the organisms depended on the size and location of the clusters of matter used. Like I explained before, the turbulent program of life had to accommodate its ability to the size of matter it was subduing to give it life. In other words, the size and nature of the process that formed the organisms also limited the extent of the detail and expression of some functions that the turbulent program of life could confer to clusters of matter it encountered. However, when I deepened the morphology and physiology of organisms, I realized that the size is not the only limiting factor. In fact, sometimes, some smaller living organisms are more complex than larger or bigger ones. Reflecting on this using my knowledge of

CHAPTER 21: FORMATION OF MACROMOLECULES

turbulence, I felt like the nature of the biological turbulence which took place during the genesis of life on Earth played a key role in the distribution or acquisition of body parts (cells, tissues, organelles, organs, apparatus, etc.) by living things according to the characteristics of their precursors and the environment they were born without forgetting their mission. What I explained for nucleic acids can also apply for other macromolecules.

21.3. Protein formation

Although proteins are polymers of amino acids, I felt like the first proteins made were not made using the same systems of protein metabolism found in organisms today. For the synthesis of proteins today required the existence of other proteins first. Enzymes involved in the replication and transcription of DNA into RNA, translation of RNA into proteins work at a very high speed, suggesting that biological systems were made to work very fast. For instance, in an instant, some cells can detect stimuli in their environment and respond to them, by something launching the expression of genes capable of producing compounds corresponding to the appropriate response. Without this ability of organisms to produce proteins fast and respond to some stimuli, life in some environments or under some conditions may have already been stalled by harsh stimuli.

As they were being formed, the initial proteins in the first organisms were also folded according to the conditions of their formation, the sequences of amino acids in their structures. When proteins are not folded, they cannot properly fulfill their functions, particularly when they are enzymes or are involved in interactions with other biomolecules. For decades, the scientific community has tried to understand the law that controls protein folding. I remember that my biochemistry teacher for graduate studies in the USA used to say that the person who will properly explain and predict protein folding will automatically receive a Nobel Prize. In those days, I did not know why such a task was so difficult. It was during my 12 years of investigating the origin of the universe and of life that I finally understood why little is still known about protein folding. For I came to realize that protein folding is highly connected to turbulence, a research field less understood before my groundbreaking discoveries of the mysterious code of the formation of the universe and life, all of which are embedded in turbulence on various scales. I am confident that some of the foundations I laid for turbulence will tremendously help many people to better understand life and some mysteries associated with it such as protein folding and other crucial biochemical challenges. Not knowing much about turbulence and the real origin of life, some people think that protein folding can be addressed by just focusing on mathematical modeling and bioinformatics. Therefore, huge research budgets have been connected with protein folding.

Proteins are polymers of amino acids of course, but the function of a protein is not just based on the primary structure of its amino acids, which is defined by its chemical composition. The functions of proteins highly depend on their quaternary structures and their interactions with other compounds and biomolecules in their environment. For after the translation of proteins, morphological and structural modifications intervene to predispose the protein to fold and interact with other biomolecules in specific ways. Just by tasting the meat of cattle, chicken, and fish, we can understand that although they are all meat, they are different for diverse reasons, including to offer human beings different tastes.

To sum it up, the original proteins in organisms were not translated from RNA but were directly formed by the biological turbulence under the influence of the turbulent program of life. Even the DNA needed to form protein exists in association with other proteins with which they form nucleosomes, chromatin, chromosomes (according to the size and level of compacting) that scientists can confirm with a microscope. Those who ignored these realities concerning the formation of protein wasted their time and resources trying to invent ways to explain how the first protein could have been formed from certain molecules. Some scientists spent their career trying to understand or believe how and why (but in vain) the first RNAs were allegedly formed by proteins. For some of those scientists, RNAs such as ribozymes, which have catalytic activities would have allegedly been formed before proteins. But in organisms, ribozymes cannot fully work independently without being associated with other macromolecules, suggesting that these types of RNAs were not formed in a world which existed before that of the proteins and other macromolecules.

According to the law of chirality, most proteins in nature are left-handed. This chirality is called levogyre in contrast to the right handedness which is called dextrogyre. The handedness of the biological macromolecules (e.g. DNA, proteins) may be aligned with the process which formed them. The direction in which proteins were naturally formed in the beginning and even today by the translation machinery may explain their chirality. This may also explain why most natural chemicals or complex compounds are left stranded (left-handedness). In contrast, proteins artificially produced by human beings are usually left stranded, therefore illustrating how artificial processes invented by human beings cannot perfectly replicate natural ones. I also wondered why more people are right-handed than left-handed? The sum of the forces applied to human limbs could affect the ability of the hands to work in certain direction. Hence, the right hand (which natural movement is aligned with a counter clockwise movement) is more powerful for most human beings. That may be why most people use their right hand more than their left hand. It may be a matter of balance of force(s) or signals translated to the brain, which, in the end, controls a lot of movement of many living things. As a comparison, if the right-handed people can be considered as prograde, the left-handed ones can be viewed as retrograde, and vice versa. While it may appear thoughtful that running fields are usually set for athletes to run counterclockwise (for it matches the natural tendency of movement), it is possible that right-handed people may have an advantage. There are many things that are geared toward right-handed people. Some are everyday things that people see and use such as: scissors, measuring cups, rulers, power tools (e.g. circular saws), watches (some left-handed people may have difficulty winding some watches, for some of them have to take off their watch first). Other examples include most musical instruments (e.g.

CHAPTER 21: FORMATION OF MACROMOLECULES

guitars), zippers on clothing, cameras (imagine trying to press the button to take the photo with your less dominant hand and not making it shake), some school desks, some kitchen utensils like hand held can openers, credit card readers, the number pad on a computer, pencils, and pens with words printed on them (when left-handed people hold them in their hand the words on the pen are in the wrong position to read). The doorknob on some doors and the way the door opens, the left and right click buttons on some computer mouse can be backwards to some left-handed people. On some coffee mugs, the print design is sometimes on one side and when used by a left-handed person the side with the cute print faces away from them or even reads backwards if there are words on it. Even the way books and magazines open up and turn pages is geared toward a right-handed person as left-handed people have to reach across the pages to turn the page, spiral notebooks and 3 ring binders (the metal spirals/rings are in their way when writing). It seems to me that there is balance or general tendency of the movement of things inside human beings. Far from judging people as wrong and right, or better and worse based on their handedness, my efforts here are just to try to explain the origin of things.

Before I close this chapter, I would like to point out that lipids were also formed according to processes that put together their chemical elements. Some lipids were associated with protein and other macromolecules to form membranes, which protect some body parts to have specific compartments in which reactions are vital for their existence and functions to take place. But for the sake of time, I will not delve into the formation of lipids here.

CHAPTER 22

WHAT IS THE ACCURATE, SIMPLE, STRAIGHTFORWARD WAY TO USE MODERN SCIENCE TO QUICKLY IMPROVE YOUR UNDERSTANDING OF THE FORMATION OF ORGANELLES, CELLS, TISSUES, ORGANS, AND APPARATUSES ... SO YOU CAN SAVE TIME AND MONEY?

22.1. Organelle formation

Organelles are biologically organized structures delimitated by a membrane that helps maintain a certain environment inside of them suitable for specific reactions aiming at performing one or more vital functions. All organisms do not have the same number of organelles. In the chapter on universal pool of abilities, I introduced some organelles present in organisms:

- amyloplast
- chloroplast
- cytosol
- endoplasmic reticulum
- Golgi apparatus
- lysosome
- mitochondria
- nucleus
- peroxisome
- plasma membrane
- ribosome

CHAPTER 22: FORMATION OF ORGANELLES, CELLS, TISSUES, ORGANS, AND APPARATUSES

- vacuole
- etc.

As of today, organelles contain many macromolecules. For instance, all organelles have a membrane, and membranes contain at least lipids, proteins, and carbohydrates. A question worth asking can be where do the macromolecules found in a membrane come from? As of today, some macromolecules can be transported from one location in a cell to others. For instance, proteins made in the ribosomes can be transported to other locations in a cell. However, the macromolecules in the first organelles were not all transported from one place to the others, but some were made in situ. In other words, when the precursors of organelles were being shaped, their internal constituents went through a split-gathering process to yield them all. Then, after the formation of the first specimens or each organism with their reproduction, replication, multiplication systems ready to go, organelles of the descendant organisms have been and continued to be formed according to the metabolic and other biological (including physiological, biochemical) processes known to support life today. For all metabolism systems related to all biochemical compounds or processes in a living organism to work, the entire body may as well have been needed to be formed beforehand. In other words, if the metabolic pathways known today must have been what provided the biomolecules that were required to form organisms, the later could never have been formed. For one must again fall into the trap of the origin of the egg and the chicken if one is trying to explain the origin of the biomolecules and all other components of organelles using a prior metabolism of all of them. The metabolic systems were also formed. Hence, I think that the initial processes that formed the organelles were above metabolic pathways, which, themselves, have had their own origin or formation.

The shape of the vesicles of the Golgi apparatus made me think about structures formed in turbulence. Likewise, the nucleus is organized as a primary body (e.g. nucleolus) surrounded by secondary bodies, which are some of the materials inside of it. Mitochondria also look like ellipsoidal bodies formed in turbulence. I could go on and on to talk about each of the organelles, but, for the sake of space, I will just summarize that all organelles were formed according to the turbulent ambiance surrounding their precursors. Once the precursor of each organism started to be formed, it could have up taken chemicals and other things from their environment (almost just as seeds and eggs can grow today and form adult bodies), split-gathered into the precursors of its body parts, which also split-gathered into their components as body fluid flowed through them and affected their shaping and all other characteristics according to the limitations imposed on the turbulent program of life as I explained earlier.

Most organelles cannot perform much of their functions when isolated from the organisms they belong to, suggesting that they are meant to work together to maintain the life of the bodies they all share. Some can work in one individual but cannot work in another individual even of the same species. Although some people may tend to classify the organelles into which one is more important than the other or which one is not important at all, it is imperative to mention that they work together as belonging to one body and share the pain and joy so much so that, if one organelle is broken, sick, damaged, or even removed, all of the others can feel the pain to some extent. For the health of an organism denuded of one organelle is never the same as that when all organelles are fully functional. Likewise, when all organelles are working fine, the entire body also works at its maximum level. Likewise, the system or apparatus to which the organelles belong to was formed under circumstances that does not need a chronological agenda to be explained. In other words, trying to say which system was formed first and which one was formed last may be pointless. But following the development of organisms today from the seed, egg, and embryonic stages until they formed adult organisms can inform on some of the processes that the first specimens of most organisms could have gone through as they were formed for the first time. Considering the way organelles work together, I felt like, for them to exist and start working together as they do, they must have been formed together in a short amount of time.

As the organelles were being formed, macromolecules were formed inside of them just as smaller bodies or smaller systems of bodies being formed inside of much larger systems of bodies. Transits or pores were made through the membrane to allow the passages for substances across them. According to the ambiance of their formation and the expression of the turbulent program of life inside their precursors, some organelles were dominated by diverse chemical elements just as some of them were made to host or gather certain types of molecules or compounds more than others. According to the size and environments where the organisms and their body parts were formed, some forms of life ended up having certain organelles than others. Hence, all organelles are not found in all organisms. For instance, prokaryotes lack many organelles found in eukaryotes. More specifically, prokaryotes do not have the following: chloroplasts, chromosomes, endoplasmic reticulum (ER), mitochondria, nuclei, and nuclear membranes. This is not because the prokaryotes lost these organelles, but because the precursors of prokaryotes were unable to split-gather into those organelles. Likewise, eucaryotes do not have those organelles because they captured them, but because the precursors of eukaryotes were split-gathered to form those organelles. The specifics of the split-gathering of the precursors of organelles according to the precursors of the organisms affected the characteristics of these organelles and the differences and similarities between them across all organisms. Similar biological turbulent events could have triggered the formation of similar organelles, but the characteristics of these organelles depended on the circumstances of their formation and expression of the turbulent program of life. The scaling of the precursors of the organisms and their parts affected also the characteristics including the sizing of the organelles and other parts.

When I think about how celestial bodies are organized, I felt like the lack of certain organelles with prokaryotes resembles how systems of celestial bodies are not the same or do not have the same number of bodies and how some bodies are

CHAPTER 22: FORMATION OF ORGANELLES, CELLS, TISSUES, ORGANS, AND APPARATUSES

single and not organized as a system of bodies. For example, while the Earth has only one satellite, a larger planet like Jupiter has 79 satellites (as of 2022), while denser terrestrial planets like Venus and Mercury have no satellite at all. Yet, planets like Mars and Pluto, which are much smaller than Venus and Mercury, have many satellites, even more than the Earth. Still at the astronomical level for instance, the organization of the Saturnian planetary system for instance is different than that of the Jovian system, which is different than that of the Uranian or Neptunian planetary systems. Yet, each of these planetary systems has a primary planet orbited by satellites and even rings. The message I am conveying here is that, the turbulence processes that split-gathered the precursors of bodies did not organize the daughter bodies the same way, but according to turbulence laws, which, among other things, involved intermittence, differential clustering of bodies with various sizes and positions, etc. Hence, the biological turbulence that shaped the organisms and their organs or organelles did not split them equally or ensure that every organism or even cell has the same composition or organization as others. Thus, at the organism level, living things do not have the same composition. Some organisms look different than others just as some cells are different than others. For instance, unlike eukaryotic cells, prokaryotes do not have a nuclear membrane, their DNA is unbound within the cell. At the celestial body levels, stars may not have gigantic solid rocks like terrestrial planets do, yet within the same stellar system, the star and the planets were formed from the same precursor: the precursor of their stellar system. To some extent, just as the constituents of the Sun are more diluted or dispersed than those of the terrestrial planets, so also, at the biological level, the constituents (e.g. organelles) of some organisms like prokaryotes are more dispersed and less compacted than those of eukaryotes. Just as both the Sun and the planets in the Solar System were all formed from the precursor of the Solar System, yet they have different composition, so also some eukaryotes and prokaryotes could have been formed from the same precursor, which split-gathered to yield different organisms having different organelle or structural compositions. This explanation of the origin of life has never been given before because people did not understand turbulence and how it can affect the same precursor to yield different daughter bodies according to even minor changes in the initial and evolving conditions.

22.2. Cell formation

Every cell contains many organelles, which were not carried from somewhere to the cell, but which were formed inside the precursor of the cell as it was being molded to become the living cell. Based on my understanding of life, as the precursors of cells were being split-gathered, the precursor of the organelles and macromolecules in them were being formed. As the biological turbulence that shaped the precursor of the cells was taking place, the turbulent program of life was also imparting living abilities onto the body parts being made. Some organisms were made of one cell because their precursor could not be split-gathered into many cells. This was not just about the size of the precursors, but also about other characteristics including their viscosity. In other words, as much as some small precursors could have been unable to split into other smaller precursors, some relatively larger precursors could not have been able to split into others because they were very viscous. Hence, although most unicellular (single-celled) organisms are usually among the smallest ones, some unicellular organisms can be larger than some multicellular organisms. In contrast, the precursors of the bodies of multicellular organisms were able to split-gather into many cells. Today, most organisms grow to their full size after going through lengthy developmental processes, which can take months for most organisms. Some organisms must eat and take months or years before reaching their full morphological and physiological potential. For instance, it generally takes more than 12 years before a human being can reach puberty and be able to reproduce. But in the beginning, considering the complex systems of organs in living organisms and their intertwining with one another, I felt like the process that formed the first set of cells and their clustering into the first organisms must have been very fast and enabled by a process faster than current morphogenesis processes.

The precursor of some unicellular organisms may have derived from some smallest precursors of organisms as the precursor of the organisms in the ecosystems were being biologically split-gathered. This same process can explain why unicellular organisms are so abundant. Although it took more energy to gather the precursor of multicellular organisms into one body than to gather the precursor of unicellular organisms, the abundance of unicellular organisms is not a matter of lack of energy, but of how the precursor of organisms in each system was split-gathered. My work on the turbulence of the precursors of the systems of celestial bodies proved that, when the precursor of a system of bodies split-gather to yield daughter bodies, the number of small daughter bodies is usually higher than that of large daughter bodies. For instance, in the Solar System, the number of small celestial bodies is higher than that of large celestial bodies. In the planetary system, there are more small satellites than large satellites. Furthermore, I showed that the intermittence of the processes that formed clusters of individuals of different forms of life can also explain why many smaller ones are found between larger ones. Hence, bacteria are present everywhere on the planets.

Because they have different histories and are shaped by different processes, all cells are not formed equal. The content of cells depends on the level of expression of the turbulent program of life inside the body made by the biological turbulence. Just as an airplane engine cannot fit a vehicle, and just as a vehicle engine cannot fit a motorcycle, and just as a motorcycle engine cannot fit a bike, for each transportation means has its own requirements and characteristics, so also, the clusters of organelles found in the forms of life varies according to their nature. Just as the body parts of a motorcycle are not sufficient to power a vehicle, so also the set of organelles that power a bacteria may not be sufficient to power a mammal for instance. Furthermore, the energy required to move a train having hundreds of wagons or railcars is much higher than that required to move a vehicle that is the size of a wagon. Hence, the energy needed for some unicellular organisms is much

CHAPTER 22: FORMATION OF ORGANELLES, CELLS, TISSUES, ORGANS, AND APPARATUSES

smaller than that of multicellular ones. The size, the complexity, and the diversity of other characteristics of the organelles from one kind of organisms to the others were calibrated according to the amount of matter used to form the precursor of that body and also according to the mission, location of that precursor during the biological turbulence that formed other organisms in its environment.

As I carefully reviewed the shape of most cells, I saw the footprint of turbulence. Even the shape of bacteria suggested to me that the flagella is like the turbulence Zones 4 and 5, while the cytoplasmic region is like the most turbulent zone (what I called turbulence Zone 3, the most developed turbulence). Some bacteria look like a baguette, with no extrusion zone similar to Zone 3, meaning that most of these kinds of bacteria could have been formed under a limited biological turbulence.

22.3. Tissue formation

Tissues are groups of cells with the same general function(s). In other terms, cells are organized into tissues. During the development of an organism, cells multiply, then are organized into tissues and then in organs, etc., but I felt like during the formation of the original specimens or the first individuals of each organism, tissues were formed from a precursor of tissue. During the formation of organisms, the precursors of tissues were differently differentiated. The precursor of some tissues could have been organized into 4 types of tissues: connective tissue, epithelium, nervous tissue, and muscle tissue. The way these precursors were shaped, connected with one another, and functioned was at the root of the different functions that their daughter bodies can perform. In the same organism, the same cell (with the same DNA) can be differentiated into different tissues because of the changes that occurred in their precursors. These changes occurred not only at the morphological and physiological levels, but also at the genetic level as different genes were expressed and different shapes and functions were acquired by the precursors of the tissues. Although these changes can be observed today during the development of organisms, which can take months (e.g. 9 months of pregnancy for human beings), during the formation of the first individual of each species, meaning at the beginning of life, the process was very quick and could have happened instantly according to the calibration of the turbulent program of life and the environmental conditions. For the secret of understanding the formation of life is not embedded in the time or duration of the processes required for life to appear, but in the understanding of the power behind the source of life, which could have made anything to happen anytime when the conditions were met. As of today, while some organisms can take almost a year (and even more than year) to reproduce, others can procreate in a matter of hours. In other words, just as the process reproducing life can happen quick, in the same manner the process of creating life in the beginning could also have been very fast and instantaneous. It is possible that some of the processes seen today during the reproduction, embryogenesis, and all development stages of living things could have happened during the formation of the first forms of life according to each species, but at a very fast speed. If celestial bodies and systems of celestial bodies were formed in a matter of a few days (and the proofs can be seen in my books on the origin of the universe— *"Turbulent Origin of the Universe"* and *"From Science to Bible's Conclusions"*)— it is also understandable that, despite its complexity, all forms of life could have been made within days as well. The difficult task is how to explain the processes by which all types of lives were formed.

In some cases, some precursors of tissues were unable to differentiate into their constitutive cells. Hence, some organisms are not organized into distinct tissues like others. For instance, sponges do not look like other animals. Although differentiated, the sponge cells are said not to be organized into distinct tissues as the case of most other organisms (Jessop, 1970). Sponges do not feed themselves like most animals, but "by drawing in water through pores" (Sharma, 2005). These traits suggest that during the formation of organisms, some precursors of cells were unable to amass into tissues, while others did. Some of these inabilities have caused some daughter bodies (such as sponges) not to act like most animals or plants, but in a unique way, which made it difficult to classify them.

Some cells are not organized into tissues and some tissues are not organized into organs. Some animals such as some comb jellies and jellyfish, sea anemones, and corals have "digestive chambers with a single opening, which serves as both mouth and anus" (Langstroth and Langstroth, 2000). The tissues of some of these animals are not organized into organs (Safra, 2003). These facts suggest that tissues were not and are not automatically organized into organs for all organisms as they were being formed. Some conditions could have been required for cells to be organized into tissues, and for tissues to be organized into organs.

As of today, tissues found in animals can be are arranged into 4 main types:
- surface or epithelial tissues (e.g. skin)
- connective tissues (e.g. bone, cartilage, fibrous tissues, and blood),
- nervous tissues (e.g. nerves), and
- muscle tissues (cardiac muscle, smooth muscle, and skeletal muscle).

In contrast, tissues found in plants can be categorized into 2 groups:
- meristematic tissue and
- permanent tissues.

While meristematic tissues (apical meristems, lateral meristems, intercalary meristems) consist of cells that are still dividing so the plant can grow in length and thickness, permanent tissues consist of cells that are differentiated, meaning they have lost their ability to divide and are permanently positioned at a fixed place in a plant.

Permanent tissues can be divided into 2 groups:
- Simple permanent tissue (e.g. parenchyma, collenchyma, sclerenchyma, and epidermis)
- Complex permanent tissues (e.g. xylem and phloem)

Plant tissues can also be divided into 3 types:

CHAPTER 22: FORMATION OF ORGANELLES, CELLS, TISSUES, ORGANS, AND APPARATUSES

- epidermis (found at the surface of leaves and other young plant bodies)
- ground tissues (less differentiated tissues)
- vascular tissues (found in vascular tissue such as xylem and phloem)

The characteristics of these tissues can be traced back to the history of their precursors and how the turbulent program of life impacted them.

22.4. Organs and apparatus formation

An organ is an assembly of two or more tissues structurally united to specifically perform a shared function. An organ system is a collection of interconnected organs working together with a common purpose. Examples of organs are skin, kidneys, intestines, blood vessels. While some organs are filled and have less space inside of them (e.g. liver, pancreas, spleen, kidneys, and adrenal glands), others are hollow (e.g. heart, stomach, intestines, gallbladder, bladder, and rectum).

Organs were not formed just by connecting different tissues after each of them were separately formed. In contrast, the precursor of some groups of cells were clustered into different groups, which then have undergone their own growth and split-gathering, yielding the precursor of different biological components, which matured and as the precursors of the aforementioned cell clusters were growing and differentiating, it came to a point when all of their biological components reached their minimum morphological and physiological stage, meaning a level when the cell clusters could function as a whole, either at a tissue level or even at an organ level. The same logic can be applied to systems of organs and apparatuses and even to entire organisms. Saying this in a different way is that the diversity of tissues, organs, systems or apparatuses in an organism is a matter of how the precursor of that organism split-gathered and was clustered or gathered into different assemblages or compartments having specific characteristics and functions which, according to the morphological and physiological scales or levels of their inclusion into one another and their interactions, can be identified as tissues, organs, apparatuses or systems of bodies, etc. Some of those processes still occur during the development of organisms today.

Although an organ is defined as a system of tissue working together to produce a specific function, it is important to know that the tissues and the acquisition of their functions were formed and honed during the process that split their precursors into different compartments and organized its cells until the matured or "final" form of such an organ was reached. Even after an organ is fully formed and functional, it can still grow, and its cells can be renewed as old ones die while new ones are formed during a maintained process that lasts the entire lifespan of the organism in which such an organ is found. Looking back at the beginning of life, the processes I just explained could have occurred at a fast rate and the building up of every cell, organ, organelle, system, or apparatus (e.g. reproductive system, digestive system, circulatory system, nervous system, etc.) could have happened or were fully formed before the degradation of such systems of cells occurred. Because all cells, tissues, organs, systems, or apparatus of organisms are connected and can be traced back to

TURBULENT ORIGIN OF LIFE

their precursors, the coordination of all systems or apparatuses in an organism gives rise to a living thing capable of performing various functions.

Following some of the processes I elaborated above, many organs and systems of organs were formed, including:
- Circulatory system (heart, arteries, veins, and capillaries)
- Digestive or gastrointestinal system (oral cavity, esophagus, stomach, liver, small intestines, and large intestines)
- Endocrine system (hypothalamus, pituitary gland, thyroid gland, parathyroid glands, adrenal glands, and related parts such as pancreas, gonads, placenta)
- Integumentary system (epidermis, dermis, hypodermis: hair, scales, feathers, hooves, and nails)
- Lymphatic or immune system (lymph, lymphatic vessels, and lymphatic structures)
- Musculoskeletal system (muscles, skeleton including the bones, cartilage, ligaments, and tendons)
- Nervous system (brain, spinal cord, and nerves)
- Reproductive system (sex organs including the ovaries, fallopian tubes, uterus, vulva, vagina, testes, seminal vesicles, prostate, and penis)
- Respiratory system (nasal cavity, nasopharynx, larynx, trachea, bronchi, bronchioles, alveoli, lungs, and diaphragm)
- Urinary system (kidneys, urinary bladder, ureters, and urethra)
- Special systems:
  - Eye (sight)
  - Ear (hearing and balance)
  - Tongue (taste)
  - Nasal cavity (smell)

Below, I will elaborate on 2 types of systems of organs: respiratory system and digestive system. Indeed, a key component of the respiratory system is the respiratory track. The respiratory tract can be divided into 2 parts:
- the upper airways (include the nose and nasal cavities, paranasal sinuses, the pharynx, and the part of the larynx before the vocal folds or vocal cords)
- the lower airways or lower respiratory tract (include the portion of the larynx down to the vocal cords, trachea, bronchi, and bronchioles).

The pharynx is an intersection passage between the esophagus and the larynx. Air passing through the larynx goes into the trachea, which branches into the right and left primary bronchi, which at its turn branches into secondary (lobar) bronchi, which at their turn branch into tertiary (segmental) bronchi, which at their turn branch into bronchioles (which are smaller airways connected to tiny air sacs called alveoli, where gases such as oxygen and carbon dioxide are exchanged). Because of its branching structure, the lower airway is also called the respiratory tree or tracheobronchial tree. Some of the ramifications of the respiratory tree include:
- trachea

CHAPTER 22: FORMATION OF ORGANELLES, CELLS, TISSUES, ORGANS, AND APPARATUSES

- main bronchus
- lobar bronchus
- segmental bronchus
- subsegmental bronchus
- conducting bronchiole
- terminal bronchiole
- respiratory bronchiole
- alveolar duct
- alveolar sac
- alveolus

The branching of the trachea into smaller airways all the way to the alveola and even afterwards is an example of the split-gathering of the precursors of the respiratory tract. Located in the thoracic cavity, the lungs are the largest organ in the respiratory system. Considering the turbulence zones, they are formed in the turbulence Zone 3 of the respiratory system, meaning the zone where the turbulence was most developed during the formation of the respiratory system. The diaphragm (a sheet of thin muscle located at the base of the lung) can be classified as belonging to turbulence Zone 4 of the respiratory system.

I could go on and on and also explain the formation of the digestive tract using the same process of branching I used above. But for the sake of space, I will not go into those details. However, without wasting your time, I would like to say that the esophagus is like the turbulent neck of the digestive system, while the stomach is like one of the biggest bodies in turbulence Zone 3, and the intestine are like bodies formed in Zone 5. Hence the intestines are also small and long like a tail.

Now that I am done handling the formation of some of the features of organelles, cells, tissues, organs, and apparatus, I will turn to the formation of organisms as a whole. I will start with plants. Then, I will review the formation of animals before finishing with much more smaller forms of life. Later, I will also handle the formation of spiritual beings.

CHAPTER 23

CAN A SIMPLE FORMULA ACCURATELY CRACK THE CODE OF THE FORMATION OF ALL PLANTS AND REVEAL THE ONE THING THAT SCIENTISTS HAVE MISSED AND THAT HAS BEEN CAUSING THEM HEADACHES, OVERWHELM, AND BURNOUT?

You are at the right place if you really want to understand how plants were formed. Considering the abundance and diversity of plants around us and the crucial role they play on Earth, it will be absurd to write a book on the origin of life without scrutinizing some key features on plants. In this chapter, I will describe the formation of the different parts of plants for the first time at the beginning of life and also their perpetuation through reproduction today. I perceived that even the most advanced books on plant biology do not understand the processes behind many aspects of plant life, and these processes are encrypted with the biological turbulence that birthed the first plants and which, to some extent, partially continued its course during the lifespan of plants today. In other words, although all of the processes that took place during the formation of the very first plants cannot be seen today, their footprints can be decoded if the processes surrounding the morphogenesis, germination, growth, development, flowering, fructification, senescence, death, and all other stages of the lifespan of plants can be carefully reviewed in light of turbulence. Therefore, because I am the one who spearheaded the decoding of the turbulence that shaped the universe, I also felt impelled to explain the biological turbulence, which took place during the formation of all organisms, and the biological processes currently found in them, but which foundations were unknown to scientists before my work, for hidden under the umbrella of turbulence, which was misunderstood.

CHAPTER 23: PLANT FORMATION

Before detailing the formation of plants, I will introduce some fundamental terminologies that can aid in specifically addressing some key plant organs. Because of the extensive terminologies already existing in plant science, I did not have to invent many terms (like I did when I was decoding the origin of the universe) before explaining the origin of plants. In other words, to present my findings, I will use some morphological terms used in plant science.

Fig. 25: Examples of plants
Source: Dr. Nathanael-Israel Israel / Science180, www.science180.com, © 2020-present, all rights reserved.

Plants are the most dominant organisms on Earth. Their mass altogether is much more than the mass of all other organisms combined. Just by walking around your neighborhood, you will see more various types of plants (Fig. 25) than animals or other forms of life. Most plants can be categorized either as grasses, vines, shrubs, trees, and other categories.

While plant anatomy generally deals with the internal structure of plants, plant morphology usually refers to their external or physical forms. Most plants have vegetative structures and reproductive structures. The vegetative (somatic) structures of some plants consist of a shoot system (composed of stems, leaves, inflorescence, fruits, and seeds) and a root system. Not all plants flower. All flowering plants do not look alike. Flowering plants having seeds enclosed in their fruits are called Angiosperms. The reproductive structures of angiosperms consist of structures such as flowers and fruits.

23.1. General processes of the formation of plants

Some developmental processes of plants are similar to the processes that initially formed them. Under a phytoturbulence (a term I coined to express the turbulence of the precursors of plants), a cluster of matter from the soil sprouted in specific combinations or arrangements, and started growing under the influence of the turbulent program of life. The biological turbulence split-gathered matter into shapes resembling plants. At the same time, the turbulent program of life was imparting into the aforementioned clusters of matter different abilities according to their size, location, and other characteristics I will explain later. Various plants started to be formed according to their location, the amount of matter clustered to form their bodies, and the level of expression of the turbulent program of life. Just as the same genome can yield different types of cells today, so also, although the same soil was used to form some plants, the characteristics of these plants differed according to the aspects of the turbulent program of life that were expressed or not in the bodies of their precursors, which were shaped by the biological turbulence. In the following segments, I will explain how each of the main plant organs was formed, namely, the roots, leaves, trunk, branches, inflorescence, flowers, fruits, and seeds as well as all of their parts. In general, plants can be divided into two components:

- Root system forming the roots and their parts and the
- Shoot system consisting of the leaves, stems, flowers, and fruits.

Below, I will explore each of their parts. As plants were being formed, precursors of some organs were split-gathering, while others were elongating under the influence of many factors, including fluid flow in the cell layers being formed. As of today, the growth rate of plant organs (e.g. roots, shoots, and leaves) is said to be controlled by hormones (e.g. auxin, cytokinin, ethylene, and gibberellin), but during the formation of the first plants, all of those hormones were not the driving forces for some of the first growth that occurred, for these hormones were progressively made as the original plants (meaning the mothers of all plants according to their families) were being made. For there was a driving force behind how matters were split-gathered to form living organisms. In other words, I do not mean that all plants come from a single plant, but that each plant according to its type or kind (e.g. family, species) once had a mother which was the first individual of that kind to be formed. Instead of one individual, many individuals of the mother plant of each species or family could have been formed. Latter environmental and genetic components played a role in the diversification, breeding, spreading, and variation across some plant families, species, cultivars, and the like.

CHAPTER 23: PLANT FORMATION

The direction of the division, multiplication, or growth of cells caused some biologists to differentiate between primary growth (which includes all elongations) and secondary or lateral growth (which is about all kinds of growth in diameter). The branching of the precursors of the parts of organisms and the direction of the fluid flow, elongation or size increase can explain the direction of the growth and the types of growth of plants. The types and forms of growth seen in organisms today are encrypted versions of some processes that took place during the formation of the first specimens of the mother organisms.

As of today, plant physiologists have shown that the movement of molecules between the soil, plants, and the atmosphere is affected, among other things, by the atmospheric pressure and water availability in soil. Because the atmospheric pressure is also affected by the atmosphere itself, the movement of molecules in plants can therefore be affected by the turbulence in their environment. Aquatic plants are surrounded by water, yet, they do not grow indefinitely, suggesting that water availability is not the limitation of these plants, and also likely for most plants.

Unlike what the mainstream traditional scientists say, land plants did not evolve from some green algae, but they were fully formed according to specific procedures I will detail soon. The formation of plants is not a simple chemical and mechanical reaction. Plant diversity is the consequence of the diversity of the turbulence that shaped the characteristics of the plants and assembled their body parts. As you will see very soon, many features are repeated throughout plant bodies. The reiteration or repetition of forms of basic models in subunits found in plants could have been produced by the ability of plants to memorize their turbulent "heritage" and to replicate the corresponding architecture throughout their lifespan as new parts are born, while some old ones are replaced during the growth, branching, and production of structures.

23.2. Germinating seeds

When a plant like peas is planted, the first part to emerge from the seed is the embryonic root, termed the radicle or primary root (Fig. 26). It permits the seedling to anchor itself in the ground and start taking up water. Afterward, emerging from the seed is an embryonic shoot comprised of 3 main parts: the cotyledons (seed leaves), the hypocotyl (segment of the shoot under the cotyledons), and the epicotyl (segment of the shoot above the cotyledons). The shoot does not emerge the same way for all types of plants. For some plants (e.g. beans and papaya) which type of germination is called epigeal germination, the hypocotyl stretches and forms a hook, pulling instead of pushing the cotyledons and apical meristem through the soil all the way into the air (Raven et al, 2005). In contrast, for hypogeal germination (case of peas and mango), the cotyledons stay beneath the soil, where they eventually decompose (Sadhu, 1989).

Fig. 26: Sweet sorghum and corn seedlings

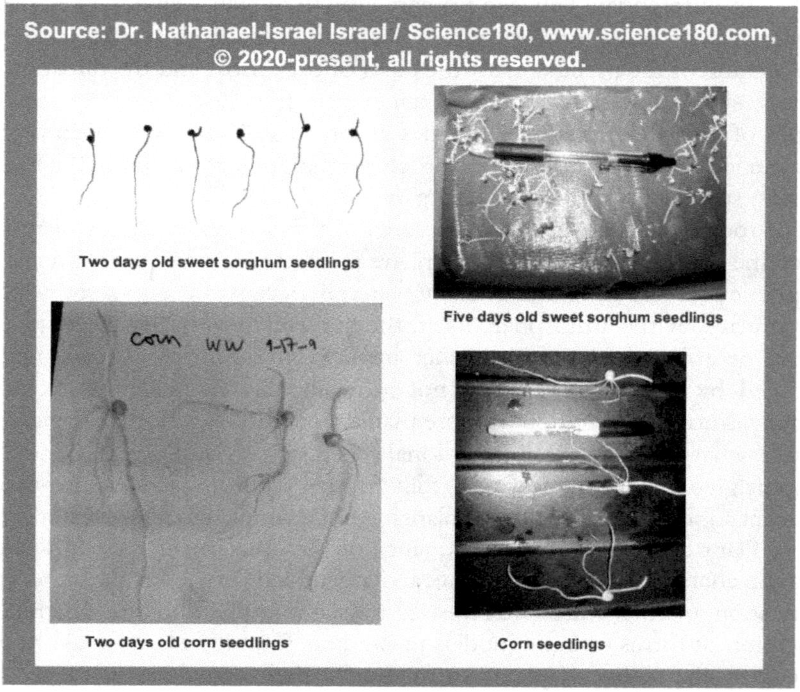

Source: Dr. Nathanael-Israel Israel / Science180, www.science180.com, © 2020-present, all rights reserved.

Two days old sweet sorghum seedlings

Five days old sweet sorghum seedlings

Two days old corn seedlings

Corn seedlings

23.3. Root system

The root system consists of the primary root and all other roots, which ramifies or branches out of it. On January 3, 2022, it appeared to me that, just as, when a plant germinates, the first root to emerge from a seed producing plants is the radicle, so also during the formation of the roots of the first plants, a precursor of the primary root similar to a radicle could have emerged first, then expanded very fast so other parts of the roots could be added accordingly. Just as no organism today reaches its complete formation instantly, but must go through growth processes, so also during the formation of plants, a precursor of the root part grows very fast as other parts could be added to it to form the complete plant including the fruits. In other words, when the turbulent program of life started acting on the precursor of plants, a precursor of root emerged, and as nutrients flow through it, it grew, branched out, and form all the root components, while the shoot system was also being formed.

In previous chapters, I explained why roots can be classified as primary bodies, while the aerial parts of a plant are like secondary bodies. This can also explain why plants can live for a while without their leaves or shoot system, but they cannot do so without their roots or root system. Some secular studies suggested that roots originate and develop from an inner layer of their mother axis (e.g. pericycle), while stem, branches, and leaves start to develop from the cortex, which is an outer layer. I will now say a few words about bulbs and rhizomes. The precursor of the aerial

CHAPTER 23: PLANT FORMATION

part of some plants did not go grow a long stem before forming fleshy leaves at their base also called scales. That is the case of bulbs which are generally short stem connected to fleshy leaves or leaf bases. Examples of such include onions (Allium), garlic, many other alliums, lily, iris, tulip, and amaryllis. For some plants, the stem grows underground and forms rhizomes (Fig. 27). Also called creeping rootstalks, rootstalks, rhizomes are stems of plants that run underground usually horizontally and which look like roots, and are able to produce new shoots to grow upwards. Examples of rhizomes are ginger, galangal, turmeric, fingerroot, bamboo, lotus root, antique spurge plant, turmeric rhizome, etc., most of which have culinary values. Some rhizomes can thicken and become stolons or stem tubers as is the case for potato (a modified stolon). More examples of roots are illustrated in Fig. 28 and Fig. 29.

Fig. 27: Rhizomes of a switchgrass I studied for my PhD. (Photo credit © Nathanael-Israel Israel)

Roots are found underground, while stems are usually found aboveground. Also called root collar or root neck, the root crown is the place where the root and stem meet; it is not always plainly visible. Although plants usually have more than one root, the primary root is usually the first root that develops from the germinating seed. Also called branch roots, secondary roots emerge from the primary one.

Fig. 28: Turnip from my 2021-2022 winter garden (Photo credit © Nathanael-Israel Israel)

CHAPTER 23: PLANT FORMATION

Fig. 29: Roots of plants
Source: Dr. Nathanael-Israel Israel / Science180, www.science180.com, © 2020-present, all rights reserved.

23.4. Meristems

Meristems consist of undifferentiated cells (meristematic cells) capable of cell division. Meristematic cells can develop into any type of tissues and organs that occur in plants. These cells continue to divide until they get differentiated and then, they lose their ability to divide. There are three types of meristematic tissues: apical (at the tips), intercalary or basal (in the middle), and lateral (on the sides). At the meristem summit, there is a small group of slowly dividing cells clustered in a region commonly called the central zone. Cells of this zone have a stem cell function and are essential for meristem maintenance. The proliferation and growth rates at the meristem summit usually differ considerably from those at the periphery.

Based on the knowledge I gained from the turbulent tree of the precursor of the Solar System, I felt like, at the meristem and at any other growing points, layers of matter exist and some are generally laid parallel to others, one on top or on the side of the other, in the direction of their growth although they can also look as if they are pointing toward the sky or the vertical for some plants. For instance, for plants that grow vertically, layers of newly formed matters are oriented toward the vertical or the direction of growth. For branches that are growing parallel to the surface or the soil, the layers of newly formed matter can be oriented as if globally perpendicular to the vertical. At the meristem level, the top or upper layers are the youngest ones, while the bottom ones are the oldest ones, and they can be programmed into becoming any type of tissue, including leaves, flowers, etc. Once a bottom layer separates from the rest of the meristem layers, it could continue its growth according to the memory it got from the stack of growing meristems before separating from them. The rest of the layers of matter in the meristem could continue their flow or growth as matters are pushed into them from the combination of what is pulled from the soil and from what is formed as organic compounds during the photosynthesis or the means of nourishment of the plant.

Stem cells are for animals what totipotent cells of meristems are for plants. For as you will see later in this book, stems cells can become any type of animal cells. For they are not differentiated yet.

The process by which meristematic cells emerge, multiply, grow, divide, and differentiate into other types of cells suggested to me that, it is not impossible that, in the beginning of the formation of life, the same "seed of life" or type of "initial biological substance" could have initiated the formation of life if the right "ingredients" were met. In other words, the same substance or biological principle could have sufficed to split-gather a type of initial biotic matter to form living things. Put another way, developmental processes in meristems also suggested to me that, in the beginning, something could have been molded into various forms of life according to a program that respected many factors, including the environment. Later in this book, I will better elaborate on how the scientific evidences about meristems hold clues that can advance our understanding of the formation of life.

23.5. Stem and trunk
Before stating how stem and trunk were formed on the very first plants, I will first explain how current developmental processes of the reproduction and multiplication of stems and trunks give clues to their origin.

23.5.1. Stem or stalk of plants
The main stem of most plants is vertical. However, for some plants, it is not. For instance, some ferns do not have vertical stems, yet some of them have rhizomes. Some plants (e.g. vines) even grow along the surface of the soil or matter that

CHAPTER 23: PLANT FORMATION

support them. The posture of a plant can be boiled down to how the precursor of the matters used to form the original specimens or mothers of these species and/or families of plants could have gone through a biological turbulence that prevented the precursor of the shoot system from growing or flowing too much before yielding secondary bodies, which here are branches holding leaves. Had the precursor of the shoot system been able to "travel" longer before starting to split, the original mothers of these kinds of plants could have had an aboveground stem or vertical stem, which could stand, provided the turbulence inside of it could allow it to support the rest of the body parts. In other words, how turbulence shaped the precursors of some plants prevented the formation of strong stems, which can stand like a trunk or a solid organ capable of holding the aerial organs of the plants. The composition of the stems being formed also played a crucial role. When the stem is weak, it cannot stand, hence it is forced to vine on the ground or around the stems of other plants. Thus some stems cannot grow vertically, but must fall on the ground and/or grow horizontally and become an underground stem like rhizomes. In some cases, almost no stem was formed at all, hence some plants are stemless.

Fig. 30: Corn at flowering stage (Photo credit © Nathanael-Israel Israel)

Just as some plants do not have a horizontal stem, so also among the plants that have a vertical stem, some do not have secondary stems, meaning branches. For instance, except a few cases like that of bamboo, most monocots do not have secondary branches. Likewise, some plants called sessile do not have a stalk. Some of these plants have their flowers and leaves directly connected to their stem or peduncle, therefore lacking a petiole (which is a stem connecting a leaf to a stalk) or pedicel (stem holding the inflorescence). For example, the leaves of most monocotyledons (e.g. corn, wheat), do not have petioles. Next, is a picture of corn, in which secondary branches and petioles are absent (Fig. 30).

When I recalled the structures of the celestial bodies, the ability of some plants to have petiole or pedicles, or to have a stem, stalk, branches, is connected to the nature of the turbulence which birthed them. Sometimes, the precursors of the aforementioned organs did not receive the required turbulence capable of forming them before the organs located downstream of them could be formed. For instance, a leaf lacking a petiole is like a satellite that has a very small semi major (which is almost nonexistent). What I am explaining here for plants can also apply to other organisms such as fungi, which fungal fruit bodies can lack a pedicel. Later I will elaborate more in the segment on leaves and inflorescence.

23.5.2. Trunk of trees

The trunk is the upright massive main stem or main vertical axis of a tree. The central trunk is sometimes called the main stem, baseline stem, mainline stem, or the master stem. While most trees have one trunk, some called polycormic have numerous, strong vertical trunks. The trunk is like the main secondary body of the tree around which tertiary bodies such as branches are formed. I used the term secondary body to describe the trunk because, like I already explained, I considered the main root as the primary body. But with respect to all of the body parts on the shoot system, the trunk is like the primary body. Just as some primary bodies have only one secondary body around them, while others have many secondary bodies, when the precursor of the cells that would become a trunk were being formed, some were split-gathered into many layers which then split from one another and led to the formation of many trunks. In other words, some trees have many vertical trunks because of how the precursor of their trunks was split-gathered very early before yielding its branches. I understand that some hormones could also be at play, but I think that the split-gathering of a precursor into many branches was also under the influence of a biological turbulence. For example, during the growth of trees, branches sprout out of the main trunk at the same position and all around the trunk because of how the turbulence that produced them led to their split-gathering at the same time and position. As you will see in some of the next pictures, the trunk did not grow much before the branches appeared. In other cases, as explained in Fig. 31, the trunks grew very much before branches appeared.

CHAPTER 23: PLANT FORMATION

Fig. 31: Trunk and branching of trees
Source: Dr. Nathanael-Israel Israel / Science180, www.science180.com, © 2020-present, all rights reserved.

Trees are usually divided into 3 main parts: crown, trunk, and roots. Tree trunks usually consist of 5 layers:
- outer bark (Fig. 32)
- inner bark (phloem)
- cambium cell layer
- sapwood, and
- heartwood

Fig. 32: Bark of a pine tree. (Photo credit © Nathanael-Israel Israel)

Fig. 33: Trunk showing dead tissues. (Photo credit © Nathanael-Israel Israel)

All of those body parts today are formed as plants develop. But in the beginning, they were formed by a biological turbulence. Unlike most animals, which tend to replenish and get rid of most of their dead tissues, many trees do not always get rid of their dead tissues, hence in many old trees, living tissues are usually found in the outside layers, while the inside of the tree's trunk is made of dead tissue. Part of this is due to the fact that the oldest parts of trees are always on the inside. Fig. 33 is a picture of a cut tree which inside is dead tissues turned into a hole. Animals do not

CHAPTER 23: PLANT FORMATION

behave like that, and the moment animals cannot get rid of old tissues, some of them can become cancerous and/or lead to various kinds of diseases.

23.5.3. Why are the trunk, branches, leaves, and all other parts of planets connected and not disconnected like the celestial bodies in a system?

Although celestial bodies were also formed under the influence of turbulence, almost as plants were also formed through a turbulence (but a biological one), celestial bodies are "disconnected" from one another while plant body parts are connected. I put "disconnected" in quotation marks because, in reality, even celestial bodies are not as disconnected as human beings may think. The distance separating celestial bodies is huge because of the size of their precursor and the scope of the turbulence that birthed them. However, the turbulence which formed living things are not as strong, but more advanced in many ways. In the case of a trunk, branches, and leaves, as their precursors were being formed, they stayed connected because the forces which could have distanced them further and even split them was not sufficient to disrupt their unity. Furthermore, as each of the body parts was splitting from the remainder of the other, the "escape speed" was not strong enough to cut them from the rest. For the flow of fluid through each organism also calibrated the speed with which the entire organism must grow and also the speed of its split of the body parts as they were formed and compartmentalized. The connection between the trunk, branches, and leaves also allows plants to stay alive and satisfy their needs (food, water, and other nutrients), some of which come from the root. If the trunk, branches, and leaves were disconnected, some parts of the plants would die. For plants to be dynamically alive and fulfill their mission, a way was designed to allow them to do so. If plants had to be a disconnected system of body parts orbiting around other parts as satellites orbiting a planet and planets orbiting a star, they could not live well for all of their bodies could not be rooted or connected to the ground from which they get some nutrients.

23.5.4. Hardest trees are not the largest ones

Many of the hardest trees grow slowly and are quite small in stature, suggesting that their bodies are not produced by a very strong turbulence. In contrast most of the soft plants are bigger, suggesting that they were produced after a strong turbulence. For when the turbulence is strong, larger, and less dense bodies are usually produced. Even if you refer to celestial bodies, according to their types, the largest ones are not usually the densest ones. For plants, the Janka hardness test is generally performed to assess wood hardness. As of 2021, the hardest wood in the world is believed to be lignum vitae (also called guayacan, guaiacum, Pockholz) which is made from the Guaiacum trees such as *Guaiacum officinale* and *Guaiacum sanctum*, all of which are native of Bahamas, Jamaica, and other Caribbean countries. Other woods made from hard trees exist across the globe, but which are much less strong than the ones mentioned above include: African blackwood, hickory, red oak, and yellow pine. According to the Guinness Book of Records, Black Ironwood (*Olea capensis*), also African Ironwood, is the heaviest wood in the world. Native of Australia, *Allocasuarina luehmannii* is an ironwood tree known for its exceptional hardness. That tree is more than 5 times harder than white oak. It is even harder than aluminum. All of the trees producing the hardest wood are not the largest or biggest trees. They are neither the tallest ones. All of the largest trees I am aware of are not the densest. Baobab trees which are some of the largest in some African countries are not made of a dense wood at all. Likewise, most of the trees that have very large leaves are not usually the densest, but most of the densest trees I am aware of tend to have small leaves. This does not necessarily imply that all dense trees must have small leaves. The biological turbulence of the precursor of plants explains why the largest trees are usually less dense than the smaller ones which produce the hardest wood.

23.6. Branches

Trees are ideal examples of how the branching of the precursors of bodies worked. Branches are also defined as lateral divisions, subdivisions, outgrowths, ramifications, bifurcations, or offshoots that diverge or derive from a main body or from the main portion of a structure. Branches are outward extensions from a trunk or from a central point. For trees, a branch is a structural member connected to, but not part of, the central trunk. Branches usually ramify, separate, split, arise, derive, proceed, or spring from the main or central structure. A branch coming directly from the trunk is usually termed a bough, limb, or an arm. The terminus of a branch is usually called twig, sprig, branchlet, spray, surcle, surculus, or ramulus. A branch found under a larger branch is called an underbranch. The area where a trunk splits into two or more boughs is called a crotch or fork. Some people may have never heard of the word crotch referring to a plant, but to where a man and woman's reproductive parts are. For the crotch is like a branching point. As it ramifies, a stem of a plant such as a trunk diverges into branches, which at their turn diverge into increasingly smaller branches, and so on and so forth until the ramification or branching ends. Branches are believed to be born from a process that starts with the growth of a bud into a twig. However, I think that the general process of the formation of branches is connected to the biological turbulence of the fluid layers in the cells of the stems.

During the split-gathering of the fluid layers of the precursor of the stems of a plant, there was a time when the stack of fluid or cell layers that were not split yet and some layers that started splitting looked like a crotch or fork. The branching pattern, which is more visible with plants, can also be found even with nonliving things. When I considered the precursor of all the celestial bodies orbiting the Sun as a trunk, the precursor of each planetary system after its split is like a branch along which the precursor of the satellites looking like under branches.

Before I proceed any further, I would like to make a disclaimer concerning my usage of the term "branching". Anywhere in my books, when I talk about branching of precursors of organisms or branching of something, I did not mean that anything evolved from one kind of organism to become another kind (just as evolutionists

CHAPTER 23: PLANT FORMATION

postulated), but I mean how a precursor of a body (living or nonliving) could have split-gathered and birthed daughter bodies, or reorganized into a shape, or form, which its mother precursor did not have.

Just as a turbulence partially controlled the branching patterns of the precursors of celestial bodies into different daughter bodies according to their position in their mother precursors, so also at the biological level, body parts on various scales (macromolecules, organelles, cellules, tissues, organs, and organisms) were shaped (according to their position) by the biological turbulence that took place during the formation of the original mothers of these organisms. The genetic code and all other processes that generate or gather matter for the bodies (including through salvage, eating, and drinking) provide the raw materials on which biological processes act to define the shape of the original organisms and that of their progeny throughout the ages. In other words, the genome is not the only system that contains hereditary information passed onto descendants. Hence, organisms have different topologies or branching patterns of their components beyond genetics. To put it another way, because of the variation of the biological turbulence which shaped them, the characteristics of the body parts of organisms are not the same from one type of individual to those of another kind. For instance, the size (e.g. length, height, and width), organization, and structure of cells, organs, organelles, and apparatuses vary across the family, species, and even individuals of the same species according to their age, environments, etc. As of today, the variation of environmental conditions and the genetic makeup of organisms lay different initial conditions upon which the biological mechanisms controlling life act to form branches. But during the formation of the very first plants, plants did not have their genetic material fully set before their branches were formed. But as matters were being biologically split-gathered, the branching patterns were formed under the influence of turbulence. The separation of fluid layers is a kind of branching, which causes a branch to develop differently than a trunk, just as secondary bodies develop differently than their peers which branch out later. Many aspects of turbulence can explain the ways that plants sprout and arrange their branches and leaves into diverse patterns (phyllotaxis). Even the way that branches and leaves spiral around their primary stem is an expression of the fluid movement in their precursors. For instance, it takes a different number of branches to be formed on a trunk before a pattern can be molded and/or repeated. The same rule affects also leaf arrangement on most plants. Since the days of Fibonacci, also known as Leonardo of Pisa (1170-1250), and even long beforehand until today, people of various origins have tried to explain the patterns observed in plants, but many of them have failed because they were unaware of the turbulence that is at the base of most natural patterns.

At one point during my investigation, I wondered why, on the same plant, branches are smaller than the trunk, and why leaves are smaller than the branches. Below is my thought on that observation. Indeed, all branches on a tree do not have the same size. Some are smaller than others. The trunk itself to which primary branches are connected decreased in diameter as the distance from the ground or root crown increased. In other words, the diameter of a trunk close to the ground is generally higher than that toward the apex of the same plant. Likewise, the size of the branches generally decreases from the node where they connect to the trunk to the apex of the branch. However, in some cases, the highest diameter of some trunks and branches are not found near the base, or node, but somewhere in the middle of the trunk and in the middle of the branch, specifically around a position I would call the turbulence Zone 3 of the trunk or of the branch. I will later explain a similar feature for leaves in which the largest areas are not near the petiole nor near the apex, but somewhere in the middle. Just as fluids of the precursors of the secondary bodies of the celestial bodies were organized into layers after they escaped the precursor of the primary body (see my book *"Turbulent Origin of the Universe"* for details), so also, the fluids of the precursor of a plant could have been organized into mostly vertical layers (for vertical plants) and horizontal layers (for plants that crawl on the ground). As the precursor of a plant starts forming or growing in height, its cell layers start splitting, giving rise to different groups of cells, which then split according to the impact of the turbulent program of life, which imparted different abilities to the plant organs. Because branches on a tree are formed as layers of cells that split-gathered from the remainder of cell layers in the meristems, their size cannot be larger than the size of the stack of layers itself. Hence, branches are always smaller than the trunk. Likewise, secondary branches are smaller.

'Science180 Academy' Success Strategy
SCIENCE180 SEMINARS

People whose awareness is raised by Science180 usually ask me to go deeper or they wonder "what's else?". That is one of the reasons Science180 trains them through strategic work sessions (during seminars or training sessions) that transfer customizable skills and solutions to them. Science180 Seminars are client-centered and tailored to strongly engage the clients so they maximize the discovery of and the tapping into new opportunities, and exponentially outperform their expectations. Science180 offers customizable seminars that can be labeled as a colloquy, conference, consultation, discussion, forum, keynote speech, lecture, lesson, meeting, symposium, summit, study group, tutorial, workshop or working section accordingly on any topic related to:
- Universe-origin for scientists and mathematicians, philosophers, laypeople, and the general public
- Universe-origin or universe creation for believers
- Life-origin for life scientists, for all other scientists, and for believers
- Chemical-origin for scientists
- Universe-origin seminars for children
- Universe and life-origin for pseudepigraphic believers

As you contact us with your needs, we can customize your program accordingly. Learn more at Science180Seminars.com.

CHAPTER 23: PLANT FORMATION

23.7. Leaves

As illustrated in the images below, plant leaves come in various shapes. A careful analysis of the diversity of leaves and the insight I have from turbulence helped me to uncover the process of their formation. Little by little, I will walk you through how I unearthed the genesis of leaves.

23.7.1. General description of leaves

Before addressing the formation of leaves, I will first describe them so you can be familiar with some terminologies I will be using later. Generally located above ground (and sometimes underground) and dedicated to photosynthesis, leaves are considered the primary lateral appendages or attachments of the stems of vascular plants. A complete leaf of flowering plants usually has most of the following components (see Fig. 13):

- Apex (the tip of a leaf)
- Bud (young stem tip capable of producing a stem, leaf, or flower located between a stem and a leaf)
- Lamina (flat and laterally-expanded portion in the blade of a leaf)
- Leaf margin (the edge or boundaries of the leaf)
- Leaflet (separate blade of a compound leaf)
- Midrib or midvein (primary or central vein of a leaf blade or leaflet
- Petiole (leaf stalk connecting a leaf blade to a stem at the node)
- Rachis (main axis of a pinnately compound leaf, meaning having two rows of leaflets)
- Secondary vein (vein emerging from the primary vein)
- Sheath (portion of a grass leaf surrounding the stem)

The characteristics and the presence or not of the aforementioned leaf components varies from one plant to the other according to their species, families, and other classification criteria. Below, I will review some of those characteristics and how they can be explained by the biological turbulence. Beforehand, let me define a few more terms. Indeed, a node is a term used to describe the point of insertion of a leaf, bud, and even branches on a stem. The space separating 2 consecutive nodes is called internode. For the precursor of some plants, the precursor of the main stem did not have to grow much before the precursor of leaves split-gathered from it. In some cases, the internodes are so short that they may even appear as nonexistent. That is the case of acaulescent leaves (and also acaulescent inflorescence) which emerge from the ground as if they have no stem. Such clustering of leaves (with very short internodes crowded together) is usually known as rosette. Rosette leaves or rosette inflorescence can be formed on the surface of the soil but also occasionally higher on the stem.

More than one leaf can sometimes emerge from the same node. For example, whorl is a term used to describe 3 or more leaves or branches or pedicels arising from the same node. Even flower parts (e.g. sepal and petals) can also form a whorl. Fig. 34 to Fig. 38 illustrate some types of leaves.

Fig. 34: Leaves of plants
Source: Dr. Nathanael-Israel Israel / Science180, www.science180.com, © 2020-present, all rights reserved.

CHAPTER 23: PLANT FORMATION

Fig. 35: Leaves of plants
Source: Dr. Nathanael-Israel Israel / Science180, www.science180.com, © 2020-present, all rights reserved.

Fig. 36: Leaves of plants
Source: Dr. Nathanael-Israel Israel / Science180, www.science180.com, © 2020-present, all rights reserved.

CHAPTER 23: PLANT FORMATION

Fig. 37: Leaves of plants

Source: Dr. Nathanael-Israel Israel / Science180, www.science180.com, © 2020-present, all rights reserved.

Fig. 38: Leaves of plants

Source: Dr. Nathanael-Israel Israel / Science180, www.science180.com, © 2020-present, all rights reserved.

All leaves do not have the same size. While leaves of most flowering plants (angiosperms) usually have all leaf components, for other kinds of plants like conifers (e.g. pine trees), the leaves are like a needle, stay green most of the time (evergreen), and usually have a single vein. Fig. 39 and Fig. 40 are images of pine trees showing needle-like leaves.

CHAPTER 23: PLANT FORMATION

Fig. 39: Pine tree. (Photo credit © Nathanael-Israel Israel)

Fig. 40: Pine tree (Photo credit © Nathanael-Israel Israel)

Looking at some conifers on December 16, 2021, suggested to me that their shape is like that of a body which precursor did not have Zones 1 and 2, but just Zones 3, 4, and 5. For the largest part of the cone of most conifers is located toward their base and beforehand, there are not much branches which can be put in Zones 1 and 2. This feature reminded me about the trend I saw in my research on the Solar System in which I noticed a turbulent leaf-like shape of the Plutonian satellites, which started with Zone 3 but not Zones 1 and 2.

On the same plant, all leaves are not the same. Some are larger than others. For instance, a cataphyll (also called a cataphyllum or cataphyll leaf) is a small or modified version of a leaf sometimes unable to perform photosynthesis, but able to accomplish other particular functions on the plants that bear them. Examples of cataphylls comprise bracts, bracteoles, and bud scales, or scale leaves. Bracts are any kind of leaves connected to an inflorescence. Some leaves even connect directly to the trunk. In other words, leaves can mostly appear anywhere on a plant. The small amount of the precursor of a cataphyll and the nature of the turbulence that birthed it can explain their size and limited functions. For, the smaller the precursor of a leaf, the lesser the probability that all of the features found in true leaves could find space and act properly. Hence, the reduction of the size of cataphyll affected their function as well. This partial correlation between the size of leaves and their function also indicates how the turbulent program of life was limited by the size of the precursors of some forms of life.

23.7.2. Arrangement of leaves on the stem

Known as phyllotaxis, the pattern or arrangement of leaves on the stems of a plant can provide key information about some of the processes involved in the formation of the plants that bear them. The size and shape of a leaf has a footprint or fingerprint of the turbulence that its precursors went through. In the case of a basal leaf for instance, the precursor of the stem did not grow much before the leaves arose from the base. In contrast, for cauline leaves, the precursor of the stem grew through the ground or soil and reached the atmosphere before the precursor of the leaves appeared, hence cauline leaves arise from aerial stems. Sometimes, the emergence and split-gathering of the precursor of the leaves and branches appeared in such a way that 2 or more leaves, branches, or flower parts can attach at the same node on the stem. For instance, when 2 leaves emerged from the same node, the leaves are termed opposite. This appearance could have been caused by a precursor of leaves split gathering or yielding two opposite layers of cells, which differentiated into 2 leaves.

As the precursor of some leaves were growing or elongating, they also rotated, turned, or twisted. Depending on the growth rate, rotation, or twist that the precursor of the stem went through before the precursor of the leaves emerge, different angles were formed between 2 consecutive leaves. For example, for decussate leaves, successive pairs of leaves were rotated 90° from the preceding. The precursors of whorled or verticillate leaves emerged or branched out from the stem of their plants at 3 or more points of a node. Likewise, instead of appearing on the stem successively one at each node, some leaves appeared all together around a node to form a rosette. All of these arrangements of leaves depended on how the

CHAPTER 23: PLANT FORMATION

precursors of the leaves were split-gathered to yield many daughter bodies leading to many leaves at some nodes.

23.7.3. Division and shape of the blade

While the blade of some leaves is divided, for others it is not. Divided blades also come in many shapes and arrangements. For example, while simple leaves have an undivided blade (meaning that the lamina is not divided), compound leaves (e.g. beans) have completely subdivided blades, which leaflets are separated along a main or secondary vein. The ability of the lamina of a leaf to be divided is a consequence of the intensity of the turbulence of the precursor of that leaf. Just like some celestial bodies born under the influence of a weaker turbulence did not have a large system of bodies (consisting of a primary body surrounded by secondary bodies), so also when the turbulence of the precursor of some leaves was weak, the corresponding lamina did not have the ability to produce all components of a fully developed leaf. Under other circumstances, the layers of matter of some precursors of blades split into several layers, which branched out and then split again, leaving the lamina arranged into many subsets of leaflets. In other words, according to the turbulence intensity of their precursors, some compound leaves even yielded leaflets, which at their turn also have petiolules, which was a miniature version of how the leaf itself is connected to the stem via a petiole. Hence, the blade of some plants is undivided while that of others is divided.

The various ways that the precursors of the leaflets emerged from the precursor of the rachis or from the rachis itself led to their division into various leaf compounds:

- palmately compound for which the leaflets radiate from the end of the petiole (e.g. the fingers of a palm or a hand) (Fig. 41).
- pinnately compound according to which the leaflets are arranged along the main vein
- odd pinnate in which the existence of a terminal leaflet vs. even pinnate in which the terminal leaflet is lacking
- bipinnately compound for which the leaflets (termed pinnules in this case) are organized along a secondary vein emanating from the primary vein
- trifoliate or trifoliolate which are pinnate leaves having 3 leaflets
- pinnatifid which leaflets are not completely separate

Fig. 41: Example of palmately compound leaf. (Photo credit © Nathanael-Israel Israel)

According to the turbulence that birthed them, leaf blades come in various shapes. At this point, using some botanical jargon, I will review a few types of leaf blades. Acicular leaf blades are slim and pointed resembling the shape of a needle. Lance-shaped or lanceolate leaves are long and wider in the middle. The wider region may be near to the turbulence Zone 3 of their precursors. Linear leaves are long and very narrow. Some of the densest and hardest plants have linear leaves.

The tip of leaves does not always have the same size as the bottom. For instance, for oblanceolate leaves, the top of the blade is wider than the bottom. Obovate leaves are shaped like a teardrop with the stem attached to the tapering point. For some of these leaves, the precursor of the blade may have escaped the precursor of the main axis of the leaf with a high speed, which may have caused it to almost be cut off from the stem, hence the larger part of the blade is developed outward of the stem. Flabellate blades are semi-circular or fan-like. The blade of orbicular leaves is circular. Ovate leaves are shaped like an egg. Reniform leaves are shaped like a kidney while rhomboid leaves look like a diamond. Sagittate leaves are arrowhead-shaped while spatulate leaves are shaped like a spoon. Trifoliate leaves are divided into 3 leaflets. Truncate leaves have a squared off end. The richness and the diversity of the shape of the leaf blade points at the huge variation of the biological turbulence which shaped the leaf blades.

23.7.4. Leaf margins or edges

The nature of leaf margins was defined by how the precursors of the veins branched out and how far they had to travel before not being able to do so anymore. In some cases, matters which were passed onto the precursor of a vein did not go very far before stopping its elongating. But, upstream or downstream, other precursors of veins could have traveled much farther, even branching out into other veins, and

CHAPTER 23: PLANT FORMATION

the repeat of such a scenario can make a leaf margin uneven. Furthermore, the speed with which the precursor of a layer of fluid/cells branched out from a primary vein defined how far its daughter cells may have gone before stopping to elongate or take certain kinds of shape. In other terms, the flow of fluid in the stack of cells in the precursor of a vein and all of the bodies connected to it could have affected the length of the resulting vein. I also felt like the smoothness and the roughness of the turbulence in the precursor of a leaf could affect the smoothness and roughness of its margin. Finally, the viscosity of the fluid (e.g. sap) in a growing leaf can also affect how fast it elongates or not. In other words, if at some point during the formation of a leaf, certain biomolecules are formed at a higher concentration than in other circumstances, the elongation of some cells may be limited and, in the end, the margins of such a leaf can be affected. I have found some of the scenarios I mentioned above to explain the leaf margin with the calibration of the distance separating celestial bodies. Therefore, some of them may still stand even with biological turbulences. Now, I will give some specific details about some types of leaf margins.

Ciliate leaf margins are fringed with hairs. Hairs found on some leaves and other organs may also be produced from the trickling down on a cascade of split-gathering of biological matters from the highest level all the way to some of the smallest levels. Hence, no leaf hair is larger or bigger than the vein that feeds it. Rough biological turbulence in leaves may lead to the formation of tooth-like leaf margins, each tooth bearing smaller teeth. On leaves which edges are toothed, the nature of the teeth varies. In some cases (e.g. serrate leaves) the edge looks saw-toothed. In the case of lobate leaves, the indentations are able to reach the center of the leaves. For a sinuate leaf, the edge looks like waves, which form an undulate leaf, the wavy edge is shallower than the edge of sinuate leaves. In some cases (spiny or pungent leaf), the edge of the leaf has sharp points like thistles. Why all these features? To answer that question, I turned to some trends I found with the formation of celestial bodies (see my book "*Turbulent Origin of the Universe*").

Indeed, when I was studying the turbulence of the precursors of celestial bodies, I noticed that, at one point, when the fluid layers in a precursor is rough, the resulting fluid ligaments lead to the formation of bodies of diverse sizes, while when the turbulence in a fluid ligament is smooth, mono-diverse daughter bodies are formed, meaning bodies having almost similar size. For instance, if I could split some highly indented leaves into pieces according to the location of the teeth at the end and the structures they formed, bodies of various sizes would be formed just as diverse sizes could be observed with daughter bodies of a rough ligament. This suggested to me that the indentation, undulation, ciliation, crenation, denticulation, serrulation, sinuation, lobation, and all other configurations of leaf margins or edges (all of which are well known to botanists) may be highly connected to the roughness and smoothness of the fluid ligament of the precursor of the leaf under the influence of the biological turbulence that birthed them.

TURBULENT ORIGIN OF LIFE

23.7.5. Leaf apex (tip)

Just like I did for leaf margin, the configuration of leaf apex or leaf tip can be explained using turbulence. For instance, acuminate leaf apex (meaning long-pointed apex extended into a narrow and concave way) can be formed if, near the end of the leaf, the precursor of the primary vein continued growing or elongating without branching out secondary veins at the same speed. In that case, the elongation of the leaf occurred more dominantly in the direction of the primary vein than into the direction of the secondary vein. Hence, a long-pointed apex. In the case of obtuse leaves, the tip is well rounded or blunt, while for truncate tips, the ending is abrupt and flat.

23.7.6. Organization of leaf veins

Also called nerves, veins of leaves start from the petiole and ramify across the entire leaf according to various patterns termed venation. Veins in leaves can be easily seen even with the naked eye without needing a microscope. Here, my goal is not to explain the vascular functions of leaf veins, but to link their organization to the biological turbulence that occurred during their formation, even during the formation of the first plant specimens or original plants.

The venation of plants depends on their types. For instance, two types of venations are generally encountered with angiosperms (flowering plants which have seeds enclosed in their fruits):

- parallel venation mostly found with monocots, which veins are parallel and
- reticulate venation (in which the veins look like nets) usually found with dicots (Fig. 42).

Fig. 42: Soybean leaf showing a reticulate venation. (Photo credit © Nathanael-Israel Israel)

Below is a cassava leaf (Fig. 43) showing a primary vein and secondary veins.

CHAPTER 23: PLANT FORMATION

Fig. 43: Cassava leaf. (Photo credit © Nathanael-Israel Israel)

Also known as first-order vein, the primary vein of a leaf enters the leaf from the petiole. Then, it branches out to produce the secondary or second-order veins, which at their turn can ramify to yield tertiary or third-order veins, and so on and so forth until the veins could no longer ramify or divide. This organization of veins is due to the split-gathering of the precursor of the primary vein as the turbulence in the blade was developing. In fact, during the formation of the veins of leaves, as soon as the precursor of the leaf emerged from the stem, just as the precursors of the other leaf components, the precursor of the primary vein was split-gathered from the precursor of the leaf. As the precursor of the leaf was growing and going through a biological turbulence, after travelling over a certain distance related to the length of the petiole, the precursor of the primary vein went through enough turbulence that allowed it to branch out secondary veins, while the remainder of the primary vein continued its growth in the same direction as that of the petiole, but away from it. As the remainder of the precursor of the primary vein continued its journey as the precursor of the entire leaf was still going through its turbulence (on different scales according to the component of that leaf), another secondary vein branched out, and so on and so forth until the remainder of the primary vein could no longer grow much further. At that point, the morphological formation (of the original leaf) and/or growth (of the leaf today) "stops" or slows down, while physiological development continues until all the veins and all other components of the leaf mature. Afterwards, the leaf can stay on the plant as long as possible before senescing, dying, falling one day, or being cut off. The distance between 2 consecutive secondary veins depended on the distance that the precursor of the primary vein had to travel before its turbulence could have had enough resources and yielded enough turbulence to branch out the precursor of the next secondary vein. The ability of the precursor of the secondary vein to ramify into the precursor of tertiary leaves and so on and so forth could have been defined by the energy and intensity of the turbulence passed onto the precursor of the secondary veins by the precursor of the primary vein and so on and so forth. In other words, if the precursor of the primary veins did not pass onto the precursor of the secondary vein enough energy, turbulence, and other resources, the precursor of the secondary vein could not have done the same for the precursor of the tertiary veins, which could branch out of it. What I explained above also can be applied to how branches were formed from the trunk of plants in the beginning, and, to some extent, even today. The nature of the split-gathering of the precursor of the veins according to the position of the vascular system contributed to defining the vein patterns. After the formation of the first specimens of plants, the turbulence and all other transmissible proprieties of the plants were kept in the hereditary stock of the plants, which, according to me, is not just contained in the genes only. For, as I explained in this book, there is a turbulent code that plants and other forms of life transmit to their descendants. But, because it was not well known by the scientific community until my groundbreaking discoveries on turbulence, these details were not understood by scientists before. My work on the origin of life and on the origin of the universe will enlighten many and open the doors for several breakthroughs, which I hope some scientists will not orient toward destroying life, but toward the advancement or betterment of causes aiming at improving lives across the globe.

In plants which leaves have parallel veins, the primary veins are organized in parallel and generally at the same distance from each other for the majority of their length before converging towards the apex of the leaves. That is the case for most grasses including Johnson grass which I pictured next (Fig. 44).

Fig. 44: Leaves of Johnson grass (Photo credit © Nathanael-Israel Israel)

The parallelism of veins could have been caused by the inability of the precursor of their leaf to split-gather into other major veins, but to just continue flowing in the same direction it emerged from the precursor of the petiole. On October 5, 2021, as I was wondering why gramineous leaves are not very large in the middle or the region I would call Zone 3 on most leaves, it appeared to me that the turbulence of the precursors of gramineous leaves were not very developed, hence, its

CHAPTER 23: PLANT FORMATION

turbulence Zone 3 was not large enough to give them the large region found in the most turbulent zone, Zone 3. This also explains their venation being parallel. The higher concentration of silicon in gramineous leaves may have also worked against the development of turbulence in the precursors of their leaves.

As for leaves having a reticulate venation, a single primary vein connected to the vasculature of the petiole usually serves as the anchor of many secondary veins, which emerge from it and spread toward the margins of the leaf, which are the limits after which the turbulence of the precursors of the leaves could not pass. Matters or structures at the end of a vein are like the outermost bodies of the precursor of that vein. Just as secondary bodies are usually smaller than their primary body, so also secondary veins are narrower in diameter than their primary veins, while tertiary veins are narrower than secondary veins and so on and so forth. In other words, the reduction of the size of the veins as they ramify from the primary vein all the way to the smallest veins is related to the reduction of the turbulence that took place in their precursors as they were being formed.

23.7.7. Characteristics of petioles

Also called leaf stalk, the petiole is the stalk portion that connects a leaf blade to a stem. All plants do not have a petiole. While petiolate leaves have a petiole, sessile or apetiolate leaves lack a petiole. For instance, the leaves of most (if not of all) grasses (Poaceae) lack petiole. Pictured next are leaves of grain sorghum (Fig. 45) and sweet sorghum (Fig. 46), lacking a petiole.

Fig. 45: Grain sorghum leaves (Photo credit © Nathanael-Israel Israel)

Fig. 46: Sweet sorghum leaves. (Photo credit © Nathanael-Israel Israel)

The length of the petiole can be related to the distance that the precursor of the leaf had to travel away from the stem it emerged from before its turbulence could be sufficient enough to birth the leaf. In other words, apetiolate leaves were formed in turbulent conditions that did not allow the precursor of the leaf to travel much farther away before forming the leaf. According to the intensity of the turbulence of their precursors, some petioles can become very big such as in the case of celery, artichokes, and other edible crops grown mainly for the succulence of their petiole. Finally, stipules are outgrowths that appear on each side of a petiole of some plants. While some plants have stipules, others do not.

To close this segment of petiole, I would like to recall that, on the scale of the leaves, the petiole is a turbulent neck. Comparing this turbulent neck with some trends I found with the celestial bodies, I noticed that the same thing that caused some celestial bodies to be very close to their primary bodies and others to be very far, while some primary bodies do not even have a noticeable secondary body at all, can also explain leaf features.

23.7.8. Position of branches and leaves affects their size and shape

As I carefully explored plants around me, I realized that, even on the same plant, all leaves and branches are not the same. Some are larger than others. As I interpreted this fact through the glance of turbulence on December 28, 2021, I felt like the variation of leaf shapes on the same plant, which causes some leaves to be smaller or differently shaped than others, is also the result of how turbulence (even

CHAPTER 23: PLANT FORMATION

biological turbulence) is influenced by its position with respect to a primary source. For instance, on a leaf, turbulence can be smaller at the base or near the petiole and at the apex of the leaf than around the middle of the leaf (in the region I would call turbulence Zone 3). Sometimes, on the same plant, lobed leaves can be found at the base of the plant, while unlobed leaves can be found at the top of the plant. Likewise, the size of the branches at the base and at the top of plants is usually smaller than that of the branches located around the middle, which can be around turbulence Zone 3 (the zone of the most developed turbulence on the scale of the aerial organs). This variation of shape and size of branches and leaves contributes to giving to plants their overall shape, wherein the base and the tip are small while somewhere in the middle the plant looks larger.

The variation of the biological turbulence even on the scale of the same plant can also explain why copies of the same organ are not always the same during the lifetime of plants. Because the biological turbulence occurring in them is not well developed yet, leaves toward the apex of a plant are different from the well-developed leaves on the same plant, and from one branch to the others, differences can also be observed. In the end, due to the impact of the memory they have of the biological turbulence that birthed the leaves, even after they mature, their shape and size are not always identical from one leaf to the other throughout the entire plant.

23.7.9. Why are some leaves shaped like a heart?

On October 5, 2021, around 7 AM, as I was walking home after dropping my children at their school, I was looking at some plants along the way, and suddenly, I came across one which leaves looked like a heart. It is like what the botanists or plant biologists could call an obcordate or an inversely heart-shaped (Fig. 47):

Fig. 47: Leaves shaped like a heart.
(Photo credit © Nathanael-Israel Israel)

Immediately, my mind turned toward a similar shape I found in some graphs I did (around September-October 2021) about the relationship between semi major axis and radius of Plutonian satellites. In the chapter on leaf-like shapes, I already explained how I made this radar type graph (Fig. 48).

CHAPTER 23: PLANT FORMATION

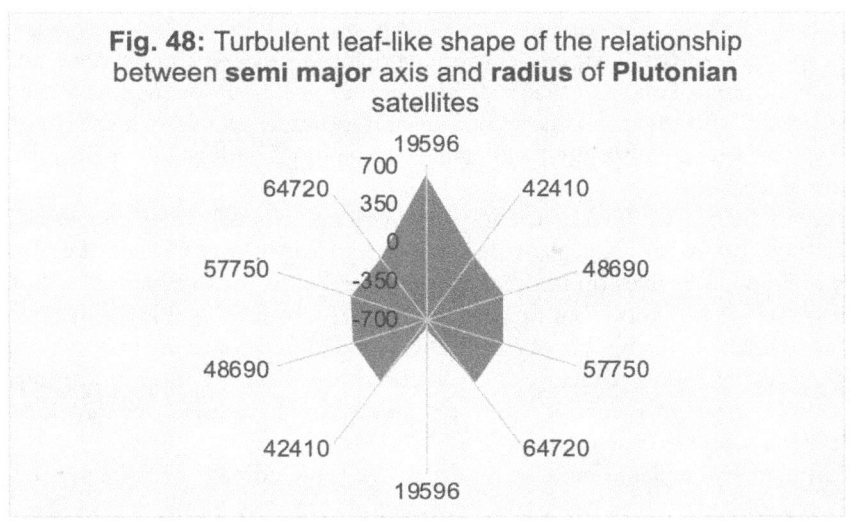

Fig. 48: Turbulent leaf-like shape of the relationship between **semi major** axis and **radius** of **Plutonian** satellites

I immediately felt like leaves shaped like an inverse heart are products of a biological turbulence similar to what the precursors of the Plutonian satellites (for instance) could have gone through: no Zones 1 and 2, but Zone 3 came first and then Zones 4. The lack of Zones 1 and 2 made the region near the branch look larger than any other region. If there were Zones 1 and 2, the larger region would have been toward the middle of the leaf. As a reminder of some of what I said in *"Turbulent Origin of the Universe"*, the turbulence of the precursor of the Plutonian satellite was not strong. Hence, the innermost satellite for instance is very close to the primary planet. Most plants which have leaves looking like an inverted heart are usually large but not very hard. This made me think that some leaves that are like an inverted heart may have been originally formed in weak or less strong turbulence ambiances. Hence the turbulent neck of these leaves is followed by a Zone 3 but no Zones 1 or 2. Some of these kinds of plants (which have obcordate or inversely heart-shaped leaves) usually have longer petiole, suggesting that their precursor could have travelled over a long distance before being able to form the leaves.

23.7.10. Leaves know and memorize their history and are one of the "lowest" levels of turbulence on a plant

From the shape of a tree as a whole, to the shape of the trunk, to that of the branches, leaves, fruits, and other organs or parts of a plant, it appears to me that each of the aforementioned plant parts (i.e. trunk, branches, leaves, fruit, and seeds) memorized the history of the turbulence of their precursors. That is why they maintained the ratio of the size, position, and even composition of their parts according to the turbulence zones that their precursors or "parents" have lived in. Hence, leaves are smaller toward the apex (Zone 5) but larger midway of the lamina (Zone 3). This and other things I observed from turbulence remind me that, to properly understand how the universe and its contents are functioning today, it is very important to first properly comprehend the history of the universe and the things it contains. To a larger extent, the memory or history of the turbulence that birthed things can explain some fractals found in nature as some of them are built on processes that are imprinted with successive split-gatherings under the influence of turbulence.

Human beings or their gathering into various groups are products of their history, and if they forget, neglect, deny, or reject their history instead of embracing it and building their lives on it, they can cause problems for themselves and others. All histories are not good, but no one can fully accomplish its mission if he or she decides to throw his or her history away 100% and just embrace the history of others 100% or craft a new one that is not somehow based on the old one. Immigrants who change their environment can testify about how hard it is to change locations and learn or adapt to new cultures.

Turbulence in a plant occurs at different levels, including the root, the trunk, the branches, secondary branches, and extending all the way to the leaves. As I was reflecting on these scales of turbulence which can be seen from the outside of plants, it appeared to me on September 29, 2021 that, as far as external features are concerned, the turbulence in leaves is weaker than that in the trunk, branches, and roots. Hence leaves can appear from any part of the shoot stem.

'Science180 Academy' Success Strategy
SCIENCE180 CONSULTING

Because Science180's trainings, seminars, or strategic work sessions (through which it transfers skills and training solutions) are great, some customers want to go even deeper on a long-term, sustainable basis. That is where Science180 Consulting, one-on-one consulting, and mentoring (that some people may prefer calling coaching programs) comes in. That is where Science180 can truly change people's behavior on a long-term basis according to their specific needs. With Science180 Consulting, you will discover and understand the deep secrets of the formation of the universe, life, and chemicals around you. Hear Dr. Nathanael-Israel Israel's personal selection and teaching on key topics that will help you break the code of the universe formation and functioning. All strategically designed to enlighten you, guide you to navigate and filter the massive data collected on the universe and its content so you know how to answer the world's most challenging origin questions, remove any scientific and philosophical cataracts that may be blocking you, and help bring you many steps closer to your best life today and forever. Science180 Consulting will train you, transfer unconventional skills to you and change your behavior so you go deeper. To get started today or to learn more, go to Science180Consulting.com.

CHAPTER 23: PLANT FORMATION

23.8. Inflorescence and flowers

The Earth is filled with diverse flowers, and before delving into their genesis, I will first present some examples of flowers (Fig. 49 - Fig. 52).

Fig. 49: Inflorescence of plants
Source: Dr. Nathanael-Israel Israel / Science180, www.science180.com, © 2020-present, all rights reserved.

Fig. 50: Inflorescence of plants

CHAPTER 23: PLANT FORMATION

Fig. 51: Inflorescence of plants

Fig. 52: Inflorescence of plants
Source: Dr. Nathanael-Israel Israel / Science180, www.science180.com, © 2020-present, all rights reserved.

23.8.1. General processes of the formation of flowers

While some plants have just one flower on an axis, others have many flowers and the united cluster of flowers is called the inflorescence. In this segment on inflorescence, I will focus on how I think turbulence influenced the way flowers were formed in the beginning and even today. Some of the conditions that prevailed the first-time flowers were formed at the beginning may no longer be found on Earth today, but some of the processes involved in the formation of the first flowers are still present and observed by most flowering plants today.

Just as all other bodies I have studied so far, flowers were also born after a certain precursor of a body was split-gathered, then went through a series of complex processes until their development was complete. In the following segments, I will address some of those processes. As of today, no flower has all of its flower parts formed instantly at the same time, so also, when plants were being formed for the first time, all of their parts were not formed at the same time. No matter how small

CHAPTER 23: PLANT FORMATION

the duration of the formation of a plant may have been, all of the flower organs were formed according to an order, which respects some turbulence laws. Indeed, as the precursor of a flower was being formed, it reached a point when its biological turbulence allowed the branching out of the precursor of the sepals (which together would form the calyx). As the remainder of the cell layers of the precursor of the flower continued its journey, it branched out, split-gathered or yielded the precursor of the petals (which together would constitute the corolla). The collection of the calyx and corolla is called the perianth, which also had a precursor. After the precursor of the perianth produced its daughter bodies (the precursor of the sepals and the precursor of the petals), the remainder of the precursor of the inflorescence continued its journey until the precursor of the androecium (i.e. the whorled collection of the stamens, which are the male reproductive units) and the precursor of the gynoecium (i.e. whorled collection of the carpels, which are the female reproductive units) were produced each at its turn. Still on its journey, as the precursor of the flower was elongating or growing, it yielded the precursor of the ovary and stamen, each at its turn.

As of today, stamens have 2 parts: a stalk called filament at the top of which is the anther, where pollens are produced. Considering what I know about turbulence, I felt like when the precursor of the stamens of the original plants (and even of plants today) were being formed, it travelled a certain distance (related to the length of the filament) before reaching enough turbulence capable of emerging and forming the precursor of the pollen at the top of the anther, which is the end of the elongation of the remainder of the precursor of the flower.

As of today, carpels consist of an ovary, a stigma, and usually a style, which all together form a structure called the pistil. The ovary is a hollow structure where ovules are produced. The stigma is a tube leading to the ovary, which is the end of the pistil, an organ which receives pollens. The pistil consists of one or more carpels fused together. The gynoecium (female reproductive organ of a flower) consists of one or more carpels. To form all of these flower parts, a biological turbulence took place in the precursor of the female reproductive organ. Indeed, during the formation of flowers, after the precursor of the male part of the flowers emerged, the remainder of the precursor of the flower continued to grow, and yielded the precursor of the gynoecium, which went through a biological turbulence before yielding its daughter bodies as I previously explained. Indeed, as the precursor of the gynoecium branched out, it started growing, after traveling for a certain distance which is related to the length of the style (a tube-like organ), its turbulence developed enough so that the stigma could be formed at the end of the pistil. In other words, the length of the filament (of the male reproductive organ) and the length of the style (tube connecting the stigma to the ovary) are like a turbulent neck, meaning a consequence of the biological matter formed between a primary body and its innermost secondary body and/or the length of travel required before a more developed turbulence occurred downstream.

As the precursor of the flower was growing and split-gathering into the precursor of different organs, all of these precursors continued to go through their own turbulence until they matured or gained their final shape, function, and other characteristics. By the time the entire flower could be considered matured, seeds (which I will handle later in this chapter) were formed and matured inside the inflorescence.

23.8.2. Pedicel

The stem that attaches a single flower to the inflorescence is called a pedicel. Considering what I already explained earlier, the pedicel is the turbulent neck for a single flower. This means that, like I have explained for other organs, the length of the pedicel can be related to the distance travelled by the precursor of a single flower after it split from its mother precursor. In other words, after the precursor of a single flower split or emerged from the stem (or the precursor of the stem) holding the inflorescence (or the precursor of the inflorescence), it took some time for it to elongate and go through a turbulence as it was growing before reaching a point where the flower was fully formed. For the flower is a product of a coherent biological structure formed out of the biological turbulence of the precursor of the inflorescence. In the following picture, the pedicel is the stem that is on top of the pumpkin (Fig. 53).

Fig. 53: Pumpkin showing a pedicel (Photo credit © Nathanael-Israel Israel)

23.8.3. Peduncle

When the inflorescence of a plant has many pedicels, the later are held together or are connected to a stem or branch from the main stem of the inflorescence called peduncle. In other words, the peduncle is a turbulent neck of an inflorescence having many pedicels. The position of the pedicels on a peduncle is related to the time that their precursors split-gathered from the precursor of the peduncle to

CHAPTER 23: PLANT FORMATION

produce a pedicel. In other words, as the precursor of a peduncle was growing, precursors of pedicels split-gathered or formed from its branching out and then after going through their own turbulence, these precursors of pedicels yielded pedicels, which in the end are connected to the peduncle, which reaches its maturity once all of the pedicels attached to it are fully formed.

The precursor of some plants did not have a chance to grow their stem much above the ground before the peduncle of its inflorescence arose from the ground with a few or no bracts at all, except the part near the rachis, leading to the formation of a scape (also defined as a long, leafless lower stalk coming directly from a root).

23.8.4. Rachis and types of inflorescences

Bracts are some kinds of small leaves or cataphylls present in the peduncle. Rachis is the main axis of an inflorescence above the peduncle. While the peduncle is denuded of flowers, flowers are present of the rachis. In other words, rachis is the portion of the inflorescence that bears the flowers or to which the flowers are connected. On the scale of an inflorescence having many flowers, the rachis is like the trunk from which emerge the single pedicels holding individual flowers. In light on this, I felt like, as the precursor of the inflorescence was being formed, the precursor of the rachis may have first emerged, and then, branched out to yield the precursor of individual flowers. As the precursor of individuals flowers were being formed, the precursor of the pedicel was the first part to be formed, then after it reached a certain point, the other flower parts were produced as their corresponding genes were being expressed and other biomolecules produced and/or mobilized to the places where they were needed. Considering what I explained above, the precursors of compound inflorescences (inflorescence having branched stems) having a peduncle to which pedicels are connected to may have gone through a more developed turbulence than the precursors of simple inflorescences (unbranched inflorescence), which are just made of one single flower, not a complex network or system of flowers. I would not be surprised if some fully-developed compound inflorescences produce relatively bigger fruits than some less developed simple inflorescences.

Simple inflorescences come with many shapes. Some are indeterminate (e.g. racemose), while others are determinate (cymose). For instance, a raceme is an unbranched inflorescence which flowers are connected to the axis via short floral stalks acting like a pedicel. The unbranching of the raceme can be explained by the nature of the turbulence of its precursor. In some cases, flowers connect to the main axis of the raceme without the mediation of pedicel: that is the case of the type of flower called spike. The length of the pedicels is not always the same for all flowers. For instance, for a racemose corymb (a flat-topped unbranched inflorescence), outer pedicels are progressively longer than inner ones. In other cases, not only is the floral axis short, but also all of the pedicels have the same length as in the case of the type of raceme inflorescence called an umbel. Many other types of simple inflorescence exist.

The classification of compound inflorescences is based on that of simple inflorescences, with the difference that the simple flowers are replaced by racemose. The branching of the stems of compound inflorescences can be traced back to the successive split-gathering of the precursors of the main stem found with those types of flowers starting from the peduncle all the way to the outermost pedicel of the compound inflorescence.

23.8.5. Genes partially controlling floral organ development

Since the discovery techniques related to gene expression, efforts have been made to understand the genes controlling the development of flowers. To make a long story short, a set of genes described in the flowering model called the "ABC model of flower development" have been advocated and are still commonly accepted by the scientific community. According to that model (Bowman et al., 1991; Haughn and Somerville, 1988), 3 genes interact combinatorically to yield the proteins responsible for the floral organs. The letters A, B, and C used to name the models stand for the functions of these genes. Based on a study on Arabidopsis thaliana and Antirrhinum majus, the ABC model says that, A genes are responsible for the whorl of sepals (which all together form the calyx). The interaction between A and B genes are said to be responsible for the formation of petals (which all together form the corolla). Furthermore, B and C genes are responsible for the formation of stamens (which all together form the androecium) while the C genes alone control the formation of carpels (which altogether form the gynoecium).

Flowering in plants is partially controlled by genes, for it has been shown that when some genes are repressed, some plants were unable to produce all of their flower parts. The sequence of the genes controlling the ABC model are not always the same from one organism to the other. Studies done on Arabidopsis, petunias, and other plants suggested that some of the genes controlling function A are called: APETALA1, APETALA2, SQUAMOSA, LIPLESS1, and LIPLESS2. Genes showing type-B function include: APETALA3, PISTILLATA, DEFICIENS, and GLOBOSA. Some of the genes exhibiting type-C function are: AGAMOUS and PLENA. However, unlike what most scientists have thought until my work on the origin of life and the universe, I do not think that flowering is solely controlled by genes. For, the genes help provide some of the raw materials or chemical compounds needed to form certain parts of the flowers indeed, but the way that the proteins expressed from those genes are mixed with other biomolecules to form structures found in flowers is responsible for the shapes of flowers. Despite what may be said about the functions and origin of the aforementioned genes, I do not think that, left alone without the influence of turbulence, they can produce the types of organs found in flowers. One thing is to have construction equipment and materials, but another thing is to follow a design to build things a certain way. For example, the same sand, concrete, and cement can be used to build different types of buildings depending on many factors including the design, the builder, the environment, etc. Likewise, genes alone cannot explain floral architecture, but

CHAPTER 23: PLANT FORMATION

attention must also be given to the biological turbulence through which the biomolecules expressed from the genes were mixed according to the (internal and external) environment of the plants. My expertise with turbulence taught me that the design and organization of floral organs have been and are still under the influence of a biological turbulence. For I demonstrated that the same precursor of celestial bodies can birth various celestial bodies according to the difference of the turbulence it went through.

> ## 'Science180 Academy' Success Strategy
> **SCIENCE180 MASTER CLASS**
>
> Hear the greatest scientific and philosophic lessons from top scientists, philosophers, thinkers, and public figures who have realized historic mistakes they made in life (concerning the origin of the universe, life, and chemicals), and that they corrected thanks to the historic discovery of Nathanael-Israel Israel, the world's first 180Scientist who founded Science180 and who is known as the one who truly decrypted the universe origin for the first time. In their own words, these renowned personalities share with the world key lessons they have learned in life and how people can learn from their experiences to improve lives instead of repeating their mistakes that many people still ignore at their own perils. To learn more, contact us at Science180.com/contact.

23.9. Fruits and seeds

Once an ovule is fertilized by a spermatozoid, an egg is formed, and then it grows inside the ovary of the plants. Progressively, all of the other floral organs die and leave room for the fruit to develop. I showed examples of fruits in Fig. 54 to Fig. 56. Because many good books were already written on fruits, I am not going to dwell on them here. What I would like to point out though is that fruit shapes are also a consequence of turbulence. A careful look at any fruit would reveal the existence of the turbulence zones I also amply discussed. The intensity of the turbulence and other characteristics of its initial and developing conditions can explain the diversity of the proprieties of the fruits formed.

Fig. 54: Fruits of plants
Source: Dr. Nathanael-Israel Israel / Science180, www.science180.com, © 2020-present, all rights reserved.

CHAPTER 23: PLANT FORMATION

Fig. 55: Fruits of plants

Fig. 56: Fruits of plants
Source: Dr. Nathanael-Israel Israel / Science180, www.science180.com, © 2020-present, all rights reserved.

When the fertilized ovules of some plants (e.g. gymnosperm and angiosperm) ripen, seeds (considered as embryonic plants) are formed in the process as part of the reproduction system. In other words, a seed is a ripened ovule. A fruit is a ripened ovary or carpel containing seeds (e.g. apple, pear, corn, pomegranate, tomato, pepper, cucumber, beans, peanuts, and zucchini). A nut is a kind of fruit, not a seed.

Fruits are usually organized into layers. Formed from the ovary, the outer layers (called pericarp) are usually comestible. In other words, the pericarp is formed from the ovary wall, and can become fleshy (e.g. berries), or form a hard-outer cover (e.g.

CHAPTER 23: PLANT FORMATION

nuts). Seeds are usually found inside the pericarp. The pericarp is usually organized into 3 layers: exocarp (also called epicarp), mesocarp, and endocarp. The epicarp is the outer layer while the endocarp is the inner layer, while the mesocarp is the layer in between. Sometimes the envelope of cell layers covering the outside of some seeds can develop into a fleshy organ, which can end up being bigger than the seed inside. The amount of flesh in the pericarp and the thickness of the layers are also used to define fruits. For instance, cereal grains (e.g. wheat, corn, and rice) are fruits which walls are thin and attached to the seed coat, therefore making those edible grain-fruits worthy to be called a seed as well.

For flowering plants (i.e. angiosperms), the seeds are enclosed, whereas for gymnosperms, the seeds are naked. In other words, gymnosperms are plants that have seeds unprotected by an ovary or fruit; examples of such include conifers and cycads. It is important to notice that some plants (e.g. ferns, mosses, and liverworts) do not have seeds, not because they are primitive as some theories claim, but because since the beginning of life, they never had seeds and they will never have seeds. For a plant that never had seeds cannot develop into a plant that will have seeds. The ability to produce seeds resulted from the expression of the turbulent program of life, which imparted different abilities to organisms. The ability to produce seeds cannot be given to plants which do not reproduce by seeds after their formation was completed. This implies that the precursors of seed producing plants could have gone through a more developed turbulence than the precursors of plants that do not produce seeds. Seeds are usually covered by a coat.

When I considered all plant organs, I felt like fruits are the ultimate level of gathering together in plants. After fruits are formed and mature, the plants bearing them either die (e.g. annual plants) or have to restart another cycle the next season (e.g. perennial plants). In plants, the fruit layers that branch out do not always collect together into one body as done by the celestial bodies, but they are spread. The time they finally collect together, they give fruits. In other words, fruits are the ultimate perfection or completion of the gathering of fluids or matter in plants. Likewise, flowers are a stage of major turbulence leading to the final stage of fructification and production of seeds.

'Science180 Academy' Success Strategy:
SCIENCE180 INTERVIEW REPORT (AKA SCIENCE180 INTERNET-TV-RADIO INTERVIEW REPORT)

Science180 Interview Report is the newsletter to read for guests and unconventional show ideas at the intersection of science and faith. Indeed, many hot questions are still unanswered on the road leading to the correct understanding of the origin of the universe, of life, and of chemicals. But most people don't know where to find the accurate answers to those challenging questions. What if with one simple call you can accurately answer all of those questions. You need to get in touch with or interview Dr. Nathanael-Israel Israel on your show, radio, tv, podcast, and even website, or invite him for a live presentation at your organization if your audience can benefit from any of the following show, talk, speaking, or interview ideas:

- Can a single variable play a crucial role in cracking life-origin?
- What is the master key to crack the origin of life?
- What is the little-known variable that hides a key to understand something unique about the origin of celestial bodies?
- How to deal with the fear of not knowing the origin of life?
- Can you be really free from doubt about the life-origin?
- Can you be really free from doubt about God's existence?
- Can mathematics and science collide to accurately explain the formation of the universe, of life, and of chemical particles?
- Can we mathematically prove that the formation of the Earth was completed on the 3rd day of creation like the Bible says?
- Can we scientifically demonstrate without a doubt that the Moon and the Sun were really formed on the 4th day of creation like the Bible says?
- Can mathematics and science rescue Christians in their efforts to rationally prove the existence of God, the Creator?
- Did the Quran and any other religious book make any gigantic error about the universe creation that any scientific formula proves the Bible got right?
- Can most Christian leaders refuse to take a stand on 6 literal days of creation and expect atheists and freethinkers not to argue that God is simply unnecessary?
- Why the secularist world doesn't care much if Christians and their leaders believe in Evolutionism, but they actually care much if they don't believe in the billions of years process?

CHAPTER 23: PLANT FORMATION

- What is one thing experts in turbulence need to do from now forth if they want to reap the fruits of the turbulence mysteries you have decoded?
- Does the Bible scientifically teach anything about the universe-origin and life-origin that most people including Christians ignore?
- Can anyone really be scientifically 100% sure and prove that God created life?
- Can we explain the formation of life and of the universe through natural processes without evoking evolution and Big Bang?
- Is it a waste of time to attempt to prove the Biblical creation using science or historical investigation?
- Can anyone scientifically prove that God created the universe without talking about the Bible
- Is accurately understanding the origin of chemical particles a choice?
- Why many people are abandoning wrong creationist theories that compromise with Darwinism and Big Bang?

I know you may be tempted to answer these questions by yourselves, but avoid landing yourself on wrong paths that caused some people to lose contact with reality, it is better to get the accurate answer from the know-how expert, Dr. Nathanael-Israel Israel, the author of many books on the origin of the universe, of life, and of chemicals, and the standout expert who accurately decoded the scientific formula that forces science to bow to the truth. You can invite Nathanael-Israel Israel to your organization asap to get his original answers to this question. If you would like to register to Science180 Interview Report to periodically receive show ideas and opportunities related to the origin of the universe, of life, and of chemicals particles, please visit Science180Interviews.com for more details.

CHAPTER 24

THE ONLY SCIENTIFIC STEP-BY-STEP PATHWAY YOU NEED TO ACCURATELY DECODE THE FORMATION OF ALL ANIMALS AND GET THE POWER, FREEDOM, AND BOLDNESS TO TAKE ADVANTAGE OF NEW OPPORTUNITIES

Animal species are different by many characteristics that scientists have used to elaborate ways to classify them. Although scientists came up with various classification names, in this book, I will hang on to 5 classification labels: fish, insects, birds, mammals, and reptiles. Like tailors and masons build different clothes and houses respectively even while using the same raw materials, so also the turbulent program of life that shaped the livings formed various organisms even out of the same raw materials. In other words, from the same soil or water, different types of animals were formed using the same raw materials, but sewed or put together differently.

Just as the formation of plants, the formation of animals can also be attributed to processes involving a biological turbulence. During the formation of the organisms other than plants (which I think were formed before the other forms of life) a biological turbulence broke out in the waters and in the upper part of the Earth (meaning near the surface). The biological turbulence split-gathered some chunks of soil and water into bodies of different forms of life, not just animals. Like I already explained in the chapter on intermittence, during a split-gathering, bodies of various sizes are usually formed and small bodies are generally found between large bodies. During the formation of the organisms, animals were most of the larger organisms between which many smaller ones were formed. In other words, when the soil and waters used to form organisms were being shaped, the precursors of animals were formed as the precursors of other forms of life (excluding plants, for

CHAPTER 24: ANIMAL FORMATION

they were already formed) were being formed. Among the precursor of animals, the precursors of the largest mammals were the largest ones. Other precursors that were also being formed were precursors of fish, insects, birds, and reptiles. Across the precursors of all ecosystems, precursors of all of those types of animals were being shaped by the biological turbulence upon which the turbulent program of life acted. The processes that formed each type of animal shaped their organs. Because plants and animals are completely different forms of life, the turbulences that shaped them were different. In many of the previous chapters including the chapters on turbulent neck, belly, tail, arms, legs, I introduced how some animal organs could have been formed. In some of the previous chapters, I also explained why many classes of animals were formed, I will not revisit all of those concepts here, but I will just give examples of animals, and then, I will illustrate how some of them were formed. I will start with fish.

24.1. Fish

Fish (Fig. 57) are usually defined as aquatic, vertebrate, gill-bearing animals that lack limbs with digits. Gills or branchia are very important for some fish to extract oxygen from water. A fish's body is divided into 3 parts: a head, a trunk, and a tail. The tail is usually posterior to the anus, meaning located downstream of the anus. The study of fish is called ichthyology. As of 2021, the Encyclopedia Britannica reported that about 34,000 fish species have been identified (Parenti, 2021).

While some fish have bones, others do not. For example, the skeleton of some fish (e.g. so called Chondrichthyes) is made up of cartilage, not bones. In this regard, sharks and rays have a cartilaginous vertebral column. With its gigantic size, sharks could have been thought to be made of very strong bones. Its cartilaginous vertebral column testifies of how the turbulence that some the precursors of some very big bodies went through did not allow them to have a very dense body. Because, bones are stronger than cartilages, organisms that have a cartilaginous vertebral column are not as strongly built as those which vertebral column is made of bones. Ignoring the origin of fish, some people think that bony fish derived from cartilaginous fish, which they perceive as ancient fish. But the fact is that all kinds of fish were formed almost at the same time, but the biological turbulence that shaped them allowed some to have a bony skeleton and others a cartilaginous skeleton. In other words, bony fish did not evolve from cartilaginous ones, but they were born out of a more developed biological turbulence. To some extent, I observed similar processes with celestial bodies where large bodies are usually less dense or less hard than smaller bodies.

When I think of what happened during the formation of the celestial bodies, I recalled that, despite its massive size, the Sun and the other stars in the universe are not as dense as some of the bodies that orbit them. Generally speaking, the larger a body, the denser it is. Even at the level of living things, I noticed that the flesh or meat of fish is not as thick or hard as that of some mammals. In fact, the meat of most fish can easily be peeled off flakily or like filets, therefore testifying of the layers of matter that were used to form those fish, but which are not as much intertwined and tough as the flesh of some mammals. I felt like fish meat is one of the softest or tenderest animal meats. Some people have testified that the meat of some carnivores including lions is very hard.

Fig. 57: Fish

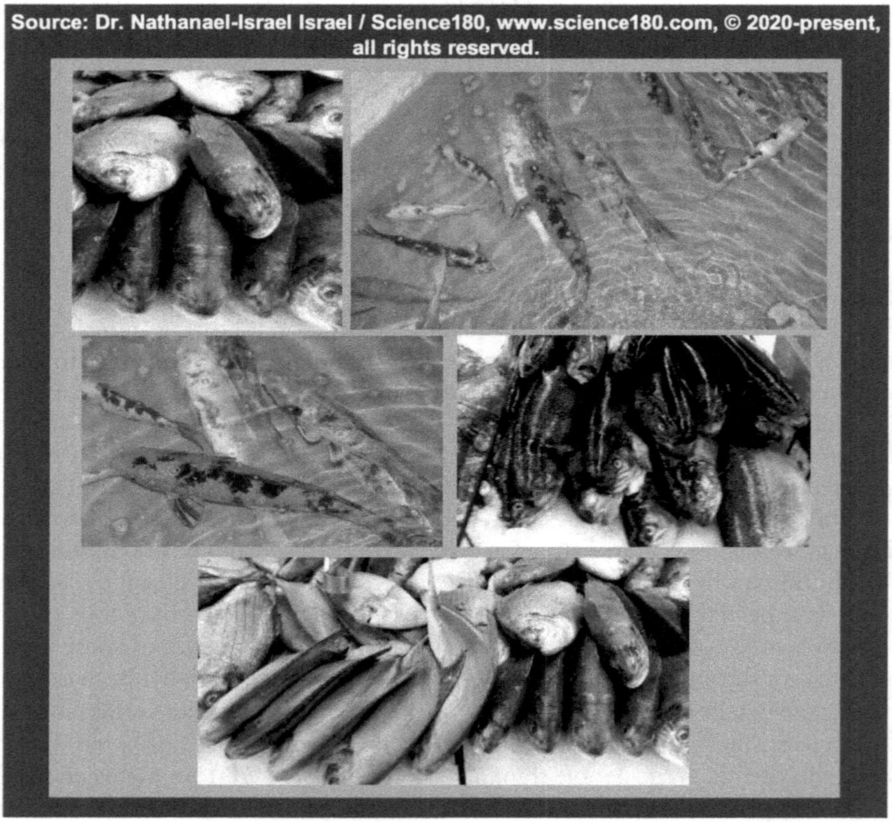

Source: Dr. Nathanael-Israel Israel / Science180, www.science180.com, © 2020-present, all rights reserved.

Although most people will consider most animals living in water as fish, it is important to mention that amphibians, reptiles, and even mammals that are able to live in water are not fish. For instance, frogs, aquatic snakes, whales, dolphins, crocodiles, and hippopotamuses are not fish, yet some of them spend all or most of their life in water. The intermittence of the forms of life during the split-gathering of the water used to form aquatic organisms can explain the presence of other forms of life other than fish in water.

Some fish have four limbs and can appear as if they have feet, but these limbs are not legs. That is the case of lungfish, which some people think are the closest living fish resembling tetrapods. Fish move using various types of fins:
- dorsal fins (located on the midline of the back; most fish have 1 dorsal fin, while others have 2 to 3)

CHAPTER 24: ANIMAL FORMATION

- caudal fin (located on the tail or at the end of the caudal peduncle; most fish usually have a single caudal fin)
- anal fin (single fin located on the ventral surface behind the anus).
- pectoral fins (paired fins located on each side, usually just behind the operculum, and considered as homologous to the forelimbs found with other animals)
- pelvic or ventral fins (paired fins located below the pectoral fins, and considered as homologous to the back limbs found with some animals)
- adipose fin (located between the dorsal fin and the caudal fin)

I felt like fins are for fish what the ramifications of the precursor of the thorax and abdomen into arms and legs are for mammals. More specifically, the precursor of pectoral fins and pelvic fins are for fish what the precursors of the forelimbs and the back limbs respectively are for other animals that go on their feet. In other words, the precursor of some fins could have become arms and legs if the turbulence that formed fish was meant to push enough biological matter into the precursors of the fins to make them grow like arms and legs. Because fish live in water and don't need legs or arms as much as fins (which are crucial for swimming), it is a good thing they have fins instead of legs and arms, which can make swimming difficult. Nevertheless, fins of fish are unable to and can never turn into or become arms and legs, for once an organ is formed, its cells cannot be reprogrammed to form another type of organ.

While some fish (called naked fish) have no outer covering on their skin at all, and others are covered with scute, the outer body of most fish is covered with scales (thin overlapping plates of bone on the skin). In other words, although most fish have fins, many do not have scales. While some scales are made of bones, others are not. For instance, the scales of cartilaginous fishes are not bony. The precursor of fish that have scales and fins could have gone through a more developed biological turbulence than that of those lacking them. Likewise, the precursors of fish that have bones in their skeleton may have gone through a more developed turbulence than those of fish which have a cartilaginous skeleton. Hence, I personally think that bony fish with bony scales and fins are more developed, more advanced, and more complete than all other kinds of fish.

Now that we have discussed fish, let's head to the air and look at birds.

24.2. Birds

Birds are vertebrates (backboned organisms) that have feathers (also called plumages) and toothless beaked jaws. As of today, more than 10,758 bird species have been identified (Clements, 2007). Examples of bird species include: albatrosses, boobies, bustards, chickens, cormorants, cuckoo roller, cuckoos, ducks, eagles, falcons, flamingos, grebes, hawks, herons, hoatzin, hornbills, hummingbirds, ibises, kingfishers, loons, mouse birds, nightjars, ostriches, owls, parakeets, parrots, passerines, pelicans, penguins, petrels, pigeons, quetzals, quails, rails, sandgrouse, storks, sun bitterns, swifts, trogons, tropicbirds, turacos, vultures, waders, and woodpeckers.

Because of the variation of the turbulence that shaped them, wings and bodies of birds come in various sizes. Some birds have large wings, while others have small wings. According to the Encyclopedia Britannica, the greatest wingspan (up to 3.5 meters) was recorded on the wandering albatross (Storer, 2021). Although all birds have wings, not all birds can fly. For example, despite their massive size as compared to that of other small birds (e.g. bee hummingbirds which size is about 5.5 cm), penguins are not able to fly in the air, but they are able to dive through the water, and while on land, they walk. While they are unable to fly, emperor penguins are known as the best divers, capable of diving at depths of 483 meters, as reported by Encyclopedia Britannica (Storer, 2021). The previous source also reported that some birds (e.g. falcons, ducks, geese, and pigeons) can fly 60-152 km/hr. According to the same source, while in a dive, a peregrine falcon has flown at more than 320 km/hr. The ostrich is unable to fly, but it runs faster than many other birds. It can run more than 70 km/hr. The inability of ostriches and penguins to fly taught me that, to some extent, a balance must exist between the weight of birds and their ability to fly.

Fig. 58: Feathers (Photo credit © Nathanael-Israel Israel)

Birds have various types of beaks, not because they have evolved, but because they were formed as such. Birds have no teeth but have wings usually covered with feathers (which are considered as outgrowths of the epidermis). Feathers (Fig. 58) consist of a central shaft called rachis, from which emerge paired branches (called barbs) on each side. Called calamus, the base of the rachis is unbranched (meaning

CHAPTER 24: ANIMAL FORMATION

has no branch) and is the portion that usually goes beneath the skin. The branches that emerge from the barbs are called barbules. The outer barbules on the barb have hooks (called hamuli). This organization of feathers points to a system of bodies under the influence of turbulence. I explored this kind of branching with plants. Indeed, compared to a tree, the rachis of the feathers is like the trunk, while the barbs are like the principal branches. The calamus are like the segment separating the primary body and the innermost secondary body. The barbules are like the secondary branches. Coming up are some photos of feathers.

I also think that wings are for birds what the forelimbs or forefeet are for mammals. Applied to humans, wings of birds are like human's arms. Therefore, I felt like wings are produced from the branching of the precursors of the thorax and abdomen of the precursors of birds just as the precursor of hands and arms branched out from the precursor of the thorax and abdomen of the precursors of most mammals. Fig. 59 to Fig. 61 exemplify some types of birds.

Fig. 59: Birds
Source: Dr. Nathanael-Israel Israel / Science180, www.science180.com, © 2020-present, all rights reserved.

CHAPTER 24: ANIMAL FORMATION

Fig. 60: Birds
Source: Dr. Nathanael-Israel Israel / Science180, www.science180.com, © 2020-present, all rights reserved.

Fig. 61: Birds

Source: Dr. Nathanael-Israel Israel / Science180, www.science180.com, © 2020-present, all rights reserved.

Like I introduced in the chapter on turbulent neck, the long neck of some animals (e.g. herons and ostriches) can be explained by the relatively longer distance the cell layers of the precursor of their torso had to travel away from the precursor of the head before developing or going through the type of turbulence required to form larger structures as those found in Zones 2 and 3. Another way of saying this is that, as layers of cells were being collected together or arranged as the embryo was growing, the precursor of the thorax and abdomen (of birds having a long neck) had to be moved over a relatively long distance before being formed. In the end, the long distance travelled by the precursor of all the body parts downstream of the head was encrypted in the long neck (e.g. Fig. 62).

CHAPTER 24: ANIMAL FORMATION

Fig. 62: Neck of a duck and a heron (Photo credit © Nathanael-Israel Israel)

According to Encyclopedia Britannica, birds are more related to reptiles than to mammals (Storer, 2021). Because of some similarities between them and reptiles, birds are wrongly thought by evolutionists to have evolved from reptilian ancestors. Although I do not support the evolution theory of birds, I felt like the turbulences in the precursors of some fish, reptiles, and birds had some similarities. For instance, the texture of fish meat is more related to that of birds and reptiles. People who have eaten snakes have testified that it tastes like fish. Because the raw material used to form fish is water, I think that birds and snakes also could have been formed using water. Many snakes live in water, suggesting that their origin may also be well related to water. But this does not mean that birds, fish, and reptiles descended from one another or that they are related.

Now that we have checked out the birds, let's talk about insects, some bird's favorite food. During part of my writing, I frequently sat outside to get some sunshine and was surrounded by birds searching for insects to eat on the ground of my backyard. To enjoy the birds a little more, I bought bird feeders and hung them up to the trees. Birds, squirrels, and raccoons enjoyed the bird food, while I enjoyed watching them treat themselves to some free food. Some pigeons that I caught and released became so accustomed to me that they ended up building nests by my car garage.

24.3. Insects

Insects are hexapod invertebrates that have a chitinous external skeleton called exoskeleton (hard outer covering made generally of chitin), a three-part body (head, thorax, and abdomen), 3 pairs of jointed legs (hence the name hexapod), compound eyes, and a pair of antennae. Insects generally have one pair of legs on each of the three segments composing their thorax and one or two pairs of wings still located on the thorax. The digestive, respiratory, excretory, and reproductive systems are found on the segments (usually 11 for most of them) of the abdomen. However, some insects such as spiders have 4 pairs of legs, a body of 2 segments (i.e. cephalothorax and abdomen) but not 3 as most other insects and they have no wings and no antennae.

More than a million insect species have been described (Chapman, 2006). In other words, over 50% of all described organisms described on Earth are insects. Examples of insects (Fig. 63) are ant, aphid, assassin bug, bedbug, beetles, bumblebee, butterfly, cockroach, corn borer, cricket, dragonfly, earwig, flea, flies, grasshopper, hornet, horse fly, house fly, ladybug, mosquito, moth, stinkbug, stonefly, sucking louse, termite, thrips, tsetse fly, walking stick, and wasp. Although they look like insects, the following terrestrial arthropods are not classified as insects: centipedes, millipedes, scorpions, woodlice, mites, and ticks. Although some people may think that all insects have wings, it is important to underline that some insects called wingless insects (also known as apterygotes) do not have wings. Winged insects (also known as pterygotes) are the ones with wings.

The matter that was split-gathered to form the apterygotes did not yield wings just as on the scale of the celestial bodies, some precursors were unable to split and have satellites. In other words, the clusters of matter that were supposed to split-gather to form wings did not go through that process when the apterygotes were being formed. This also implies that the ability of some insects to have wings is an advanced step in the completion of the processes that formed insects. But this does not mean that winged insects are superior or descended from wingless insects.

For insects, the turbulent trunk can be assimilated to the (dorsal and ventral) tracheal trunk, whereas the bones can be the (thoracic and abdominal) spiracles. In other words, the precursor of the tracheal trunks is for the insects what the precursor of the spine and vertebrae are for bony animals. The head, thorax, and abdomen could have been formed according to turbulent processes like those that formed similar body parts of other animals I already discussed.

CHAPTER 24: ANIMAL FORMATION

Fig. 63: Insects
Source: Dr. Nathanael-Israel Israel / Science180, www.science180.com, © 2020-present, all rights reserved.

24.4. Reptiles

Reptiles (Fig. 64) are vertebrates such as alligators, amphibians, caimans, crocodiles, dinosaurs (extinct), geckos, iguanas, lizards, snakes, tuatara, turtles, and tortoises. Their skin is mostly covered with scales, which are made of keratin (a structural fibrous protein). Some reptiles possess limbs (e.g. crocodiles and lizards), while others are limbless (e.g. snakes). Their length varies according to their species. For instance, while the length of some crocodiles are over 7 meters, some pythons are more than 10 meters long.

Although most amphibians are classified as reptiles, some literatures classify them as a different class of animals. Such is the case of frogs, which, like human beings, have no tail. In addition to their nostrils, frogs and many other amphibians breathe also through their skin.

Encyclopedia Britannica reported that "reptiles mostly reproduce sexually and their sexual / reproductive activity happens through the cloaca, the single exit/entrance at the base of the tail where waste is also eliminated" (Zug, 2021). The lack of a separate hole / orifice for defecation and reproduction is another example of how simple the organization of reptiles is as compared to that of more advanced organisms such as mammals.

The only mode of locomotion for some reptiles is to crawl, creep, slide, bipedal, undulate, or push backwards against their supports. Encyclopedia Britannica reported that reptiles do not have the ability for the "rapid sustained activity found in birds and mammals" (Zug, 2021). In other words, reptiles are not as enduring and rapid as birds and mammals. The lack of arms and legs of snakes for instance can also work against their potential. This may also be explained by the fact that reptiles may have been born from a biological turbulence similar to that of Zone 5. In general, the more some organs or body parts are missing from some organisms, the more those organisms can be limited in certain functions performed by organisms that have those organs.

The giraffe-necked reptile called Tanystropheus (that I already introduced in the chapter on turbulent neck) could have been a reptile like snakes or lizards, but as the precursor of all of its body parts downstream of the head was being formed, it flowed or elongated or produced many cells over a relatively long distance before it went through a more developed turbulence worthy of yielding the body parts downstream of the head. By the time the body parts in its abdomen were being formed, the ability of the cells (which were being formed) to keep multiplying and forming long and big body parts was reduced. Hence a long neck and a very small thorax and abdomen.

Although they are very gigantic and look like mammals, dinosaurs are generally classified as reptiles. Fossil records suggested that some dinosaurs could have been up to 30 meters long. Some people may even wonder why and how dinosaurs were formed in the first place. As I showed in previous chapters, when turbulence occurs in a system, beyond a certain intensity, both large and small bodies are formed. The size of the large bodies is affected by the intensity of the turbulence and some characteristics of the matters of their precursors. The availability of the matters for the precursors of the forms of life was not a limiting factor for the biological turbulence, for water and soil were hugely available on Earth during the formation of organisms. The way the biological turbulence was calibrated or scaled to make the organisms formed with the opportunities and constraints that their environment would impose on them is one of the key factors that limited the maximum size of organisms on Earth. Just as in the turbulence of the precursor of the celestial bodies, large bodies are always fewer than small bodies, the number of the largest organisms formed on Earth were smaller than that of the small organisms. Hence, small organisms like bacteria could be found everywhere on Earth, yet dinosaurs could

CHAPTER 24: ANIMAL FORMATION

not and the few that were formed did not last long before disappearing.

Fig. 64: Reptiles
Source: Dr. Nathanael-Israel Israel / Science180, www.science180.com, © 2020-present, all rights reserved.

Now that we have covered the reptiles, let's take a look at the animals that are warm blooded, and which most people enjoy seeing.

24.5. Mammals

Derived from the Latin word mamma (meaning breast), mammals are vertebrate having mammary glands (which is the milk-producing gland in females). In other words, mammals feed or nourish their young or children with the milk from their mammary glands generally called breast. Encyclopedia Britannica added that mammals are characterized by a "neocortex (a region of the brain), fur or hair, and

3 middle ear bones" (Jones, 2011). Most mammals have 4 limbs (usually referred to as legs and arms), but some aquatic mammals have no limbs, but fins, which look like limbs. In the case of bats, the forelimbs are structured into wings, making bats the only mammals that truly fly. Humans have 2 legs and 2 arms. Like most mammals, human beings have a head, neck, trunk (i.e. thorax and abdomen), 2 arms (with hands and fingers), and 2 legs (with feet and toes).

Like I introduced in my writing on clean and unclean animals (see chapter 28), the hoof of some mammals is divided, while that of others is not. I explained why I think the division or separation of the hoof indicates the advancement level of the completion of the biological turbulence that formed those animals. Hence, some of those animals which hooves are divided and which also chew the cud are considered as more advanced and more qualified for certain higher usages (e.g. food and/or sacrifice) than others that don't.

Mammals are the only kind of animals that generally have hairs. However, some mammals such as whales (which some people may refer to as fish) do not have hairs. Other characteristics of mammals is the direct connection between their jaw and skull, the existence of a diaphragm between their heart and lung from the abdominal cavity. Finally, mammals are the only organisms which mature red blood cells lack a nucleus.

All mammals known as of 2025 give birth to their children except for a few species of monotremes (egg-laying mammals). For instance, mammals like platypus and echidnas of Australia lay eggs and do not carry a pregnancy like most other mammals do. In other words, except for a few mammals that lay shelled eggs, most mammals bear their young in their mother's womb during their pregnancy. Hence mammals are generally called viviparous, meaning their females bear live young. While in about 9 months, human pregnancy usually arrives to its term and babies can be born, the pregnancy of some mammals lasts just a few months. Most mammals have a placenta in which the embryos spend most of their time in the mother's womb before the mother births the children. Some mammals like marsupials give birth to immature children, which have to complete their development in their mother's pouch where they are fed via their mother's nipples. To some extent, I see breasts as a ramification of the thorax of mammals. Their presence with mammals and absence with fish, birds, reptiles, and insects is a sign of the advancement of some of the processes which formed mammals. Nevertheless, this does not mean that all mammals are more advanced than organisms from the other forms of life.

As of 2018, about 6,495 mammal species were identified (Burgin et al., 2018). The most massive animals on earth are found among mammals. Mammals are numerously dominated by rodents, and they are found on all continents. Some mammals are semi-aquatic, meaning they can partially live in both water and on land. Examples of these are cane rats, hippopotamuses, and otters. Aquatic mammals such as whales live in water most of the time and they do not have legs.

Examples of mammals (Fig. 65 – Fig. 86) include apes, bats (which some people may think are birds), bears, cats, cows (Fig. 66), coyotes, dogs, elephants (Fig. 69),

CHAPTER 24: ANIMAL FORMATION

goats, hippopotamuses, hogs, horses, kangaroos, leopards, lions (Fig. 83), marsupials, moles, monkeys, opossums, pangolins, panthers, rabbits, raccoons, rhinoceroses, rodents (e.g. cane rats, mice, porcupines, rats), seals, sheep (Fig. 71), shrews, whales, wolves, and human beings (which are the highest quality of organisms on Earth). Although some primates (e.g. apes and monkeys) look like human beings, they are not as completely formed as the latter, nor do humans descend from them. The similarities between some humans and primate features are due to the similarities of some of the processes that shaped their precursors. Again, when I talked about the precursor of human beings, I do not mean a being that became human beings, but I mean a cluster of matter upon which the turbulent program of life acted to form human beings.

The diversity of the characters among mammals as well as among the other forms of life is due to the abilities imparted on them by the turbulent program of life and to the limitations imposed on them by the characteristics of the organs that were formed from their precursors. From personal experience, I have come to realize that, even a tiny variation of some chemicals in living things can significantly shift their behavior. Some carnivores may display more aggressiveness than herbivores just because of the makeup of their bodies and the variation of their chemical composition for instance in their brain and elsewhere, which altogether forged the animals to act in certain ways. Hence some carnivores (lions, leopards, cats, dogs, weasels, bears, and seals) can display some behaviors that will never be seen with herbivores like cows, deer, goats, and sheep etc.

As announced in the chapter on turbulent tail, the tail of some mammals is just a body formed under the circumstances of turbulence Zone 5, the least turbulent zone. In other words, tails are among the last organ to be formed during the formation of mammals. Furthermore, the entire body of some reptiles (like snakes) looks like a tail, which points at how less developed was the biological turbulence that shaped the precursor of most reptiles, particularly those having no leg or arms but which crawl on their belly only.

Fig. 65: Camels
Source: Dr. Nathanael-Israel Israel / Science180, www.science180.com,
© 2020-present, all rights reserved.

CHAPTER 24: ANIMAL FORMATION

Fig. 66: Cows
Source: Dr. Nathanael-Israel Israel / Science180, www.science180.com, © 2020-present, all rights reserved.

Fig. 67: Deer
Source: Dr. Nathanael-Israel Israel / Science180, www.science180.com,
© 2020-present, all rights reserved.

CHAPTER 24: ANIMAL FORMATION

Fig. 68: Dogs
Source: Dr. Nathanael-Israel Israel / Science180, www.science180.com, © 2020-present, all rights reserved.

Fig. 69: Elephant
Source: Dr. Nathanael-Israel Israel / Science180, www.science180.com, © 2020-present, all rights reserved.

Fig. 70: Giraffe
Source: Dr. Nathanael-Israel Israel / Science180, www.science180.com, © 2020-present, all rights reserved.

CHAPTER 24: ANIMAL FORMATION

Fig. 71: Goat
Source: Dr. Nathanael-Israel Israel / Science180, www.science180.com, © 2020-present, all rights reserved.

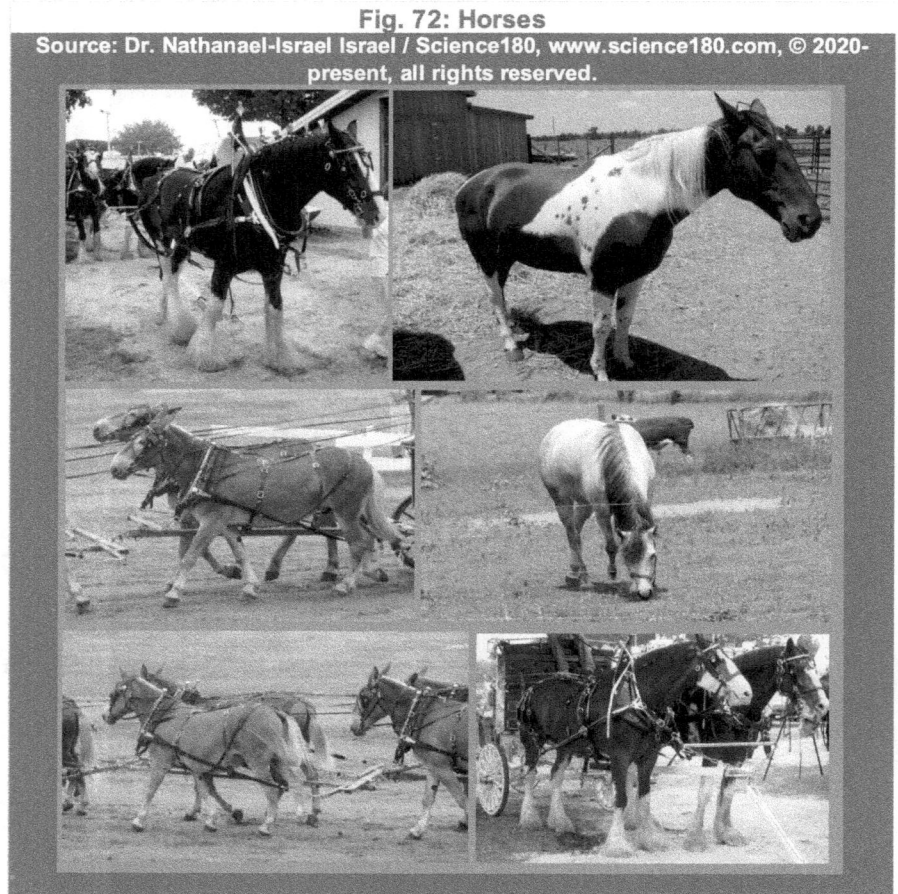

Fig. 72: Horses
Source: Dr. Nathanael-Israel Israel / Science180, www.science180.com, © 2020-present, all rights reserved.

CHAPTER 24: ANIMAL FORMATION

Fig. 73: Tiger
Source: Dr. Nathanael-Israel Israel / Science180, www.science180.com, © 2020-present, all rights reserved.

Fig. 74: More ruminants

CHAPTER 24: ANIMAL FORMATION

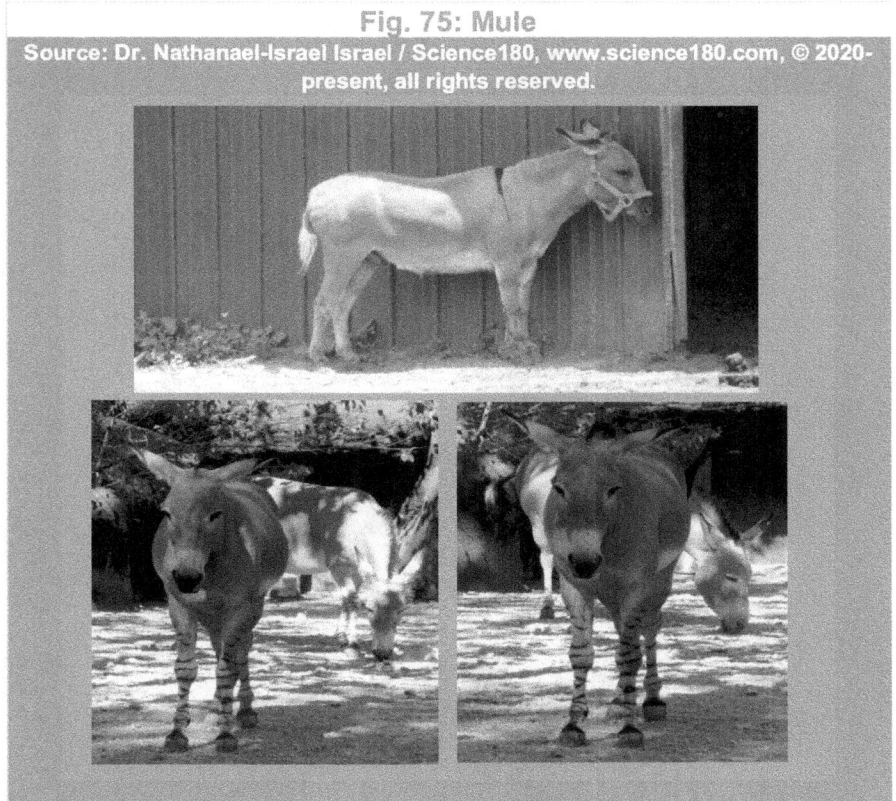

Fig. 75: Mule

Fig. 76: Rhinoceros
Source: Dr. Nathanael-Israel Israel / Science180, www.science180.com, © 2020-present, all rights reserved.

CHAPTER 24: ANIMAL FORMATION

Fig. 77: Squirrels
Source: Dr. Nathanael-Israel Israel / Science180, www.science180.com, © 2020-present, all rights reserved.

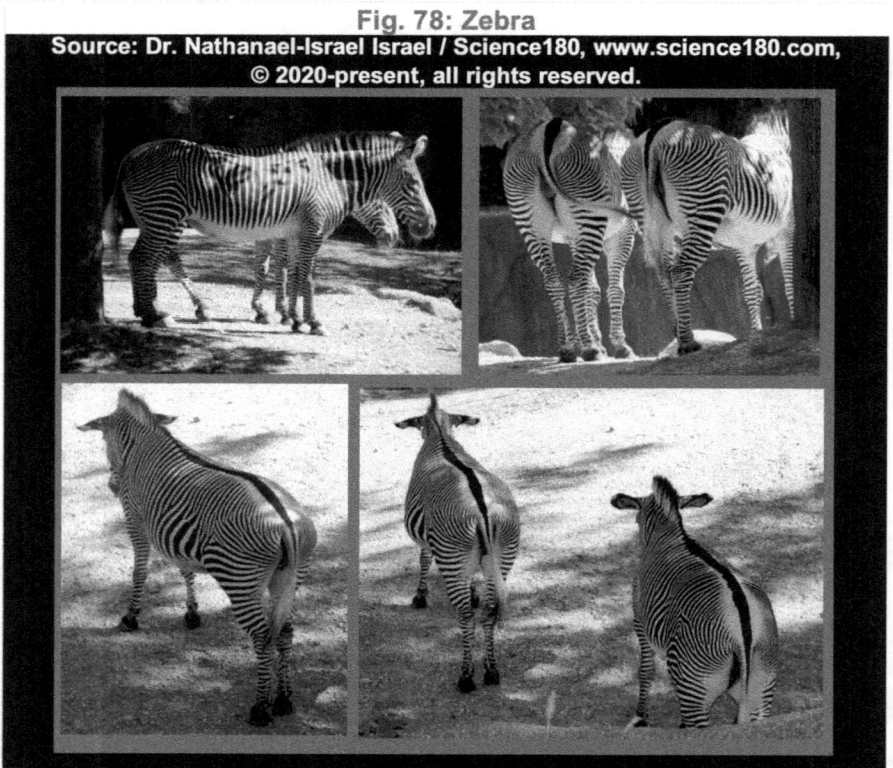

Fig. 78: Zebra
Source: Dr. Nathanael-Israel Israel / Science180, www.science180.com, © 2020-present, all rights reserved.

CHAPTER 24: ANIMAL FORMATION

Fig. 79: Chimpaze

Source: Dr. Nathanael-Israel Israel / Science180, www.science180.com, © 2020-present, all rights reserved.

Fig. 80: Llama

Fig. 81: Koala

Fig. 82: Rabbit

CHAPTER 24: ANIMAL FORMATION

Fig. 83: Lion and lioness

Fig. 84 Leopard

Fig. 85: Grizzly

Fig. 86: Lamb

At this point, I would like to say a few words about human beings, which are also classified as mammals. With all his abilities, hierarchical position, and dominance of all other organisms on Earth, human beings cannot and should not be considered as a mere animal. Although we share some features with some wild animals, I don't think we were formed by just a biological turbulence. A certain biological turbulence was used to first shape the body of human beings. Then, the abilities conferred to him by the turbulent program of life were significantly different than the ones imparted into the other forms of organisms. Hence, human beings can for instance think and do other higher intellectual activities not found with other organisms. Using their mind, human beings have been able to research, discover, invent, and improve many things in their environment and even beyond. Like I will detail later, all human activities were not, are not, and will not be good. Some human

CHAPTER 24: ANIMAL FORMATION

actions degraded the environment and are complicating the existence of life on Earth. Nevertheless, human beings have the power to dominate the Earth. They may not be able to confront some big carnivores (e.g. lions, leopards, and tigers) face to face in the bush, but they have managed to capture these big animals and put them in captivity. However, except for some carnivores that have killed human beings, no carnivore has been able to submit humans to its ecological niche yet or put humans under its control yet. Human beings do not have a long neck like some giraffes and dinosaurs, but they were able to invent ladders and even lifts with motors they can use to reach higher places. Human beings do not have wings like birds, but they have invented airplanes and helicopters, which can fly faster than the fastest bird. They do not know how to swim like some fish, but they have invented boats, ships, and submarines which can take them very far much more than the fins of fish can take them. Human beings do not know how to hibernate like some reptiles and fish, which can stop their activities when bad weather comes and restart them when conditions are favorable, but human beings are able to invent things that allow them to cope with heat (e.g. air conditioning), cold (heaters), or even to travel from one location to another not like some artic birds which take months just to migrate toward the equator when the cold starts to hit, but humans can do so very fast. In less than one day, human beings can reach anywhere on the globe. They may not be able to accommodate the regular changes in the environment like most wild animals do by controlling the activities in the membrane of their skins, but human beings can sew and put on various kinds of clothes according to the weather. They are not able to dig holes in the ground with their fingers and mouth like some rats and other animals do, but they have invented ways to design and build various kinds of houses, even reaching heights that some animals cannot climb. This book alone does not have enough space to detail everything human beings have improved and damaged on this Earth and beyond by their efforts to improve their life, which for many reasons (that I will review later) turned against them, and against many other organisms and nonliving things on the planet and beyond. I can go on and on about listing more things that human beings have accomplished with their knowledge and its application to invent things throughout the ages, and particularly more recently in this technological era, during which certain things that could have been classified as spiritual or supernatural are being done using technology, which seems to be "advancing" at a pace which limit sounds difficult to predict.

To make a long story short, human beings are the highest quality of existence of the organisms on Earth. All of the theories that have tried to explain the origin of life have put human beings as the last organisms formed in the series of the formation of living things. Considering life on Earth, human beings are undoubtedly the highest ranked organism. Although some people may keep claiming that there is "no yardstick of highness or of ranking organisms", the abilities of human beings as compared to those of other organisms also support the idea that man is the crown of creation or formation of life. For no other organism on Earth is as abled as human beings. Although many people will not believe in it, the manifestation of human life is beyond the natural, hence I will also revisit the formation of human beings later in this book, adding some philosophical and religious connotation to it. Though some may perceive angels as higher or more powerful, a thorough investigation will show that, angels are inferior to men, although most human beings have not lived by the highest standards of life worthy of their rank among the living. But the day is coming when some human beings will manifest themselves as they really are or as they should have.

Just as environmental conditions affect the ecology of organisms, so some of the morphological traits and behaviors of some human beings are affected by their geographical location. For instance, the height, diet, culture, and habits of most human beings are highly affected by their location. On top of the influence that the environment can exert on them, human beings are able to acquire some attitude and characteristics by their own choices and lifestyle. For instance, some gluttons may become obese, while those who are starving may become skinny. In other words, on top of the shape their environment could have imposed on them, some people can become obese if they eat beyond certain limits. Likewise, people from certain countries are very short, while those from other nations are tall. Because my goal for this book is not to review all biometric features, I will not address any more characteristics of humans according to their geography or culture. But my point is that many things that humans have been doing have not been defined only by natural laws (as they applied to most wild animals, which have not innovated anything since their beginning), but also by human efforts to alter their life for good or bad, willingly, or unwillingly.

Before I close this chapter, I need to quickly mention an observation I made concerning the abilities of organisms, particularly human beings and even of nonliving things. Indeed, I have observed that, most things in nature are moving and anything that stops moving dies or shifts from one form of existence to another. Precursors of all bodies (living and nonliving) were formed under "pressure". The aggregation and impartation of conformation of everything in the universe could have been "painful" or "uncomfortable" to the precursors of bodies in the universe. Even after their birth, human beings are shaped by some of the problems they go through. They keep moving to solve their problems, and meanwhile, they live and perform living functions. When they will stop moving, they will generally not last long before dying. For life and existence are tied to movement. Even celestial bodies and chemical particles are bound to keep moving within the limits set by the parameters of their formation, else they will cease to exist, which they cannot by themselves. Hence, everything in the universe is in motion. From the living to the nonliving, the extent to which something is molded by its environment defined its characteristics and, sometimes, the harder the molding, the more refined the products that came out of that process. I have seen some human beings manifesting characters under pressure that they have never express beforehand. The more some people think to get out of some problems, the more creative they can be. I am not saying that problems are always good things, but they can bring the best out of some people (who have good attitudes). All living organisms may be able to respond to

CHAPTER 24: ANIMAL FORMATION

some changes by expressing certain genes, but on top of that, human beings have an advantage of being able to use their mind beyond the ultimate level of "any" other organism on Earth. However, in their arrogance, some human beings have elevated themselves above their "limits", while minimizing the potential and intrinsic abilities of other living things on Earth and beyond.

'Science180 Academy' Success Strategy:
SCIENCE180 MODELS OF THE ORIGIN OF THE UNIVERSE AND ITS CONTENT

Science180 Models consist of all the theories elaborated by Nathanael-Israel Israel regarding his ground breaking discovery on the origin of the universe and its content including all forms of life and chemical particles. These theories are detailed in various books written by Dr. Nathanael-Israel Israel encompass the following:

- **1. SCIENCE180 MODEL OF COSMOLOGY**, also called Science180 Cosmology, Science180 Model of Cosmology, Science180 Cosmological Model, a scientific theory that explains Science180 to the scientists. Discover the details of this model in Nathanael-Israel Israel's book titled *"Turbulent Origin of the Universe"*. In that book, you will also unearth the new physics that will revolutionize science forever and land you into a zone of original ideas that improve lives nonstop regardless of your expertise.

- **2. SCIENCE180 CREATIONISM**, also called Science180 Model of the Creation of the Universe and Life by God, a scientific theory that presents the origin of the universe in a biblical language. If you want to learn more about how to scientifically prove the Biblical account of the creation of the universe and the existence of God in a way that makes the head of God deniers to spin faster than a DJ's turntable, then get Nathanael-Israel Israel's book titled *"Reconciling Science and Creation Accurately"*.

- **3. SCIENCE180 MODEL OF THE ORIGIN OF CHEMICAL PARTICLES**, a scientific theory that explains the origin of chemical particles with the perspective of Science180 Turbulence. If you want to professionally learn how to transform the true knowledge of the origin of chemical particles into insights that significantly add value to your life in less time, successfully establish you as a symbol of freedom, power, creativity, and originality in your field of expertise, get Nathanael-Israel Israel's book *"Turbulent Origin of Chemical Particles"*, THE ultimate how-to guide for great people wanting to correctly decode the origin of the chemicals and positively transform their lives. Get this celebrated book today. Don't wait!

CHAPTER 24: ANIMAL FORMATION

- **4. SCIENCE180 MODEL FOR THE GENERAL PUBLIC** (which explains the origin of the universe and life to the general public in a language that laypeople can understand). Find out more in Nathanael-Israel Israel's book called *"From Science to Bible's Conclusions"*, a scientifically verifiable, bestselling book to finally get the accurate, jaw-dropping answer that has been rationally shaking believers, skeptics, and freethinkers. Get this very popular book today.

- **5. SCIENCE180 MODEL OF LIFE-ORIGIN**, or Science180 Model of the Origin of Life, a scientific theory that explains the origin of all forms of life using turbulence. To unlock the step-by-step pathway to decode the origin of life and get the power, freedom, and boldness to detect, correct, and remove all misinformation, ambiguity, and misleading claims and theories surrounding the life-origin and take advantage of the opportunities that an accurate understanding of the life-origin creates, get Nathanael-Israel Israel's book titled *"Turbulent Origin of Life"*.

- **6. SCIENCE180 MODEL FOR CHILDREN**, a children's version of the theory of the origin of the universe and life in a language that 7-12 years old children can properly understand. To know the proven formula that helps children to easily answer their huge universe-origin and life-origin questions with confidence, humor, and joy, get *"How Baby Universe Was Born"*, the pragmatic book that has been causing children to belly laugh and thank those who offered it to them.

- **7. SCIENCE180 MODEL OF PSEUDEPIGRAPHA**, a deep explanation of the secrets of the origin of the universe and life revealed a long time ago, but hidden from the general public. To discover how the only one ancient blueprint has the reliable power to help you to accurately decrypt the spiritual origin and history of everything in the universe, get Nathanael-Israel Israel's book called *"Origin of the Spiritual World"*. In it, you will discover deep rejected secrets that have prevented humankind from unearthing the beginning of the universe and know how to properly use the lost and rejected scriptures to articulate the process by which the universe was formed, so you can use that insight to improve your understanding of the Bible, innovate in your domain of interest, and improve your life perpetually.

- **8. *SCIENCE180 MODEL OF THE PROOF OF THE EXISTENCE OF GOD*,** a theory that ties together most of Nathanael-Israel Israel's discoveries that scientifically prove the existence of God. With Nathanael-Israel Israel's book "*Science180 Accurate Scientific Proof of God*", you will surely know the only way to scientifically know if God exist, and if so, which of the thousands of beings worshipped across the globe is the true God. In that book, you will also discover the errors in the scientific and religious theories (about the origin of the universe, life, and chemicals) that are putting you at a high risk you will never recover from if you don't quickly and confidently learn how to rationally take control over threats lurking at the edge of your efforts to understand the universe and life today.

- **9. *SCIENCE180 THEORY OF EVERYTHING*,** (also called the theory of all theories), ties together everything in the universe into a single theory. Checkout Science180.com to learn more about the incoming book that covers this extremely important topic.

CHAPTER 25: FORMATION OF FUNGI, PROTISTS, ARCHAEA, BACTERIA, AND VIRUSES

CHAPTER 25

THE EASY YET ACCURATE THEORY THAT SCIENTISTS TRUST TO QUICKLY UNDERSTAND THE FORMATION OF FUNGI, PROTISTS, ARCHAEA, BACTERIA, AND VIRUSES SO THEY CAN BECOME FULFILLED THOUGHT LEADERS IN THEIR FIELD OF EXPERTISE

Fungi, protists, archaea, and bacteria are generally smaller than most plants and organisms not by chance. For, just like I explained for the formation of celestial bodies, when a turbulence is acting on a cluster of matter, it can form bodies or various forms and shapes including a few large ones separated by several small ones. In the case of the formation of organisms, the biological turbulence that acted on water and solid to form the bodies of living things including the precursors of plants and animals as the precursors of larger bodies separated by the precursors of smaller bodies like fungi, protists, archaea, and bacteria. However, in some places, the turbulence was not developed enough to even form any plant or animal, but small bodies like fungi, protists, archaea, and bacteria. In some locations, all of these bodies could not even be formed, for the turbulence taking place could not accommodate them. Hence, sometimes, only a population of fungi, protists, archaea, and/or bacteria were formed alone and separated from other forms of life. The abundance of fungi, protists, archaea, and bacteria in the biosphere is because they required a less developed turbulence to be formed and their precursors were abundantly formed as more "developed" organisms were being formed. Hence, if a comprehensive inventory of all forms of life could be done, the number of species of fungi, protists, archaea, and bacteria would be far beyond that of plants and animals. Even with celestial bodies, I noticed that in most systems, there is usually fewer large bodies than small bodies, meaning that there are more small bodies than large bodies.

The presence of some protists, archaea, and bacteria in higher organisms (e.g. plants and animals) can also be explained by many reasons including:

- The precursor of some protists, archaea, and bacteria were formed when the higher organisms (plant and animals) were being shaped;
- The precursors of some protists, archaea, and bacteria were incorporated into the clusters of matter of the precursors of more "advanced" organisms;
- The precursor of some protists, archaea, and bacteria were formed inside the precursor of "higher" organisms as the latter were going through the processes of their genesis.

In other words, plants and animals were not just formed after fungi, protists, archaea, and bacteria came into existence and vice versa. For each form of life was shaped almost at the same time that others were formed. However, considering their position and role in the existence of other forms of life, plants could have come into existence first. To make a long story short, unlike what the conventional secular scientists think and have maintained throughout the years, the smallest organisms (e.g. bacteria, protists, and archaea) were not the descendants of the bigger organisms such as plants and animals that I discussed in the previous chapters. I will elaborate on this in the chapters to come.

25.1. Fungi

Fungi have characteristics intermediate between plants and animals, yet, they cannot be considered as descendants of any of them, for they were brought to life by a unique process, that I will describe very soon. When the precursors of fungi were being shaped, their cell wall ended up having both glucan and chitin, but no cellulose was found in plants. As their hypha (elongated and filamentous structures of fungi) were being formed, they branched out or forked as plants do, yet they were made of different materials. Hence, some fungi branch like plants. However, unlike plants, the precursor of fungi was unable to form chloroplasts, which could have allowed fungi to make their own food like plants. Therefore, some fungi resemble plants, but lack key features found in plants.

As the precursors of hyphae were being shaped, some of them divided into compartments, while others did not. Hence, while some hyphae (i.e. septate hyphae) are divided into compartments separated by cross walls, others (i.e. coenocytic hyphae) are not divided nor compartmentalized. When I reflected on this with respect to what I know about the celestial bodies, I felt like what could have caused some structures like rings of celestial bodies to be unable to split-gather into large and unique bodies but to be spread all over a certain space in their system could help explain why the precursor of some fungal structures were and are still unable to split into distinct cells, but to have to be made of compartments containing many cells, which, put together, sometimes look like an umbrella (e.g. mushroom head) (Fig. 87) or a tube (which resemble the stem or trunk of a tree). When I considered

CHAPTER 25: FORMATION OF FUNGI, PROTISTS, ARCHAEA, BACTERIA, AND VIRUSES

how soft most mushrooms are, it appeared to me that the turbulence that formed them was not strong at all. Hence, although they look like plants, they are not structurally strong like plants.

Fig. 87: Mushrooms (Fungi)
Source: Dr. Nathanael-Israel Israel / Science180, www.science180.com, © 2020-present, all rights reserved.

25.2. Protists

Considering their resemblance to plants in many ways, including the ability of some of them to photosynthesize, protists may have been formed in environments where the biological turbulence was not strong enough to form plants. Knowing that some of them even live inside animals, on November 18, 2021, I felt like their precursor also did well among the precursors of animals. Hence, they are found near plants and animals and can even live in them. The weakness of the turbulence that formed them can explain why some protists do not even have chloroplasts, which could have made them to be more advanced like plants. Just as each organism is formed with a set of abilities, which go also with some weaknesses, so also some protists were formed lacking certain organelles (e.g. chloroplasts). Unlike what some people think, protists were not born with some organelles that they later lost. In other words, the fact that many protists lack some key organelles found in plants supports the idea that the processes that birthed them were not as strongly turbulent as those that birthed plants. Hence, although they can quickly colonize their environment, protists are not as gigantic as plants. Below, are some illustrations of algae (Fig. 88), which are the most dominant protist in the world.

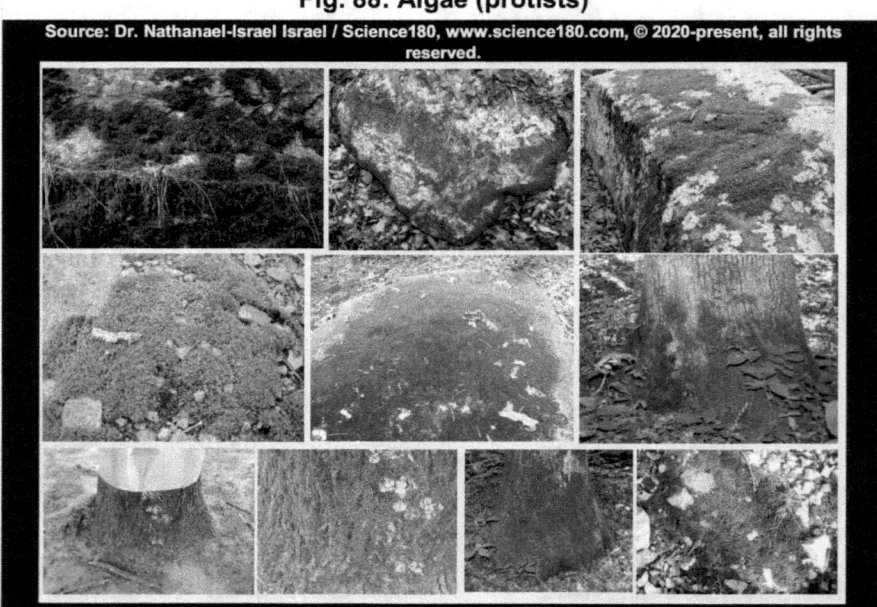

Fig. 88: Algae (protists)
Source: Dr. Nathanael-Israel Israel / Science180, www.science180.com, © 2020-present, all rights reserved.

25.3. Archaea

Derived from a Greek word meaning "ancient things", the name "archaea" was given to some organisms because the first archaea to be characterized were methanogens (i.e. a methane-producing bacteria, for instance capable of reducing carbon dioxide to methane) and because some people think that the early Earth was dominated by methane (CH_4), they thought those organisms were the primitive ones. What a mistake! For the Earth has never been dominated by methane but by nitrogen, oxygen, and carbon dioxide as of today (see my book *"Turbulent Origin of Chemical Particles"*). It is unfortunate that, every time some scientists have a glimpse at the proprieties of some organisms, they always try to link it to a chain of descendance as if every organism on Earth was really formed after a common precursor that evolved and/or as if it is relevant to always try to link every single organism to an evolutionist or evolving parent as a child that must have a parent.

CHAPTER 25: FORMATION OF FUNGI, PROTISTS, ARCHAEA, BACTERIA, AND VIRUSES

The precursors of archaea were not able to form more than one cell; hence, they are single-celled prokaryotes (i.e. organisms lacking a distinct nucleus with a membrane and certain specialized organelles). The ability of an organism to be made of many cells is the result of a more advanced biological turbulence than the ability to be made of just one cell. Therefore, archaea are formed out of a weak biological turbulence, implying that many features of the turbulent program of life were not expressed in most archaea.

The fact that archaea have characteristics intermediate between bacteria and eukaryotes suggested to me that some of the turbulent conditions that prevailed in the environments where their body parts were formed also dominated when the body part of some bacteria and eukaryotes were being shaped. To put it another way, the turbulence that shaped the precursor of archaea could have been intermediate between that which shaped bacteria and the one that shaped some eukaryotes. The precursors of some archaea and bacteria may have been formed during the intermittent split-gathering of the precursors of organisms into the precursors of eukaryotes. This also can explain why archaea are also abundant in environments where bacteria and eukaryotes are found, for some of them were formed by near similar processes. In other words, it was not that archaea were formed before or after bacteria and eukaryotes. But even in the same ecosystems, the biological turbulence that took place divided the precursor of life (water and soil) into a "default" body including the precursor of archaea in locations where the turbulence was intermediate between the one that formed bacteria and eukaryotes. Hence, the scientific community had a hard time classifying the archaea. In the past, they were classified as bacteria, because their size or shape is similar to that of bacteria, but they contain genes and metabolic pathways similar to those of eukaryotes. It should not be surprising that the cellular membranes of archaea are made of molecules different from those of other organisms. That was the stage of my thinking on November 18, 2021, as I was reflecting on the origin of archaea.

25.4. Bacteria

Because bacteria usually consist of one cell and are prokaryotes (meaning their cells lack a nucleus) made me think that they were born from a turbulence less advanced than that which formed archaea. For archaea have some eukaryotic features, meaning more advanced traits. Bacteria are present everywhere on the planet because the biological turbulence that was needed to form them occurred more broadly than that which formed higher organisms. The weakness of the biological turbulence of the precursors of bacteria can also explain why their DNA is not even compacted inside their cell, which does not even have a nucleus, whereas the nuclear DNA of eukaryotes is usually packed. For it takes a certain amount of energy and turbulence intensity to gather the precursor of DNA and or other biomolecules together tightly. Because the biological turbulence of the precursors of eukaryotes was stronger, their DNA was more compacted and they have more organelles than prokaryotes. The weak intensity of the biological turbulence of the precursors of bacteria can also explain why their organelles usually lack membranes.

When the precursors of bacteria were being molded, the turbulence was not very strong; hence, their DNA is not even compacted and wrapped inside a nucleus as is the case for eukaryotes. By the time they were formed, bacterial cells lacked a nucleus. In other words, like most prokaryotes, the precursor of bacteria did not have the ability to form a nucleus and/or the size of its precursor did not allow the turbulent program of life to express the ability related to the formation and activity of a nucleus. In the chapter on the formation of macromolecules, I extensively explained the processes that were involved in the formation of DNA.

Bacteria are present everywhere on the planet because, during the formation of organisms, their precursors were formed as the precursor of other forms of life were also being split-gathered just as I explained in the chapter on intermittence. Their smaller size (usually a few micrometers long) and the lack of membranes in their organelles result from the weak intensity of the turbulence which shaped their precursor.

According to how their precursors were split and collected together, bacteria come in various forms or shapes. In other words, the shape of bacteria was affected by the biological turbulence that took place when the matter constituting these bacteria were being gathered. For instance, the precursors of spherical bacteria (e.g. cocci, singular: coccus) may have been gathered all the way around. The turbulence of some of their precursors could have been more tightened and the matter constituting them may be denser or more compacted. In contrast, the precursors of rod-shaped bacteria (e.g. bacilli, singular: bacillus) may not have been gathered as tightly as those of spherical ones. The precursors of bacteria that are curved rods or comma-shaped (e.g. vibrio) may have spiraled to a certain degree, with the comma-shaped ones less compacted than the curved rods. Spiral-shaped (e.g. spirilla) also had their precursor go through special turbulence which conferred them to such a shape. Bacteria which are tightly coiled (e.g. spirochaetes) could have had their precursors tightly assembled together, and these bacteria may be the densest. How tightly some bacteria are coiled or how dense they can affect the ability of some medications to treat bacterial infections related to them. The diversity of the shape of bacteria reminds me of the diversity of galaxies in which some are spherical, others are spiral, and others have mixed shapes.

Finally, just as some bodies formed in a turbulence have structures resembling those molded in turbulence Zone 5, on November 23, 2021, I felt like the flagella (i.e. thin threadlike structure or appendage that enables bacteria and other organisms to move, including to swim) and the filamentous structures found on some bacteria are like structures of turbulence Zone 5. To put it another way, the flagella of bacteria are a turbulent tail. The same logic can also be applied to other organisms. In short, the flagella found on bacteria and the filamentous structures found in some protists and even fungi are like structures of turbulence Zone 5 but which were never split from their mothers. Finally, because the turbulence that formed them was weak, some bacteria do not always have an extrude or "bumpy" turbulent belly

CHAPTER 25: FORMATION OF FUNGI, PROTISTS, ARCHAEA, BACTERIA, AND VIRUSES

like some animals do.

25.5. Viruses

Because viruses are not conventionally classified as organisms, I should not have talked about them at all when dealing with life, but I felt like I should clarify a few things that may interest those who are eager to know their origin. The composition and functions of viruses point to the fact that the ability to perform functions was not intrinsic to organisms as known by science today. Virus composition (which can be simplified as DNA and/or RNA surrounded or protected by a coat protein and sometimes also by an external lipid envelope) points to a level of turbulence of their precursor which could have not even allowed more biomolecules to be assembled. Hence, although viruses could infest other organisms (Fig. 89), the turbulence that formed them and which control their existence today did not leave room for them to amass any more matter to be as large as conventional organisms. Because viruses are much smaller than bacteria and cannot perform all functions performable by bacteria, I think that the biological turbulence that shaped the precursors of viruses was weaker than that which formed the precursors of bacteria. The abilities of viruses to perform some functions found with organisms suggests that some of the processes that formed them were also found, to some extent, with the precursors of similar functions in conventional organisms. Unlike what some people think, higher organisms did not descend from viruses, and vice versa. But the precursors of viruses were formed as the precursors of organisms and their constituents were being formed. Hence, viruses exist between and within organisms.

Fig. 89: Callus caused by a virus living in the tree (Photo credit © Nathanael-Israel Israel)

TURBULENT ORIGIN OF LIFE

All viruses are not the same. The fact that some RNA molecules called viroids (not considered as viruses) do not have the aforementioned protein coat found in viruses although they share some features with numerous viruses suggests that the precursors of these virions went through a less developed turbulence than the precursors of conventional viruses. The day is coming when other forms of existence (not necessarily form of "life") will be defined by turbulence weaker than that of virions. Viroids may not have shared some of their formation processes with viruses. In other words, some processes that shaped viruses were missing in those that shaped viroids. Viruses are not ancestors of viroids and vice versa, and none of them has ever had an ancestor of any kinds, but each of them was formed after specific processes, which had some overlaps. The diversity and variation of these formation processes can explain the variety of viruses and viroids in the world. In the end, according to their shape, while some viruses are linear, others are circular, and others still are segmented. Similarly, according to the strandedness of their nucleic acid, some viruses are single-stranded, others are double-stranded, and others yet are double-stranded with single-stranded regions. The variation of all of these proprieties were caused by the processes that split-gathered their precursors and which were later maintained and reproduced by the codes of life including, but not limited to, the genetic code. Here, by precursor of viruses, I mean the clusters of matter on which the program of life enacted to form the viruses.

'Science180 Academy' Success Strategy:
SCIENCE180 ACADEMY PROGRAMS

Owned by Science180, Science180 Academy is a training, speaking, consulting, and mentoring program specialized in everything universe-origin, life-origin, chemicals-origin, and anything at the intersection of reason and faith, or science and religion.

Science180 Academy deals with different subjects according to the needs of its members or target groups. When people register to Science180 Academy, they must choose the program(s) they want to focus on so their training can be properly personalized accordingly. This is similar to how people register to a university, and take classes in a specific department matching their needs!

Science180's breakthroughs are so complex and dense that it is not realistic or good to try to explain all in just one academy, else people will be overwhelmed, disinterested, and confused by the plethora of data to handle. In other words, Science180 Academy offers a wide range of origin-related training in various domains strategically designed to allow people to choose the most suitable for their needs so that, regardless of their background or field of expertise, people can equip themselves, align their mindset, improve lives today and forever using the accurate explanation of the origin of the universe, of life, and of chemicals.

CHAPTER 25: FORMATION OF FUNGI, PROTISTS, ARCHAEA, BACTERIA, AND VIRUSES

Science180 Academy curriculum is based on 12 years of deep unconventional research that culminated with the publication of many much-admired books on the formation of the universe and its content: (www.Science180.com). The content of each Science180 Academy is strategically crafted by Dr. Nathanael-Israel Israel (who is acknowledged as the internationally-acclaimed world's authority in origin-related issues) to suit both scientists and nonscientists, religious and nonreligious people, leaders as well as followers, so they can fully decode the proofs of the formation of the universe, of life, and of chemicals they have been wanting to demonstrate or grasp.

The current programs of Science180 Academy are:

1. SCIENCE180 ACADEMY OF COSMOLOGY (Designed for all scientists who want to scientifically study cosmology, the science of the origin and fate of the universe)

2. SCIENCE180 ACADEMY OF TURBULENCE (This is a perfect fit for scientists and other experts interested in studying abiotic turbulence).

3. SCIENCE180 ACADEMY OF LIFE SCIENCES (Tailored to those who want to study biotic turbulence):

4. SCIENCE180 ACADEMY OF CHEMISTRY (Designed for chemists, biochemists, scientists, and other educated people who want to understand the origin of chemical particles)

5. SCIENCE180 ACADEMY FOR LAYPEOPLE OR THE GENERAL PUBLIC (Very fit for any layperson or "less" educated people who wants to learn (in a simple language) deep insights that even those who went to university for years were unable to decrypt by themselves, so these laypeople can be equipped to eliminate all forms of scientific and religious universe-origin prejudices)

6. SCIENCE180 ACADEMY FOR CHILDREN (This Academy breaks down origin key topics into language that children can fully understand). This is the only Science180 Academy that your whole family will like and enjoy together, and which will set children on the path of success by accurately showing them early in life the formation of the universe, and how to detect errors in theories or stories that would misguide them as they grow up.

7. SCIENCE180 ACADEMY OF THE PSEUDEPIGRAPHA AND SPIRITUAL WORLD (Only one ancient blueprint has the reliable power to help you to accurately decrypt the spiritual origin and history of everything in the universe. If you are a believer and want to delve into the prophetic, angelic, and higher order of knowledge based on the spiritual world, then this Science180 Academy is for you. This program is suitable for those who took at least "Science180 Academy of Creationism".

8. SCIENCE180 ACADEMY OF CREATIONISM (Science180 Creationism is a scientific theory spearheaded by the groundbreaking discoveries of Nathanael-Israel Israel, that scientifically explained the origin of the universe, life, and chemicals using turbulence, and that mathematically reconciled science and the Biblical account of creation for the first time in history. Science180 is different from all existing creationist theories known before 2025. Science180 Creationism reconciled science with the Biblical account of creation, including scientifically proving that the Earth was formed on Day 3, while the Moon and the Sun were formed on Day 4 of creation!). As you attend "Science180 Academy of Creationism", you will receive accurate answers to all your questions concerning the creation of the universe).

9. SCIENCE180 ACADEMY FOR FREETHINKERS & ALL ANTI-CREATIONISTS (This Science180 Academy is designed for evolutionists, anti-creationists, and all other types of unbelievers seeking to rationally explore and understand alternative arguments for creation or formation or origin of the universe, life, and chemicals from a fresh, scientific perspective).

10. SCIENCE180 ACADEMY OF LEADERSHIP-(Also called "Science180 Academy for Leaders", this program will enlighten leaders of organizations on how to solve their people problems, process problems, and profit problems related to the origin of the universe, of life, and chemicals according to their domain of expertise). With "Science180 Academy of Leadership", leaders will gain new insights so they can cast new visions and avoid focusing on screwed-up processes, products, and services related to universe-origin initiatives that need to be fixed, faced, or dealt with. Science180 Academy of Leadership will also equip leaders to address process problems related to inefficiency, gaps, missed opportunities, wasted time and efforts, too many steps, bureaucracy, useless layers between organization and customers concerning the innovation, research methodology, research, product development, strategic planning, workforce diversity in alignment with the historic Science180 breakthroughs so that they can sell more often at full price, avoid regrets in the end, open new markets focusing on real solutions, expand their products and services lines, cut useless costs and research, …

CHAPTER 25: FORMATION OF FUNGI, PROTISTS, ARCHAEA, BACTERIA, AND VIRUSES

11. SCIENCE180 ACADEMY FOR GOVERNMENTAL AGENCIES (Do you want to know how and why most nations and governments are wasting millions of dollars on universe-origin and life-origin researches they don't need … and how to avoid it? Indeed, for most developed nations, and even for some under developed countries, universe-origin projects can cost billions of US dollars and other expensive things that cannot be afforded without sacrificing crucial priorities. Even in developed countries, the impact and the return of investment of the space researches are subject of intense political and economic debates. What if your nation or institution can reduce wasteful spending on universe-origin research and life-origin research, as well as your dependency of wrong theories on the origin of the universe and life? "Science180 Academy for governmental agencies" will show you how to use the latest scientific breakthrough to better understand the origin of the universe without wasting money on what is already known or what we think we don't know, but that most scientists ignore. Having spent years accurately decoding the origin of the universe, of life, and of chemicals, Dr. Nathanael-Israel Israel delivers science-backed insight to properly understand all the processes connected to the universe formation—so you don't waste more money and time on trying to

12. OTHER SCIENCE180 ACADEMY: If you did not relate with any of the Science180 Academies mentioned above, but you are still interested in learning something specific about the origin of the universe, life, and chemicals that better fits your needs, please visit Science180Academy.com to contact us so we can discuss that with you.

CHAPTER 26

HOW TO TALK TO SMART PEOPLE ABOUT THE CHARACTERISTICS OF CELESTIAL BODIES AND THEY WILL IMPLORE TO REVIEW WHETHER THE SIMILARITIES BETWEEN LIVING THINGS IS SUFFICIENT TO IMPLY SIMILAR DESCENDANCES

Is there any scientific detail that seems bogus and unwise but that can really defend the fact that the similarities between living things don't imply a similar descendance or origin? Or do they?

As you would have noticed by now, all organisms have some features that can be (at least partially) found in other organisms. Sometimes, those features are so similar in many ways that it can be very easy to confound their sources. When different organisms share many features, the confusion can be more. Because some organisms are so similar, some people try to deduct their origin, but unfortunately, many mistakes are made while doing so. Here, I will explain what I have learned from investigating the similarities and dissimilarities of organisms with respect to their origin. Because, while studying the turbulence that formed the universe, I learned lessons that tremendously helped me to decode the origin of life, I will start this chapter by presenting what I discovered from the similarities and difference between celestial bodies.

26.1. Similarity and dissemblance between celestial bodies are not alone sufficient to deduce their descendance

During my investigation of the origin of the universe, I learned that the existence of similitude between celestial bodies is not always correlated with the similarity of their descendance. For instance, all of the celestial bodies in the Solar System descended from a common precursor (i.e. the precursor of the Solar System), which was split-

CHAPTER 26: SIMILARITIES DO NOT IMPLY DESCENDANCE OR ORIGIN

gathered into many daughter precursors, one of which ended up producing the big and hot Sun, while the other daughter precursors in the Solar System formed bodies like planets, asteroids, and satellites, etc. Some precursors produced just chemical particles. Although all of the planets in the Solar System descended from a common precursor, they do not all look the same. Some are terrestrial, while others are giant gas or giant ice. When I looked at the satellites in the Solar System, some look the same, yet they do not all come from the same precursor. For instance, some Neptunian satellites look exactly like some Jovian satellites, yet they come from different precursors: the precursor of the Neptunian satellites and the precursor of the Jovian satellites. Likewise, some Saturnian satellites resemble Jovian satellites, yet they come from different precursors. However, even inside the same planetary system, some satellites do not look the same at all, yet they come from the same precursor: the precursor of their satellite system. For instance, although all of the Jovian satellites come from the same precursor, most of them are different, some are very large, while others are very small. Furthermore, in the same planetary system, some bodies move very fast, while others move very slow, yet they all come from the same precursor, but were just molded differently. For instance, looking just at their radius, inclination, eccentricity, and rotation, it may be hard to even believe that the largest satellites in Zone 3 come from the same precursor of the satellites in Zones 4 or 5 of their planetary system. In the same planetary system, some satellites have a very huge orbital speed and others have a very small one, yet, they all descended from the same precursor: the precursor of their planetary system. The atmosphere of some planets is rich in certain chemical elements that are absent in the atmosphere of others. By the time that all celestial bodies in the Solar System were formed, they were diverse, and each has its own history, which is connected to the history of the precursor of the Solar System. But this does not mean that some celestial bodies in the Solar System descended from others.

To summarize, my work on the origin of the universe taught me, among many things, that similarities between the celestial bodies for instance are a false indicator of their origin and are insufficient to allege them having a common ancestor. If similarities alone can be considered, not all of the satellites in the same planetary system could even be considered as descending from the precursor of their planetary system. Likewise, although a primary planet is usually different than its satellites, both the planet and its satellites descended from the same precursor: the precursor of their planetary system. In other words, although the bodies in a planetary system can be very different from one another, they all descended from the same precursor of their planetary system. This example is similar to how children in a family differ and how, sometimes, people may tell you that you look like someone else, who may truly resemble your appearance, yet that someone else is nowhere near related to you or your family. Likewise, although the Sun and the bodies orbiting it all descended from the precursor of the Solar System, neither the Sun nor the bodies orbiting it (planets, asteroids, and satellites) look the same. In other terms, based on the physical appearance only, a few people can know that the Sun and all the bodies orbiting it came from the same precursor, which was split and differently shaped to yield the various celestial bodies present in the Solar System today.

In short, I demonstrated in *"Turbulent Origin of the Universe"* that the nature or characteristics of the celestial bodies does not just depend on their precursors or "ancestor" (if you prefer that term), but on the processes that shaped those precursors into distinct daughter bodies. In other terms, the processes through which precursors of bodies went through before being formed are better indicators of the history and story of their formation than any allegation based just on the physical similarities and differences between these organisms. I have discovered some of these processes by carefully studying the makeup and current functioning of these bodies. Likewise, on November 23, 2021, I came to realize that the similarity and the dissemblance between living things have fooled some people about their origin. Below, I will explain what I mean.

26.2. Similitude is not synonymous of hierarchical descendance or similar origin

In a house, many types of items can be found: couch, tables, plates, pots, refrigerator, stove, electronic devices, bed, office appliances, kitchen appliances, dresses, jewelry, etc. These things have different forms, designs, and functions indeed, but most of them are made using matters or materials consisting of atoms. These materials were differently molded to fit their purposes. For instance, the same wood used to make a table can also be used to make a chair, couch, wooden spoon, door, window frame, etc. Likewise, matters in the universe were molded differently to form nonliving and living things. Because a chair can be made with wood, which can also serve to make a bed frame, does not mean that a bed frame descended from a table or that the chair evolved to become a bed, and vice versa.

In a library, many books can be found and they are usually made using paper and ink. But the fact that, without opening them, books look alike does not mean that they descended from one another of that some books evolved to become or birth others. Similarly, the fact that chairs, tables, beds, firewood, and fresh wood are made with stem cuts from trees does not mean that chairs descended from beds or that tables descended from chairs or from trees.

Likewise, the fact that plants and animals have some similitudes in the way their genetic code works, or the way they function physiologically does not imply that plants descended from animals, and vice versa. In other words, the similarities between animals cannot prove that they descended from one another, regardless of how thoughtful a linear reasoning on this matter may statistically sound.

Furthermore, from one nation to another, or from one continent to another, people have built various types of houses: shacks, huts, igloos, teepees, sandy or muddy houses, brick houses, concrete houses, mansions, apartment complexes, various rental houses, many commercial houses, various residential houses, complex monuments, etc. But all of these buildings have different beauty, architecture, and cost indeed, but they are all made with materials coming from the earth. Every house

CHAPTER 26: SIMILARITIES DO NOT IMPLY DESCENDANCE OR ORIGIN

has a door and usually windows. Some windows are in wood, while others are in glass or metal and some are in plastic materials. Houses contain various equipment. In short, all of the houses in the world have some similarities and dissemblance indeed, but none of these is sufficient to claim that some houses descended from the others according to their complexity or location or that the houses of the rich descended from the houses of the poor or that the beautiful houses of the rich have collapsed to yield to modest or poorly designed houses for the poor! In other words, it is unthoughtful to allege that more-advanced or more-equipped houses descended from less-furnished houses. House designs depend on many factors including the culture, the cost, and the location. A brick or concrete house built on a rock may require different materials than a house built on water or on sand, or on a muddy environment. Although they are all made by human beings, some houses built on one continent do not resemble those built on another continent. Because some houses are apparently more beautiful than others, does not mean that those houses are made with precious material or that their designers were more intelligent than others. There is more to things than their physical appearances. Beautiful houses have collapsed where apparently ugly shacks stood. Likewise, roads that appeared nice were broken or cracked by earthquakes and drought, while others withstood the weather. The beauty of a person does not say much about his or her attitude and inner being!

However, a linear way of trying to explain the origin of life and the universe is filled with similar errors that I apostrophized above. I have come to realize that, because they are unable to understand the origin of the universe and the things and beings it contains, some people have invented linear and progressive ways of postulating the formation of everything, even if astronomical timelines have to be attached to them as if time can really make things happen by itself. For instance, evolutionism is a linear effort to connect the origin of the forms of life in the world as if the seemingly "less-developed" or "less-advanced" forms of existence came from the seemingly "more-advanced" or "more-complex" ones or as if complex things and beings came from seemingly simpler ones, and vice versa.

The decomposition of organic matters after the death of living things is a reaction that returns to the Earth the matter that was taken from it to form life. Due to the processes involved in their formation and also the nature of their constituents, some organic matters (e.g. bones) take more time to decompose than others. Bones and flesh are formed within the same organisms, yet their constitution is different. Likewise, hairs are tiny, yet they can take more time to decompose than some larger organs such as stomach and liver can do. If the complexity of hairs and bones could be investigated, some people may not believe that they both came from the same organisms, which have soft parts like intestines. This is another indication that the differences between things is not enough to define their origin.

Fish were among the first animals to be formed, and the process of their formation is also responsible for why their meat is usually very soft. The relative simplicity of the formation of fish can also explain why even their bones are not as hard as that of some mammals. Fish bones can be broken more easily and can also decompose more easily and faster than the bones of most mammals. In contrast, it took a little longer for mammals to be formed, hence their meat is usually harder than that of all other animals. In other words, the hardness of the bones of most mammals can be linked to the complexity of some of the processes that formed those organisms and their parts. Likewise, the simplicity of the intertwinement of the compounds of most plants can explain why plants also easily break down after their death. Organs that are made of complex compounds (cellulose) take longer time to break down than those made of simpler ones like glucose. Woody parts of plants take longer to decompose than leafy parts. For instance, for the same plant, a trunk will decompose more slowly than the leaves. While intestines can decompose within days after death, bones and hairs can last millennia. Excavation of some tombs showed bones dating thousands or years.

Although their theories cannot really explain everything, some people have never given a real thought to the revealed explanation of our origin such as those found in some religions, some of which I addressed in this book! For to ignore the power of God and His ability to quickly form various things in the same environment using laws like turbulence is to open the doors to human reasonings and imaginations attempting to explain the complexity in the world by breaking down each component of nature and then trying to piece them together with a human mind, which is doomed to fail without acknowledging the creator, that most people unfortunately don't want to talk about, hear, fear, obey, and seek before everything else!

Hence, since 2013, meaning 12 years before the publication of the first version of this book, I came to realize that similarities between things and beings are encryptions of the resemblance of some of the processes of their formation. Likewise, the similarities of the biochemical and genetic processes of organisms are not a proof of their common ancestor, but a sign of them sharing some similarities in the processes of their formation and the components required to sustain them. Because most DNA can be transcribed into RNA, which at its turn can be translated into proteins does not imply that all organisms come from the same ancestor.

CHAPTER 26: SIMILARITIES DO NOT IMPLY DESCENDANCE OR ORIGIN

'Science180 Academy' Success Strategy:
SCIENCE180 SERVICES AND PRODUCTS YOU WILL LOVE

Because you are reading this book, you are probably very interested in answering your questions about the origin of the universe, of life, and of chemicals. Imagine you want to be trained by Dr. Nathanael-Israel Israel and his team so you can benefit from their outstanding expertise to empower yourself or your team. Or you want him to give a keynote speech, a seminar, or any other kind of talk or conference at your organization. Or you want him to mentor you or some people or team at your organization. Maybe you have critical origin-related questions that you need his help to accurately answer. You want a true expert to talk with you about the customized program or game plan that fits your needs. You want him to tailor his advice, expert feedback, and proven shortcuts to the stage of life you are in and help you get to where you want to be in your desire to properly understand the origin of the universe, life, and chemicals and harness the benefits that come with it. Perhaps you don't know how to properly get any of these important tasks done according to your specific needs or the needs and demands of your organization. That is what Science180 Academy is all about. Visit Science180.com/services for more details about how to benefit from the services that Science180 provides.

Maybe you are a leader that wants to hire Dr. Nathanael-Israel Israel and his team to train some departments at your organization. Or you want to refer them to other companies like a good dish passed around the dinner table, and you want to explore how Nathanael-Israel Israel can pay you something for that referral. Maybe you attended Nathanael-Israel Israel's speaking program, for which, without going into details, he accurately raised your awareness about how the universe, life, and chemicals were formed. Or maybe you attended his training, in which he detailed and showed you how he decoded the scientific data using various tools and certain thinking strategies that helped him and which transferred some skills to you; and now, you are interested in a long term one-on-one consulting, or mentoring program with him, so that, he delves into more details about how to use proven techniques to decode the universe (strategies for data collection, data analysis, data presentation, writing, and even tips for future research) and change your behavior on a long term basis. If you related to any of the points mentioned above, Science180 Academy is the right fit for you!

Other customizable services that Science180 provides include: Assessments, Books, online courses, posters, how-to-guides, study guides, Book publishing, Conferences, Consulting, Executive mastermind groups, Face-to-face visits, Master classes, Online courses, Podcasting, Retreats, Seminars, Speaking engagements. Training, Video programs, Virtual presentations

TURBULENT ORIGIN OF LIFE

Here are other reasons why you should choose to work with or hire Nathanael-Israel Israel and the team at Science180:

- A simple life-origin theory that made no assumption
- Accurately understand life-origin. Be happy forever!
- All the life-origin solutions you love
- Bringing people together through the power of the accurate decoding and understanding of the origin of life
- Customizable life-origin trainings with unique materials
- Discover the key variables needed to decode life-origin
- Easily understand complex life-origin equations in minutes
- Enjoy the scientifically verifiable life-origin model
- Fearless life-origin decryption trailblazer
- First-stop life-origin information center
- Get ready to understand life-origin and live forever
- Improve your understanding of life-origin with new, accurate products and services
- Life-origin problems final bus stop
- Life-origin theory that helps you fight wasteful programs
- Light in the heart of science the lamp of understanding
- Nonconformist, rule-breaker, and accurate demonstrator of the life-origin
- One-stop platform for the origin life
- One-stop-destination for life-origin experts
- Source of unconventional wisdom and knowledge on the origin of life
- The all-in-one proven & uncomplicated life-origin formula
- The best life-origin decoding experience. Only on Science180
- The best way to explain life formation
- The formula at the intersection of life-origin science and the Bible
- The go-to source for valuable life-origin information
- The most accurate, reliable, safest, best explanation of the origin of life ever
- The place where life science accurately meets Biblical creation
- The place where life-origin gets decoded accurately
- The place where the accurate interpretation of life-origin data matters
- Science180: The premier organization that scientifically decoded the origin of life accurately using natural processes excluding evolution
- The science that refreshes your mind, faith, or doubt

CHAPTER 27: BIBLICAL ACCOUNT OF THE ORIGIN OF LIFE

CHAPTER 27

DO WE HAVE TO DENY GOD TO SCIENTIFICALLY TEST WHETHER HE CREATED LIFE–OR IS THERE ANY RATIONAL EXPLANATION OF THE BIBLICAL CREATION OF LIFE THAT CAN SCIENTIFICALLY BAIL ANTI-CREATIONISTS OUT OF ANY DOUBT?

- Is a church or a pastor making you doubt God?
- Is your belief making you doubt the Biblical account of creation?
- Is your school teaching about the universe-origin making you doubt creation?
- Is science making you doubt your faith?
- Is the Biblical account of creation making you doubt science?
- Is science making you doubt God or the Bible?
- Is your secular education making you doubt God and kiss biology books?
- Is there any important demonstration of the Biblical account of the origin of life that can bail evolutionists out of a dangerous doubt?
- Do you want to stand as a lightning bolt that electrifies those who are still struggling to understand the formation of all forms of life in the universe?
- Are you looking for proven ways to push the boundaries of human abilities to properly understand the un-understandable, mysterious, supernatural, unimaginable, impossible, and unthinkable that hold people back?
- Do you want to holistically detect, correct, and remove all misinformation, ambiguity, and misleading claims and theories surrounding the origin of life?

TURBULENT ORIGIN OF LIFE

27.1. Disclaimers and disclosure of my faith

Books after books, articles after articles, theories after theories, and religious views after religious views have been written or crafted to explain the origin of life, but as you will see below, the Biblical account of life has many things that caught my attention and deserved to be mentioned in this book. Being born and raised in a Christian family, I was taught early that God is the creator of all forms of life. But I had not deeply thought about the scientific implications of the Biblical story in the book of Genesis until 2013, when a concourse of circumstances led me to ponder on life.

Indeed, since I was born until today, I never believed that human beings or any other living thing is the product of evolutionism. Furthermore, even if I was not a Christian, the scientific demonstrations I did concerning how the formation of the Earth was completed on the 3rd day of creation, while the formation of the Moon and Sun was completed on the 4th day, and fluid layers were being separated by the 2nd day (see my book *"Turbulent Origin of the Universe"*) just as the Bible said more than 3500 years ago does not and cannot allow me to stay an unbeliever of the God of the Bible. Putting this another way is to say that, the undeniable support of the Biblical creation story by scientific evidences does not allow me and should normally not allow anyone who wants or chooses to follow the scientific evidences wherever they lead to ever doubt the Bible. Therefore, I felt convinced more than ever that, because the scientific data backed 100% the creation story that Moses said occurred in the first 4 days of creation, the other parts of the story that Moses said happened on the 5th and 6th day must also be true. For, it is illogical that the story of the Day 5 and 6 of creation could be false when the story of the first 4 days told at the same moment, more than 3500 years ago, is proven to be true today. In other words, the God that Moses said created the Earth, Moon, Sun, and many other celestial bodies by the 4th day of creation MUST also be the One who created all living things by the 6th day of creation. I have already spent 12 years of intensive research to prove that there is a God. In this book on the origin of life, I did not intend to repeat that exercise again. For the record, I want to be very clear at this stage of my writing that I believe 100% that life was created by God. However, what I did in this manuscript is to use my unusual insight into the origin and formation of the celestial bodies and microscopic particles to explain how God could have created life of formed living organisms. I may not have properly explained what happened, but I think what I presented in this book will enlighten many people that it is possible to explain the origin of life without rejecting the Creator, who deserves all the honor and glory.

CHAPTER 27: BIBLICAL ACCOUNT OF THE ORIGIN OF LIFE

Another Book by Nathanael-Israel Israel:
HOW GOD CREATED BABY UNIVERSE

THE FIRST AND ONLY BOOK THAT ACCURATELY EXPLAINS EVERYTHING ABOUT THE FORMATION OF THE UNIVERSE AND LIFE IN A WONDERFUL LANGUAGE THAT ALL CHILDREN AGES 7-12 CAN EASILY, FULLY UNDERSTAND & ENJOY!

As the only universe-origin book that your whole family will like and enjoy together, *"How God Created Baby Universe"* will set children on the path of success by accurately showing them early in life the formation of the universe, and how to detect errors in theories or stories that would misguide them as they grow up. Therefore, you need to add this great, efficient, trustworthy, and cost-effective book to the strategic journey of children toward their best tomorrow.

With *"How God Created Baby Universe"*, you will:

- Have a peace of mind that children will get accurate, fit, and easy to understand universe-origin information that will produce real results in their life
- Become the leader that captures the heart of children craving for the original explanation of the formation of the universe so you can clear their way for freedom, power, technology, innovation, and breakthroughs of the future (learn more at Science180.com/children)
- Protect yourself and loved ones from wrong theories in the literature and the media by keeping children secured and empowered with the true knowledge of how the universe began
- Explain complicated secrets to children about how to locate mistakes in origin-related theories so you can save time, money, and other resources to improve their lives
- Help children to easily sort out their origin-related questions using strategies that get them to tap into deep secrets that even highly educated people ignore
- Clearly explain to children how to mathematically know without a doubt whether God created the universe as the Bible says or billions of years evolution processes formed it

Accurately explaining the complex formation of the universe and of life to children can be very hard in our modern world, but by getting *"How God Created Baby Universe"*, you will know the proven formula to help children to easily understand their huge universe-origin and life-origin questions with confidence, humor, and joy.

They will surely laugh aloud while reading this book and thank you for it! It is time to buy this pragmatic book to help the children in your life today.

Member of the American Association for the Advancement of Science, American Chemical Society, and the American Society for Microbiology, **Dr. Nathanael-Israel Israel is** a Beninese-American scientist and international consultant, who shows the world how to scientifically decode the formation of the universe, of life, and who is known as the creator of the Chemicals Turbulent Origin Formula™, the inventor of the Life Turbulent Origin Formula™, and the discoverer of the Universe Creation Formula™. Learn more at Israel120.com.

Another Book by Nathanael-Israel Israel:
HOW BABY UNIVERSE WAS BORN

If you don't believe in God or you hate God, or you don't think there is anything or anyone called God, but you want your children to understand how the universe was formed from a scientifically-proven perspective that considers the facts, then this book is for your children. Learn more at www.Science180.com/childrensecular

---

Dr. Nathanael-Israel Israel is the founder of Science180, the American organization that helps people enter the realm of true knowledge about the universe formation. In other words, he is known as the first human being to ever use modern science to give people the state-of-the-art decoding

27.2. Origin of life according to the book of Genesis

The same Bible that revealed more than 3500 years ago that God created the universe in 6 days (including the Earth by the 3rd day, the Moon and the Sun by the 4th day, which I scientifically demonstrated to be true), also said that God created all living things. Because after thousands of years, the creation story of the celestial bodies is proven true using scientific evidences, we must also seriously take the Biblical story of the formation of living things:

Genesis 1: ... 11 God said, "Let the earth yield grass, herbs yielding seeds, and fruit trees bearing fruit after their kind, with their seeds in it, on the earth;" and it was so. 12 The earth yielded grass, herbs yielding seed after their kind, and trees bearing fruit, with their seeds in it, after their kind; and God saw that it was good. 13 There was evening and there was morning, a third day. ... 20 God said, "Let the waters abound with living creatures, and let birds fly above the earth in the open expanse of the sky." 21 God created the large sea creatures and every living creature

CHAPTER 27: BIBLICAL ACCOUNT OF THE ORIGIN OF LIFE

that moves, with which the waters swarmed, after their kind, and every winged bird after its kind. God saw that it was good. 22 God blessed them, saying, "Be fruitful, and multiply, and fill the waters in the seas, and let birds multiply on the earth." 23 There was evening and there was morning, a fifth day. 24 God said, "Let the earth produce living creatures after their kind, livestock, creeping things, and animals of the earth after their kind;" and it was so. 25 God made the animals of the earth after their kind, and the livestock after their kind, and everything that creeps on the ground after its kind. God saw that it was good. 26 God said, "Let's make man in our image, after our likeness. Let them have dominion over the fish of the sea, and over the birds of the sky, and over the livestock, and over all the earth, and over every creeping thing that creeps on the earth." 27 God created man in his own image. In God's image he created him; male and female he created them. 28 God blessed them. God said to them, "Be fruitful, multiply, fill the earth, and subdue it. Have dominion over the fish of the sea, over the birds of the sky, and over every living thing that moves on the earth." 29 God said, "Behold, I have given you every herb yielding seed, which is on the surface of all the earth, and every tree, which bears fruit yielding seed. It will be your food. 30 To every animal of the earth, and to every bird of the sky, and to everything that creeps on the earth, in which there is life, I have given every green herb for food;" and it was so. 31 God saw everything that he had made, and, behold, it was very good. There was evening and there was morning, a sixth day." (World English Bible version, of the Holy Bible).

My goal in this chapter is not to argue about the things mentioned in the Biblical account of creation, which I personally believe is 100% true, but I will instead offer a few comments about the significance of the Biblical story of creation of life, and why I wrote this chapter.

Although most people may not know it or confess it, human being's ways of thinking and perceiving things are influenced by his or her worldview. For instance, people's belief can highly affect how they perceive the origin of life. This also implies that some of the biases that people have toward their understanding of the origin of the universe and of life are rooted in their philosophical (including religious) viewpoint. Therefore, any bias I may have in my explanation of the origin of the universe can also be affected by my belief. Hence, at this point, I felt it is important that I reveal my religious viewpoint and explain how it has also contributed to the way I perceive life. That way, people may also have a glimpse at where I stand religiously.

As a prophetic messianic believer, I strongly believe that God is the Creator of all forms of life. Throughout this book, my goal was not and is NOT to contradict the Bible, but to demonstrate how life could have arisen within the creation context revealed in the Bible and backed by scientific data, including turbulence that I extensively studied and which proved that the universe was really created according to the processes and timeline revealed in the Bible. I also understand that some people do not believe in God nor in the 6 days of creation account mentioned in the Bible's book of Genesis, but, as I have demonstrated in my books on the origin of the universe, God created the world in 6 literal 24-hour days indeed, and rested on the 7th day to give us an example and a sign for the timetable that is supposed to govern this world until its "end" very soon.

Describing the creation of life, the Bible uses terms such as fish in the waters, creeping organisms (which some commentors said include invertebrates), birds of the air, insects, livestock, crawling animals on the earth (e.g. reptiles including serpents), beasts, and human beings. I also realized that the animals later described in Leviticus 11:20-23 as "winged swarming animals" can be some winged insects. In general, the story of the formation of life can be summarize in 4 points:

1. Earth putting forth or bringing forth plants (Genesis 1:11-13) according to which God commanded the Earth to produce plants. The above mentioned verses can also be interpreted as God giving the previously "barren" Earth the ability to produce vegetation, and it did so on the 3rd day of creation as God commanded it.
2. Waters swarming with fish, creeping living things, and birds (Genesis 1:20-23), which was about the formation of fish and other aquatic organisms, and birds (birds were formed with water, but were commanded to multiply on the earth) on the 5th day of creation.
3. Earth bringing forth kinds of livestock, crawling animals, and wild beasts (Genesis 1:24-25) which is about the formation of mammals, insects, and reptiles on the 6th day of creation.
4. Formation of man (Genesis 1:26-27) on the 6th day of creation.

Below, I will make some comments about how the formation of these organisms can be fit together without contradicting the word of God and the scientific evidences. Here, by scientific evidences, I do not mean scientific theories that deny God, but the scientific raw materials that can be empirically collected in nature. The command that God gave to the earth to put forth or bring forth grasses (Genesis 1:11-13) was what empowered the earth to produce plants. This command was executed through a turbulent program of life acting on the earth to produce biological systems to which life was given according to their types. Likewise, the command that God gave the Earth to bring forth many kinds of livestock, crawling animals, and beasts (Genesis 1:24-25) was responsible for the formation of mammals, insects, and reptiles. In other words, the same Earth brought forth grasses with one command from God, and then it also brought forth mammals, insects, and reptiles with another command from God because these commands were not the same. This also suggests that the Earth listens when God speaks. Truth be told, I have come to realize that, even as of today, the Earth can also listen to some human beings when they speak. For the ability of the Earth to hear has not been taken away from it, but it is the ability to hear or to make it obey the instructions of most human beings that has been removed, reduced, or limited after the fall of man that most people mistakenly deny. Nevertheless, I have seen human beings giving instruction to the Earth and it listened and obeyed. I don't have space to detail how and why the Earth and the Sun listened to some ancient people and

CHAPTER 27: BIBLICAL ACCOUNT OF THE ORIGIN OF LIFE

stopped moving at their command to do so.

Finally, the command that God gave to the water to swarm with fish, creeping living things, and birds (Genesis 1:20-23) triggered the turbulent program of life to act on the precursors of the bodies so that aquatic animals and birds could be formed through the biological turbulence. The Biblical account of the formation of man helped me know that during the formation of organisms, the putting together of the bodies and their empowerment or abilities were 2 different things. Hence, death also allows the spirits of beings to leave them so the bodies can decompose and their particles return to the dust of the Earth where they were before. In general, the formation of all forms of life happened within 4 days: from the 3rd day to the 6th day.

I came to realize that the abilities given to water and soil on the Earth to produce organisms has not been withdrawn from them. Therefore, until today, life can be reproduced on the land as well as in the waters. For the earth and the water are still obeying the instruction given to them, not to just allow organisms to be produced from them just during the 6 days of creation, but throughout the timeline God allows life to continue on Earth. Throughout this book, I talked about biological turbulences and the turbulent program of life that formed organisms indeed, but at this point, let me emphasize that no turbulence or program by itself has the power, resources, and abilities to impart anything without the power given by God. In other words, all of the processes used to form or to formulate all forms of life were under the command of God. And we should be grateful to God for having made us as we are and for placing us above all other organisms on Earth. For, I have realized that some human beings have overestimated their abilities, which are the fruits of God's grace, while they ignore the grace and glory that God also gave to other organisms. For instance, if human beings can have a frank dialogue with some organisms and give them a chance to talk, we may be shocked to hear some of them arguing for their own superiority, although they are not "superior" to us. Although I personally believe, and the ecological niche of organisms can prove, that human beings are the most advanced organisms on Earth, even if we compare them with angels, I also believe that our classification of superior and inferior beings is wrong. For nothing is inferior or superior in this world where everything has a reason to exist and where some so-called superior beings can disappear if so-called inferior beings can cease to exist. Even organisms like bacteria that some people may consider inferior than human beings have complex organs such as their flagellum, to which some people have cringed to try to "crush" some evolutionary assumptions; furthermore, without some bacteria, the digestion of some foods could be impossible.

Although by November 18, 2013, I knew that life on Earth was created before the formation of the Sun and Moon was completed, I did not know how I could scientifically demonstrate the formation of the celestial bodies. As a man of faith, I just believed that God used the word to create and form everything. However, in those days, I also perceived that the same matter was not used to form all kinds of organisms. Some organisms were formed using water, while others were formed using soil, etc. For instance, plants were formed on the third day of creation using the soil (Genesis 1:11-13). That same day, they reproduced. Today, no fruit can grow and reproduce itself in one day. In fact, some fruit trees take several years after being planted before producing any fruit. Therefore, I understood that the process and/or the conditions that prevailed during the creation of plants must have been different from those reigning today, else in one day, plants containing seeds could not have been formed, while today it takes days, even months, before most plants produce seeds. Because the Sun was created on the fourth day of creation (Genesis 1:14), a fact that I also scientifically demonstrated, the type of light that plants received on the third day of creation was different from the solar light known today. In *"Reconciling Science and Creation Accurately"*, I also demonstrated that the precursor of the Sun produced the light that plants depended on before the Sun was formed on the 4th day. In other words, just as the Sun is the main provider of light in the Solar System today, before it was fully formed, the precursor of the Sun was the main provider of light in the Solar System or in what would become the Solar System. In was on the 4th day that God empowered the Sun and Moon to govern day and night.

The Bible states that birds and aquatic animals were formed using water (Genesis 1:20). Here, water may not mean pure water only, but water mixed with other chemical elements beside hydrogen and oxygen. Mammals, reptiles, and all other land animals were formed using soil (Genesis 1:24). Human beings are the only organisms about whom it was said: *God breathed on man so he became a living soul.* In *"Reconciling Science and Creation Accurately"*, I amply explained how God used His word to create the universe. Just as He did during the creation of the celestial bodies, God also used His word to create living things. But this time, His commands of forming living organisms were given to the water and soil that were already formed. I also learned some lessons from how curses and benedictions work today and how the word of God has worked to create things in the beginning. For instance, some people would argue that, using His word, God did not create the world and everything in it, including living things, but they will not argue that human beings are able to curse people, do charms or witchcraft using the power of their tongue. If human beings can respond to spoken words, why do some people think that matter could not be formed using words? If you do not know that a spoken word can change the life of a person for good but more often for bad, you may need to check with some people who are highly involved in spirituality.

27.3. Formation of the first human beings (Adam and Eve) and their mysterious lifestyle

The Bible says that human beings were the only organism that God had to make with His own hands in the image and likeness of God Himself! Unlike what most people will say, I believe that the bodies of human beings were made using some of the processes that formed other organisms. Human beings are the ultimate target of creation. Just as many bodies were formed as the precursor of a system of bodies was split-gathered, I also believe that all organisms on Earth were part of a program

CHAPTER 27: BIBLICAL ACCOUNT OF THE ORIGIN OF LIFE

that was targeting the formation of human beings as the ultimate product. For instance, to form the Earth, the precursor of the Solar System had to go through many split-gatherings after a series of branching of fluid layers until the precursor of the Earth-Moon System was split, and then the Earth and the Moon were formed. After the Earth was formed, other celestial bodies were also birthed. Without all of the other bodies formed before and after the Earth, the Earth itself could not have been what it is. Likewise, before the first human beings were formed, other organisms were formed as part of the biological turbulence that shook the precursors of the organisms on Earth. Hence, some mammals (e.g. primates) looking like human beings were formed before man. Although some similarities exist between human beings and monkeys, they did not descend from one another like some evolutionary theories assume. Humans evolve from being babies to old people but, in our lifetime, nobody has ever seen a monkey evolving into a human as some theories claim happened. Why did the monkeys stop evolving into humans? Some of the similarities between primates and human beings are due to the similarities of some of the processes of their formation. However, special attention was given to the completion of the formation of man. The Bible states that God had to breath in him so he became a living soul. Such impartation was given to no wild animal. There are some abilities that human beings have but that even angelic beings lack. Abilities that human beings have but that wild animals lack (at least we human beings don't know or can't say that they have) are spiritual qualities such as intellect and will.

Just as human beings are more advanced than all wild animals because their bodies were endowed with special bodies, it is also possible to remove those abilities and make a human being behave like a wild animal. For instance, the Bible recounts the story of King Nebuchadnezzar who hardened his heart with pride, denied the power and supremacy of God, and who was driven away from among men and his mind became like that of an animal, his hairs had grown like an eagle's feathers, and his nails like bird's claws, he ate grass like an ox (Bible's Book of Daniel 4:25-5:1-30). Then, at the appointed time, his sanity returned to him, implying that it is possible for a human being to lose his human nature and behave like a wild animal, while it is possible for a wild animal to behave like a human being if the proper programs of life can be fine-tuned accordingly. Just because wild animals have not been empowered to act like human beings does not mean that such an empowerment is impossible. For God can do anything He wants. We are who we are because of God's grace and we need to be grateful to Him who created us and placed us above creation despite our shortcomings. Although some wild animals can still cause trouble for human beings in the bush or forest, human beings are still the king of the Earth.

The Hebrew word used to signify man when God said "Let us make man" is "adam", which is a generic noun for "mankind", and which does not imply that "adam" is male. In other words, although we human beings seem to imply that in Genesis 1:26 God created only the male human being and later created the female, the Hebrew doesn't seem to mean the same. In other words, the second chapter of the Book of Genesis, which describes the formation of Eve, may just be a detail of some of the processes that occurred in Genesis 1:26. However, the Book of Jubilees (also called the Book of the Law and the Testimony) mentioned that Adam sinned after being in the Garden of Eden for seven years, and that it was not the same date he was created that he named all the animals of the field. According to the Bible, when Adam was naming the beast of the field, he did not find any mate among them, suggesting that Eve may not have been created by then. Whether Eve was created the same day as Adam or not, or whether they stayed in the Garden of Eden for a few hours or a few days, I think the main message is to believe that God is the Creator of everything and the Bible alone cannot explain everything God has done and about which Apostle John said that the entire universe does not have enough space to contain all the books that can describe what God has done.

I could go on and on in detailing the formation of human beings, but I need to stop here for the sake of space and time. I devoted other books to this subject, for certain things about the human life cannot be demonstrated scientifically but believed, and because my goal for the current book is to stick more to the scientific explanations, I did delve into too many spiritual things here, although they are very important. Those who are interested can learn more by visiting the website associated with this book (www.Science180.com), where I also publish other materials (including articles and other books) on the subject of life, and updates about the origin of the universe.

At this point, I will dive into the mysteries about the lives of Adam and Eve, the first human beings. Indeed, throughout the years, I have asked deep questions about the origin of man and the lifestyle of the first people who lived on Earth (Adam and Eve), but I did not find the answers until I discovered a book called the "Books of Adam and Eve" known to the Jews centuries before the birth of Jesus Christ. In fact, according to the Books of Enoch (which is quoted in the Bible), the Books of Adam and Eve were written by Enoch, a Biblical figure and patriarch in the seventh generation from Adam, who faithfully walked with God to the point that God took him away (Gen. 5:21-24). Although some people (even Bible believers) may not believe in everything in the Books of Adam and Eve, I felt like I still need to mention its existence here, for it may answer some questions some people may have. Although I cannot confirm the authenticity of all the scriptures in that book, I have to confess that it answered many of my questions about Adam and Eve, and while some passages were beyond my understanding, I did not let that assessment to cause me to reject a book that is said to be written by Enoch, a patriarch in the Biblical faith. Moses (that some people consider as the greatest prophet of all time) thought it was important to put Enoch's account into the Torah or the Pentateuch. Apostle Jude (the brother of Jesus Christ) also mentioned the Book on Enoch in his writing in the New Testament (Jude 14-15). Therefore, I decided to study the writings of Enoch. Considering the fact that Enoch walked and talked with God, I believe he knew quite a lot of things, and his writings were just a glimpse of these things he

CHAPTER 27: BIBLICAL ACCOUNT OF THE ORIGIN OF LIFE

recorded so that others could know and understand.

The Books of Adam and Eve was reportedly written between 200-300 BC. But, considering the author of this book, I believe it was written in the days of Enoch. In that book, I learned that, initially, human beings were made to see in the spiritual realms. But after Adam and Eve sinned, the spectrum of their eyes was changed and they could no longer see clearly in the spiritual realm as they used to. The Books of Enoch revealed that, before the fall, wild animals were obedient to Adam and Eve. Before Adam sinned, he had a bright nature and could see to heaven. Before they sinned, Adam and Eve did not need water, but afterwards, they could not do much without it. In the Garden of Eve, Adam and Eve knew neither night nor day, for they were constantly in light or in the presence of light. The Books of Adam and Eve also mentioned that, before the fall, the snake used to be the most beautiful animal, but afterwards, it became the least and the meanest of all animals, for God changed and cursed it. Due to its continual rebellion even after the fall of man, the snake was muted by God, hence, although some snakes can make sounds with the rattling of their tail, snakes do not speak or make (loud) sounds with their mouth like other animals.

Before Adam and Eve sinned, cherubs (a type of angel) used to tremble before them but afterwards, it became the reverse. Likewise, before their fall, Adam and Eve could not be burned by fire, but afterwards, fire scorched them. Before sin, Adam and Eve did not need clothes, but after sin, God had to clothe them; and clothes are a mysterious sign of the weakness and death human beings put on them. In other words, clothes that human beings put on is a prophetic reminder of the token of death on their bodies. After their fall, Adam and Eve were changed. For instance, in the beginning, Adam and Eve were not made to be able to eat as we do today. Just as animals are clean and can be eaten while others are not edible, so also among plants, some are clean and others are not (Genesis 1:29). For instance, seeds bearing plants and trees with seed-bearing fruits are the best foods. All kinds of wild animals were supposed to eat just green grasses and initially, humans were supposed to eat vegetables and some nuts. But because human bodies were altered after their fall, they even had to learn how to eat earthly food. It was around the time of Noah during the great flood that God had officially permitted the consumption of meat by human beings and at the same time, the human lifespan was reduced to 120 years (Genesis 6:3) instead of hundreds of years as it was in the days of the first generations of human beings. For instance, Adam lived for 930 years (Genesis 5:3-5). The record of the human lifespan (969 years) was obtained with Methuselah, a son of Enoch (Genesis 5:25-27).

After the fall of Adam and Eve, God also had to make their bodies suitable so that they did not die or suffer from eating earthly food they were not supposed to eat. Unlike what people think and teach, the Books of Adam and Eve said that God did not give Satan power over Adam, but Adam fell under the rule of Satan by accepting Satan's counsel, hence, Adam lost control of the Earth. Similarly, in today's time, Satan has no reign over the children of God (those who believe in

TURBULENT ORIGIN OF LIFE

Yeshua, Jesus, and live according to His word), yet if children of God give ear to and submit to the counsel of Satan, they too will fall under the rule of Satan. In other words, Satan did not receive the rulership of this world from God, but Adam gave it away by choosing to obey Satan's evil commands.

As a consequence of sins, the ground was cursed and thorns (e.g. Fig. 90) were also created to make human life miserable. Some people do not think that God will punish human beings in eternity. Yet, a lot of things in nature point at how God punishes. God is gracious, loving, and kind of course, but He is also a God of justice who cannot let sins go unpunished, while calling His people to be clean. Hence, even after the 6th day of creation, God had to create other things so His work could stand according to His power and justice. For instance, thorns on trees and plants were created after the fall of man as a way to punish him and make life on Earth more painful. For a sinful man does not deserve a life meant for a holy and obedient man. Else, God will not be right in his justice and judgment. For sins must always be punished so the demarcation between good and evil can be clear. Get more details on these topics in my books called:

- *"Origin of the Spiritual World"*
- *"Reconciling Science and Creation Accurately"*
- *"Science180 Accurate Scientific Proof of God"*

Fig. 90: Thorns of plants
Source: Dr. Nathanael-Israel Israel / Science180, www.science180.com, © 2020-present, all rights reserved.

According to certain Jewish literatures, Adam and Eve would have stayed in the Garden of Eden for about 5.5 hours before being expelled the same day they were created. Like I previously said, other references such as the Book of Jubilees pointed to at least a stay of 7 years in the garden. I used to think that 5.5 hours to 7 years was a very short amount of time until January 2021 when I realized that, in the days before Adam and Eve sinned, 5.5 hours was a lot of time. For that stay occurred before the introduction of sin and human abilities were much higher than those

CHAPTER 27: BIBLICAL ACCOUNT OF THE ORIGIN OF LIFE

afterwards, meaning the things that could take years to be accomplished today could have been accomplished in a matter of seconds before the fall of man. For in the spiritual realm, spiritual beings are not as limited by time and space as human beings are after the introduction of sin. In other words, the short period of time that Adam and Even could have passed in the Garden of Eden (whether it was a few hours or a few years) is like an eternity that was temporarily suspended or cut short so that the physical time could affect human beings until the fulfillment of the covenant that God had with them and which is connected to the passing of about 6000-7000 years before eternity resumes again! Even today, each human being has an eternal component: the spirit. In my book "*Origin of the Spiritual World*", I delved into these mysterious details.

There are a lot of things I could say about the spirituality of life, but I have to stop here and address them elsewhere. For many people (including believers) have unfortunately refused to believe in the spiritual aspects of life until some of them were destroyed by some spiritual beings such as demons well known to those aware of and/or initiated into witchcraft and other forms of occultism, while others who fail to believe in the spiritual aspect of life the right way will be severely punished one day, sooner or later by the father of all spirits, referred to as God, the creator and author of all forms of life. In other words, the search for the meaning of things, which some people have tried to accomplish using physical means only, have blocked or prevented them from understanding life and the mission assigned to all things and beings, which will never be comprehended using some secular ways of thinking. Because I cannot start detailing life without referring to its spiritual foundation, I have to point out a few things I cannot ignore for the sake of anything including that of trying to be scientifically correct and accepted. For me, being accepted and approved by God (the Originator or Creator of life) is more important than being accepted by the scientific community, which, without referring to the real source of life, will never explain the origin and fate of living things. Hence, although some people may not appreciate the significance of these mysteries, I have to point them out here, for I do not want to make the mistake of diluting too much the substance of my understanding of life, which existed before it appeared on Earth and which will also exist even after the Earth will be gone! I hope and pray this chapter will be a blessing to you and that you will open your mind to something new.

'Science180 Academy' Success Strategy
HOW TO RAISE RATIONAL CHILDREN IN OUR MODERN WORLD

In our modern secular world, and with the many things that kids are taught at school and over which parents have little control once the kids head to public school, parents have a lot to worry about. But it does not have to be that way. Universe origin and life origin scientist Dr. Nathanael-Israel Israel has discovered that, more than ever, parents have a crucial responsibility to rationally prepare their kids to have a strong worldview that properly embraces both science and faith, so their kids are not pulled on one side by the secular education and on the other side by religious belief. But how can parents and their children achieve that common goal?

Listen to this Beninese-American scientist and mathematician Dr. Nathanael-Israel Israel to figure it out. Nathanael-Israel is the author of the acclaimed book *"How Baby Universe was Born"*, an easy to understand scientific book primarily written for children age 7-12 years old to help them properly crack the code of the formation of the universe in a language they completely enjoy, and that prepares them to fight any secular or religious theory that may try to rationally drift them away from the reality of everything!

Sample questions that will get answered include the following and many more:

- How can parents use the latest breakthrough about the universe origin to rationally raise their kids?
- How can parents prepare their children from being victims of the danger of wrong theories and dogmas on the origin of life and the universe?
- What can parents do to shield their children from the influence of religious and scientific beliefs that try to enslave them in the name of reason or faith?
- Why is wrong science not the only danger of raising rational children, but wrong belief as well?
- How can we help children to positively navigate the intersection of science and faith?

Learn more at Science180.com/children

27.4. Linear thinking has prevented people from understanding the formation of the universe and life

On January 1st, 2014, long before I knew I would be writing a book on the origin of the universe and life, I left like the formation of the universe and of life was not as linear as most people think. For instance, according to the Biblical account of creation, plants were the first organisms to be formed. Animals were formed next.

CHAPTER 27: BIBLICAL ACCOUNT OF THE ORIGIN OF LIFE

But this does not mean that plants are inferior to animals or that animals are more advanced than plants. In fact, without plants, animals cannot live, for the latter need the former for food and for other usages. God did very well by creating plants before animals. Because God did not want to waste time, He did not wait for the Sun to be fully formed before He created plants on the 3rd day of creation. Those who ignore the mysteries of creation don't understand why and how plants could have been formed before the Sun. For among other things, they ignore that, although the Sun was not fully formed on the 3rd day, the plants were completely formed and operational, the precursor of the Sun was already going through changes and was already providing enough light to every plant formed on Earth on Day 3.

It is easier for some people to solve a single equation than a system of equations. The solution of an individual equation in a system of equations is not always the solution of the entire system of equations. In other words, a solution can be found for a single equation but still that solution may not fit the whole system of equations. The best and only way to solve a system of equations is to consider all the equations involved without ignoring any. Likewise, linear reasoning, which causes some people to stick with certain types of knowledge only to explain the origin of the universe and of life while ignoring other sources of knowledge, is partial and will never holistically apprehend the truth.

In the same manner, believing that a living organism comes from another one or that a chemical particle or a celestial body descended from another one, or using a similar reasoning to explain complexity in nature is like solving an individual equation of a system of equations but ignoring all other equations provided in that system. Although what I am saying about systems of equations may sound odd or silly, it is exactly what many scientists and nonscientists have been doing (according to their discipline and religious doctrines) with the truth, which they refuse to embrace from a global or holistic perspective while they cling to the little that they know in their field, but which they unfortunately magnify as the ultimate truth that everybody must abide with. I have come to understand that if some scientists could have reflected or thought about things systematically, they could have understood that the universe and the things and beings it contains were not formed by evolving over millions or billions of years, but quickly according to turbulent laws, which I spent 12 years (2013-2025) decoding so they are revealed to the world for the improvement of lives and for the glory of God, the Creator. But because they refuse to see the universe from this angle, many people (not just the scientists) have launched themselves into reasonings that led them to elaborate on trees of life (including so-called phylogenetic trees) to design gradual relationships between organisms as far as their origin is concerned. Eventually, those who refuse to view life with the perspective of God as the Creator, will be surprised of their mistake, which consequences can be dangerous. For some of those people, it will be too late to change the trajectory of their eternity that most people deny because, among other things, they cannot see past their present life around them.

While some people categorically deny God, others believe in Him but have failed

TURBULENT ORIGIN OF LIFE

to fully believe in what the Bible said that God did during creation. I have searched the literature and realized that many people have tried to explain the origin of life using the Biblical story of creation, but they explained it wrongly. Yet, those people act as if they know it all. I will not name anybody here, but some people (e.g. unbelievers) need to understand that some of the theories that some believers exposed to them were wrong. Some people argued that, because "living things are too well-designed to have originated by chance, life must have been created by an intelligent creator, which is God". Although I strongly believe that the design found in nature is after God's plan, I have been disappointed by some people who support design theories, but who unfortunately advocate for theories alleging that the universe is billions of years old and that human beings were also formed after billions or millions of years of evolutionary processes mirroring evolutionism theories, which have unfortunately denied God. In other words, although God designed the order in the world, it did not take millions of years for Him to do so, nor did it take the universe and the organisms in it millions or billions of years to become what they are today. The inability of some people to understand the process that formed the universe and life, and their desire to explain what they do not understand has caused many believers or so-called believers to come up with theories that support some scientific data (wrongly analyzed) while denying the truth laid in the Bible in which they say they believe. As for me, it would have been better for some scientists to shut their mouth and confess their inability to explain life according to the Bible rather than crafting theories that blatantly go against the word of God, but which seem to "fit" the scientific data, while they claim that they believe in God. How can some people claim they believe in God who created the world and everything in it in 6 days while at the same time the same people say that they also believe 100% in some scientific theories which claim that there is no God and that the universe was formed after billions of years of process? Why and how can people believe in such an opposite thing, while still acting as if the 6 literal days of creation can ever be reconciled with the billions of years of processes claimed by evolutionism? I was able to reunite the scientific data collected on the celestial bodies with the Biblical story of the formation of the celestial bodies, but I never tried and I will never try to reconcile evolutionism with anything related to the origin of life. For, I know it is purely a mistake, just as I have proven that the Big Bang is a mistake. Just as it is pointless to try to reconcile light and darkness, so evolutionism and the true origin of life can never be reconciled.

To summarize, I have no problem with people trying to explain the formation of life using scientific data, but I have a real problem with their theory when they start saying that it took God millions of years or even billions of years to form the universe and the organisms it hosts. For the Bible did not say so, but it said that God created everything in 6 days. I scientifically demonstrated that the 6 days of creation were literally 24 hours each. That is what I not only have believed since I was born, but which I ended up demonstrating scientifically after sacrificing my plant science career for that purpose. Therefore, those who think they are serving

CHAPTER 27: BIBLICAL ACCOUNT OF THE ORIGIN OF LIFE

God by advocating for theories that contradict the Word of God need to be very careful, for they will end up paying a high price. Before the publication of my work on the origin of the universe, a lot of confusion existed even among the believers about the timeline and the historical account of creation by God. But since my work has plainly demonstrated that God did indeed create the universe and everything in it in 6 literal days of 24 hours each, people need to be very careful. For there is no more room for excuses.

The incapacity of many human beings to properly understand the origin and fate of the universe has caused them to engage in various actions that are complicating and even destroying lives. For instance, according to the Bible, although life was formed perfect, it did not take too long before human beings in collaboration with some fallen angels brought chaos onto the Earth, troubling everything in their vicinity. Because many books are already devoted to the original sin and its consequences, I will not dwell on it here. On the biological scales, some of the anthropic actions are exterminating some organisms. Despite efforts to manage and conserve natural resources, several organisms are becoming rare and threatened or extinct. Sometimes, I am shocked at how some human beings work hard to protect wildlife, but they are not committed to improving their own life, future, and most importantly, their eternal fate.

Human beings are incapable of creating organisms, but they are inclined to destroying lives. Some destructions occur through abortion, while others are through wars, and many kinds of pain inflicted on people. Conflicts are rising every year in different parts of the globe and peace efforts are not producing expected hopes. Despite scientific progress, many problems still exist. Many diseases are being caused by some medicines. Indeed, a key lesson I learned from turbulence is that no structure formed in turbulence is indifferent to the history of all the other structures formed by their common precursor. Consequently, all bodies in a system are interconnected somehow, even if the factors connecting them are not usually well understood by most human beings, and much more when theories ignoring turbulence are the lens through which things are interpreted. Seen from the angle of turbulence, I felt like the bodies in a living organism are highly connected. If one organ or body part cannot be hurt without the other feeling some form of pain, it can be unthoughtful to try to make a drug or medication for certain parts without properly assessing how they affect other parts. Throughout the years, I have come to realize that many medications and medical procedures are not meant to consider their holistic impact on all the other organs of the body of living organisms, but mainly a limited and sometimes biased effect of the pain their inventors tend to solve. In the end, many medicines are more toxic to human beings than anyone could have ever imagined. I am not trying to put all the blame for this problem on pharmaceutical companies only, but to show how, the ignorance of the origin and functioning of life is destroying organisms and threatening the planet.

While many people still do not believe in God, the world continues its course toward an end that no human being can stop regardless of their rebellion. At the appointed time, a gathering of human beings will happen. The rapture of the saints (believers) that I addressed in another book (see my book on the age and end of the world) is a major split-gathering. It is an escape of the saints from this world. As for now, death is a partial split-gathering. Most people just don't know where the components of the living go after "death". The remains of dead people are usually buried in the grave, but there is an invisible (or spiritual) component of life which people cannot see, which leaves and goes somewhere at death. People may not believe in the end of the world, but the undeniable fact, that even the unbelievers can testify about, is that the life of many people ends every single day and most of them were surprised by their own death. Hence, it is important to be ready for death at all times. Many books and movies were made and written about the end of time or the end of the world, but no book other than the Bible gives clearer instructions as to what human beings can do to escape God's judgement. To make a long story short, the state of the world is not strong, but weak and all signs point to an imminent end as predicted by the Bible.

27.5. Forms of life before birth and after death

Based on my understanding, there are more than the 6 forms of life (animals, plants, fungi, protist, archaea, and bacteria) usually mentioned in modern classifications of life. For instance, some people may say that the life lived by God is different from that lived by human beings. The life of angels (whether they are holy angels or bad angels such as demons) is different from ours. In other words, the definition of the life of God or that of other spiritual beings like angels cannot be completely described by simply using scientific terms like cells, survival, struggles, or evolution. For God is not made of cells and He does not need to struggle. He is Almighty. He is light. He is fire. He is much more than what the whole world contains the space to describe Him. Angels are not made of earthly materials and they are not limited by the things that can definitively block some organisms on Earth. The life lived by each of the aforementioned 6 forms of life is not the same. For instance, the life of an animal is different from that of a plant. Even among animals, they do not live the same life. Likewise, all plants do not have the same life.

As I considered human beings, I realized that they have lived many types of life that they usually ignore. All human beings have lived a type of life long before they were conceived by their mother. Here, I am not talking about the time embryos spend in their mother's womb before being born, but about a life lived by human beings long before their parents were even born. For life is not just physical. Before being born through the flesh, all human beings have existed in the form of a spirit. Even if you do not want to believe in that, I hope you agree that before your mother conceived you, part of you was in the testicle of your dad and in the ovary of your mother. If the chain can be traced way back, the spirit of all human beings could be traced back to the first person who ever lived, and whom Christians believe to be Adam.

The Bible for instance says that God knows everybody long before they were

CHAPTER 27: BIBLICAL ACCOUNT OF THE ORIGIN OF LIFE

conceived by their mother. The Bible also says that all forms of life were at once in existence in the thoughts of God, the father of all spirits. The spirit which gives life to human beings is sent into the flesh of a human egg as it is being formed during the act of reproduction. The day that the spirit will depart the flesh, death of the flesh will follow. And from that moment, the spirit of man will start living another form of life different from the one lived before being incarnated into a human flesh on Earth. Then in the long term, after what I termed the "Great Human Escape" and which Messianic believers and Christians call "rapture", the spirit of some human beings will be clothed with a different garment or glory according to the choices their "owner" or "tenant" had made while living on Earth. And this is the eternal life that all human beings were supposed to be looking after on this Earth, but that they unfortunately relegate to the last position, while they live their lives as if it was not a kind a death! For the life people live today on Earth is a set of choices that will qualify them or not for the real life or eternal life, which will occur at the end of this age! During that age to come, which is an eternity (never ending), those who refuse to believe in God (the giver of life) will spend eternity in hell, which is another form of life, but a death, for it is a life lived far from God, the only source of real life. Along the way, all of the forms of life that a human being could live were, are, and will be calibrated according to their environment and mission. Hence, the life before conception is different from that of an embryo, which is different from that of a baby, which is different from that of a toddler, which is different from that of an adult, which is different from that of a very old man ready to die, which is different from the life after death that many people refuse to believe in just as they also refuse to believe in the life before birth or even before conception.

As of today, everything human beings have managed to label as life are not even living. Those who are living (can perform some physiological functions more than others) are not even living the same. For instance, the life of human beings is different than that of all other animals which is different from that of plants and that of all other types of organisms on Earth and beyond. I know that some people classify human beings as animals, and all other animals are called wild animals as if no human being is wild. But if I had to be radical, I will never call human beings animals, for they are spirts living in animalistic bodies. Human spirits are not the same just as their physical bodies are not. Some spirits are regenerated, while others are corrupted. In other words, some spirits are clean and believe in God, while others are unclean and deny God, the father of all spirits.

To some extent, the nature of the spirit living in human beings defines the nature of the actions and behaviors they can display. Spirits can enter and live in any organism and even in nonliving matter. But, according to the Bible, the Spirit of God lives only in those who believing in Him. Even so, it has not been that way since the beginning. According to the Bible for instance, the Spirit of God started dwelling in human beings after God tabernacled (dwelled) in human beings on Earth. Before the Common Era, the Spirit of God occasionally dwelled on a few anointed people and then left. But since about the Common Era, human beings are eligible to be the Temple or dwelling place of the Spirit of God, with the Virgin Mary, the only human being to host the spirit of God via pregnancy. Because of the nature and objective of this book, I will not dwell much on these religious facts. For the sake of reaching out to as many people as possible, I will not engage into details about spiritual forms of life (for I handled that in another book), but I will base my explanation of the origin of life on the scientifically acceptable forms of life. Nevertheless, the spirit is what gives life and that spirit can accommodate any form, and the shape of the containment of the life defines the kind of functions the spirit can have.

Much more, chemicals could have never been formed without a Creator who purposely designed everything for a reason. Imagine coming home from work and walking into your house to find a coffee table (that you did not buy nor move into your house) sitting in front of your couch, would you think that it just got there without anyone placing it there or would you want to know who placed it there or where it came from? If it was me, I would not believe it got there on its own although it does have legs but cannot walk; for I would be searching to find out who brought it into my house. Nevertheless, I am not going to base my explanation of life on these kinds of logic, which some religious people use without being unable to scientifically demonstrate how life originated. For one thing is to believe in a Creator, but another one is to be able to scientifically explain how life was created. Furthermore, ignoring that life after the so-called death on Earth is eternal and will be based on the bearing of the consequences of choices made on Earth regarding future life or death, some people (who denied God on Earth) think that the "grave is the sweetest place on Earth". Yet, the grave is an entrance into another world that will be sorrowful for those who failed to believe in God, our Creator. Unfortunately, most people will experience this fact at a time very late when they can no longer do anything to change the direction of their life.

27.6. Biblical definition of life and how the life after resurrection is different from the current life

So far, I have talked about the origin of life according to the Biblical account. But at this point, I would like to say a few things about the future of life after the current life people are living on Earth. For like I already mentioned in previous chapters, life does not end when people die physically. In this segment, I will also elaborate on the spiritual aspect of life that I introduced a few chapters back.

I have heard many preachers, teachers, and prophets talk, but a few really understand and generously share deep secrets about life and the spiritual world. Men are supposed to have dominion over everything on Earth. Spirits drive physical bodies almost just like people drive cars. The human life can be felt. For instance, people who are adequately sensitive can feel the life of others as they get close to them. For instance, some prophets can sense people's lives by inviting them close so they can feel them.

The life of the human flesh is in the blood, which is the mediator between the

CHAPTER 27: BIBLICAL ACCOUNT OF THE ORIGIN OF LIFE

spirit and the flesh. The soul is in the blood, meaning that human feelings, sight, and hearing for instance are in the blood. The blood has both a spiritual aspect and a physical aspect. All curses and misfortunes are placed or given to the blood. For the blood can receive even spoken words, which it transfers to the flesh. In other words, every curse that people are dealing with today was handed to them over their blood. The life at resurrection is different from the life lived today. At the resurrection, the life of the flesh will come directly from the spirit and bypass the blood which will no longer be. Indeed, today, the spirit gives life to the blood and the life that the blood has gives life to the flesh. In other words, today, the blood converts the life of the spirit and gives it to the flesh. Hence without the spirit, there is no life and without the blood, there is no life. However, at the resurrection, there is a bypass of the blood, the spirit and the flesh come together. The resurrected life will receive its life from the spirit directly. At the resurrection, the flesh becomes one with the spirit, therefore able to do things that it could not do before because of the limitation of the flesh hosting the spirit. Therefore, the resurrected flesh can even walk through walls and on water. The resurrected body is rupturable and can manifest in many forms. This life we are living today is a phase to make choices and decide who we are going to become in the Kingdom of God or not.

I have learned that gravity was given dominion over the blood, not over spirits. Hence, spirits can go or travel to places that the flesh cannot. For people to live by the spirit, their blood must stop being their source of information. They must consult the spirit and live according to the spirit, which fruits are against those of the flesh. Witches are not spiritual people, but they have mastered the human life (which is not a spirit) so much so that they take advantage of its potential to do evil things.

Until today, the Jews (to whom the Bible was originally given) still believe that everything in the universe has a soul, meaning able to feel things. This does not mean that everything that was created can feel things the same way we do, and even if they do, we may not know, and just because we don't know cannot cause us to think they don't feel anything. Just as all human beings do not feel things the same way, the abilities of the soul that the created things have differ. Since 2013, I have also come to believe that everything really has a soul, but the problem is that we human beings are not able to know or sense the feelings of everything around us or in the universe. Furthermore, in 2020, due to the preaching of some contemporary prophets, I have come to realize that everything in the universe has a spirit. But all spirits are not the same. The spirit of a man is different from the spirit of a wild animal, which is different from the spirit of a nonliving thing. The problem that some people have had to understand that every creature has a soul and a spirit is that most people think that the only expression of a soul is to have emotions like human beings and that the only expression of spirits is to be able to worship God like humans. When we hear birds singing, flowers flourishing, plants growing and producing foods in due seasons, and when we see the order in nature, we may not understand what is behind them, but we should never again think that all the created things do not have life in them. For they have their own life according to their world, missions, and the processes that formed them. Instead of trying to understand everything in nature (something humankind will never achieve), human beings should seek God, their Creator, obey Him, and ensure that they are working on things that have eternal rewards rather than on some earthly things, which will not help them in eternity.

Just as all leaves on a tree can be connected or threaded back to the root of that tree, so also all human beings in the world are connected to the first man (Adam) and the first woman (Eve). Paul explained how, at the spiritual level, the gentile believers are like branches grafted on the main branch (which is the Jewish people). Jesus also likened God to the main vine and his believers as the branches which abundant fruits make the father happy and bring glory to Him (John 15).

An internet search for the word "life" in the Bible returned to me 422 results, suggesting that the Bible talks abundantly about life. According to the Bible, after God formed the first human being (Adam), He breathed into his nostril the breath of life which gave life to Adam and empowered him to become a living soul (Genesis 2:7). Other passages suggest that man is not the only organisms that has the breath of life, but all flesh under the heavens have it as well (Genesis 6:17 and Genesis 7:15). But the life of man is different from that of other organisms. The Bible also mentions the existence of a tree (the Tree of Life) that can impart eternal life (Genesis 2:9). When sin entered Adam and Eve in the Garden of Eden, they were driven out from the Tree of Life (Genesis 3:24).

Of all the important parts or organs in the body, the blood is the only one that the Bible singles out as hosting life: "... the life of all flesh is in its blood" (Leviticus 17:14). Although human beings came up with many definitions of life, the Bible introduced the notion of eternal life that God has been giving to only those who believe in Him. Those who deny Him cannot have it despite God's love for the whole world: *"For God so loved the world, that he gave His only-begotten Son, that whosoever believes in Him may not perish but have everlasting life"* (John 3:16). In other words, the eternal life that God gives is received by believing in Him. Later, Jesus went on to specify that He is the bread of life: *"Jesus said to them, I am the bread of life: he that comes to me shall never hunger, and he that believes in me [Jesus] shall never thirst at any time"* (John 6:35). He is not just the author of life, but He is also the resurrection: *"Jesus said to her, I am the resurrection and the life: he that believes in me [Jesus], though he is dead, yet shall he live"* (John 11:25). To make a long and eternal story short, the Bible clearly says that only those who believe in Jesus Christ will have eternal life, anyone who denies Him will go to hell. In other words, any human being who does not believe in Jesus is a dead creature walking and performing physiological functions like some wild animals. All translations in this segment on life are from the Darby Translation of the Bible.

Those who know God, can operate from a different perspective and dimension. For instance, the Bible recounts that, there was a time when wild animals communicated with human beings. Even as of today, animists can make wild

CHAPTER 27: BIBLICAL ACCOUNT OF THE ORIGIN OF LIFE

animals to talk. Some witches can even communicate with nonliving things. Something must have happened to rupture the communication between human beings and wild animals. Plants were able to listen to God's instruction. The Sun and the Moon were able to listen to God's instruction about their mission of dominating the day and the night and they have faithfully carried it out until today. I have witnessed human beings commanding nonliving things (e.g. broken refrigerators, cars, cellphones, and other appliances) to work and they started working without a human mechanic or technician fixing them.

Even myself, I attended an online service during which a minister ordered people's hands to grow. The hands of many people including mine grew. Mine grew about 5 inches longer. Then upon the command to go back to its place, the hands also returned to their initial position, meaning that my hand and that of many others listened to an instruction to grow and also to another instruction to shrink. My wife was at work when that happened, but when she came home, she did not believe me at first, when I told her what happened. To test what happened to me when she was at work, I commanded her hand to grow and it did. I did the same for my oldest daughter and her hands grew. Then, I commanded the hands to go back to its place and it did. I even recorded a video of those experiences. When my mother-in-law visited us a few weeks later and I talked to her about the experience, she did not believe when she heard the story. Before her own eyes, I ordered the hand of my child to grow and it happened as I commanded. All I said was "hand, I command you to grow. Grow! Grow! Grow!" and the hands of my child obeyed. But in the process, the hands of my mother-in-law also grew, however, she did not say anything, but just "smiled". For, she does not believe in the prophetic. Since that day until this very moment, every time I speak to my own hand and command it to grow it obeys. In other words, a human body can listen to instructions if the right commands given to it are aligned with the law and mission that were given to those things. For everything in nature has a mission and if we know and align ourselves with that mission, we can command nature and it must obey. Else, people can do everything they want, but they will never have the result they expect.

Despite what some unbelievers may say or think, all human beings are passengers in the same vehicle (called Earth) going to two types of remote destinations (which are either positive or negative). Death is an exit from this world and an immediate birth into another kind of life called afterlife or after death, which is a waiting room or an antechamber to eternal death or eternal life that most people refuse to believe in or to accept for free. The main question that everybody needs to answer is whether he or she is ready to die and meet God? Meanwhile, do you know why you were born? What is your mission on Earth? As for me, after spending years trying to make it in life (first in Africa, which and where I was wrongly taught was a poor continent) and then in the US (that I was also wrongly taught was heaven on earth) and after discovering my purpose in the historic 12 years (2013-2025) during which I went through the thick and the thin, I have finally understood, embraced, and accepted the fact that I was born to holistically (e.g. scientifically, philosophically, and spiritually) decode the mysteries of the formation of the universe, chemical particles, and life so that human beings can learn and improve their life today and forever. Therefore, I encourage you to not just read the chapters of this book for book knowledge, but to first of all seek a relationship with the God who created you, and to also dearly seek to discover the mission of your life; for you were not born by chance or created without a mission. Know your Maker and serve Him so it can be well with you today and in the world and life to come!

Another Book by Nathanael-Israel Israel:
RECONCILING SCIENCE AND CREATION ACCURATELY

THERE IS ONLY ONE SIMPLE, COMPELLING, SOLUTION-DIRECTED SCIENTIFIC FORMULA ACCURATE ENOUGH TO RATIONALLY EXPLAIN HOW GOD CREATED THE UNIVERSE

"*Reconciling Science and Creation Accurately*" is a landmark book in universe-origin writing from a rare perspective by one of the most respected minds of our time. It scientifically explores the most challenging questions of all times that believers, nonbelievers, and all freethinkers are interested in: How can we rationally demonstrate, without checking our brain at the door in the name of faith, that God created the universe? How did the universe begin and what processes did God use to create it? Are these processes still operating in the universe or not? Can believers abandon wrong theories if they think it is impossible for science to literally prove the Genesis story, or if they think that science is evil and diametrically opposed to faith, or if they compromisingly embrace scientific theories that contradict the Biblical account of creation written before the scientific era? What can believers do to help the skeptics believe in the Biblical narrative of creation?

Lucky you, Dr. Nathanael-Israel Israel successfully navigated all those questions with an accuracy that both scientists and nonscientists have been applauding across the globe. After reading "*Reconciling Science and Creation Accurately*", you will confidently:

- Scientifically prove the Biblical account of the creation of the universe and the existence of God in a way that makes the head of those who deny God to spin faster than a DJ's turntable
- Know how to rationally talk to anti-creationists, evolutionists, Big Bang proponents, atheists, skeptics, and other freethinkers about the universe-formation and they will beg you to know more about God, the Creator, that they mistakenly rejected
- Discover very accurate, rare, and factual truths about the universe-origin that will save you time and money, and get you much closer to the better and joyful life you want to live today and forever

CHAPTER 27: BIBLICAL ACCOUNT OF THE ORIGIN OF LIFE

- Improve your health and faith by knowing that the existence of God can be scientifically justified using Science180 Cosmology and particularly Science180 Creationism
- Enter a new area of freedom and power by crushing the head of and breaking free from the suffocating expectations of all wrong theories that have highjacked secular and religious education, and that have held the Biblical account of creation captive for almost 3500 years
- Break free from the suffocating expectations of some forms of creationism that have sequestered the mind of some believers for a long time
- Uncompromisingly, intelligently, and scientifically explode the myth of those who, instead of literally taking the Biblical days of creation as 24-hours consecutive days, think that they were millions of years, or were representative of long ages, or that millions of years existed before them or were positioned between them
- Understand the accurate standard to interpret the Biblical account of creation thanks to Science180's breakthrough that transformed science and laid a foundational bedrock for the inerrancy of Scripture

Now that Genesis (the oldest manuscript in the world, written before science and most religions were born) is scientifically proven to be correct (Science180.com/biblical), what unstoppable, jaw-dropping paradigm shift will the discovery of the perfect alignment between science and the Bible bring into the religious, rational, and secular world today? Get this thoughtful book now to figure out what happened at the beginning, what is coming up, and why it is time to urgently rethink everything you have been told about the universe-origin so you don't eventually regret! Don't say nobody told you!

Founder of Science180 Academy, **Dr. Nathanael-Israel Israel** is acknowledged worldwide as the discoverer of the all-in-one, proven, and simple scientific formula that accurately cracked the origin of the universe, of life, and of chemicals, and that scientifically unearthed the holy grail at the intersection of science and the Biblical account of creation. Learn more at Israel120.com.

CHAPTER 28

IS THE BIBLE IRRATIONAL OR AT WAR WITH SCIENCE WHEN IT REVEALED THAT SOME ANIMALS ARE CLEAN AND OTHERS UNCLEAN OR IS THERE ANY ENCRYPTED SCIENTIFIC CODE BEHIND THE LEVITICUS 11 LAWS?

As my efforts to explain the origin of life was to be as holistic as I could, I decided to also review the origin narratives in the world religions. Like I already demonstrated in my books on the origin of the universe, the creation story (about the formation of celestial bodies) in the Bible's Book of Genesis truly concurs with scientific data. Therefore, while I was working on this book, I also decided to pay particular attention to what the Bible said about life and organisms. As I was investigating this issue, one of the things that caught my attention was a mysterious classification of animals into clean and unclean (in the Bible's Book of Leviticus) in such a way that human beings were not supposed to eat all kinds of animals:

Leviticus 11:1 "Yahweh spoke to Moses and to Aaron, saying to them, 2 "Speak to the children of Israel, saying, 'These are the living things which you may eat among all the animals that are on the earth. 3 Whatever parts the hoof, and is cloven-footed, and chews the cud among the animals, that you may eat. 4 "Nevertheless these you shall not eat of those that chew the cud, or of those who part the hoof: the camel, because it chews the cud but doesn't have a parted hoof, is unclean to you. 5 The hyrax, because it chews the cud but doesn't have a parted hoof, is unclean to you. 6 The hare, because it chews the cud but doesn't have a parted hoof, is unclean to you. 7 The pig, because it has a split hoof, and is cloven-footed, but doesn't chew the cud, is unclean to you. 8 You shall not eat their meat. You shall not touch their carcasses. They are unclean to you. 9 "You may eat of all these that are in the waters: whatever has fins and scales in the waters, in the seas, and in the

CHAPTER 28: ENCRYPTED CODE BEHIND CLEAN AND UNCLEAN ANIMALS

rivers, that you may eat. 10 All that don't have fins and scales in the seas and rivers, all that move in the waters, and all the living creatures that are in the waters, they are an abomination to you, 11 and you shall detest them. You shall not eat of their meat, and you shall detest their carcasses. 12 Whatever has no fins nor scales in the waters is an abomination to you. 13 "You shall detest these among the birds; they shall not be eaten because they are an abomination: the eagle, the vulture, the black vulture, 14 the red kite, any kind of black kite, 15 any kind of raven, 16 the horned owl, the screech owl, the gull, any kind of hawk, 17 the little owl, the cormorant, the great owl, 18 the white owl, the desert owl, the osprey, 19 the stork, any kind of heron, the hoopoe, and the bat. 20 "All flying insects that walk on all fours are an abomination to you. 21 Yet you may eat these: of all winged creeping things that go on all fours, which have long, jointed legs for hopping on the earth. 22 Even of these you may eat: any kind of locust, any kind of katydid, any kind of cricket, and any kind of grasshopper. 23 But all winged creeping things which have four feet are an abomination to you. 24 "By these you will become unclean: whoever touches their carcass shall be unclean until the evening. 25 Whoever carries any part of their carcass shall wash his clothes, and be unclean until the evening. 26 "Every animal which has a split hoof that isn't completely divided, or doesn't chew the cud, is unclean to you. Everyone who touches them shall be unclean. 27 Whatever goes on its paws, among all animals that go on all fours, they are unclean to you. Whoever touches their carcass shall be unclean until the evening. 28 He who carries their carcass shall wash his clothes, and be unclean until the evening. They are unclean to you. 29 "These are they which are unclean to you among the creeping things that creep on the earth: the weasel, the rat, any kind of great lizard, 30 the gecko, and the monitor lizard, the wall lizard, the skink, and the chameleon. 31 These are they which are unclean to you among all that creep. Whoever touches them when they are dead shall be unclean until the evening. 32 Anything they fall on when they are dead shall be unclean; whether it is any vessel of wood, or clothing, or skin, or sack, whatever vessel it is, with which any work is done, it must be put into water, and it shall be unclean until the evening. Then it will be clean. 33 Every earthen vessel into which any of them falls and all that is in it shall be unclean. You shall break it. 34 All food which may be eaten which is soaked in water shall be unclean. All drink that may be drunk in every such vessel shall be unclean. 35 Everything whereupon part of their carcass falls shall be unclean; whether oven, or range for pots, it shall be broken in pieces. They are unclean, and shall be unclean to you. 36 Nevertheless a spring or a cistern in which water is gathered shall be clean, but that which touches their carcass shall be unclean. 37 If part of their carcass falls on any sowing seed which is to be sown, it is clean. 38 But if water is put on the seed, and part of their carcass falls on it, it is unclean to you. 39 "If any animal of which you may eat dies, he who touches its carcass shall be unclean until the evening. 40 He who eats of its carcass shall wash his clothes, and be unclean until the evening. He also who carries its carcass shall wash his clothes, and be unclean until the evening. 41 "Every creeping thing that creeps on the earth is an abomination. It shall not be eaten. 42

TURBULENT ORIGIN OF LIFE

Whatever goes on its belly, and whatever goes on all fours, or whatever has many feet, even all creeping things that creep on the earth, them you shall not eat; for they are an abomination. 43 You shall not make yourselves abominable with any creeping thing that creeps. You shall not make yourselves unclean with them, that you should be defiled by them. 44 For I am Yahweh your God. Sanctify yourselves therefore, and be holy; for I am holy. You shall not defile yourselves with any kind of creeping thing that moves on the earth. 45 For I am Yahweh who brought you up out of the land of Egypt, to be your God. You shall therefore be holy, for I am holy. 46 "This is the law of the animal, and of the bird, and of every living creature that moves in the waters, and of every creature that creeps on the earth, 47 to make a distinction between the unclean and the clean, and between the living thing that may be eaten and the living thing that may not be eaten." (World English Bible version).

In this chapter, I will show why and how the clean and unclean animals as mentioned in the Bible hide a code of the characteristics of a developed turbulence or an advanced level of development of the program of life. Indeed, on the morning of Saturday, September 25, 2021, meaning on the 5th day of the Sukkot of the year 2021, around 5 AM, I lost my sleep and, from that time until about 7 AM, I just stayed on bed and my mind was flooded with so much information concerning the program of life, and how, in the precursor of some animals, such a program was not executed to its highest level, therefore making some animals clean and others unclean. In those days, getting up at 5 AM was too early for me because, like I shared in other books, from 2021 to 2022, I used to stay up until 1 AM. Most of the content of this chapter are based on the inspiration I got on that day.

In 2013, meaning 12 years before the publication of this book, I was heavily inspired about why some animals are clean and others are not. In 2015, I even did a communication on that topic at a meeting. The feedback I received caused me to deepen the subject to the point that I even wrote a separate book on it, but which I did not publish yet, for I ended up being busy with decoding the origin of the universe. Here, I will just pick a few points about this mystery that I realized contains a code about the formation of life and criteria by which we can assess a fully formed animal according to its kind. For God (the creator) who is perfect and who wants and calls for His people to be perfect cannot request them to eat the lowest quality of animals. Being holy and perfect, God also wants His people to eat the perfect animals according to their kinds and the limits imposed on things and beings on Earth. In other words, I realized that the singling out of some animals to be the best and the only ones to be eaten points to a mysterious, encrypted code of the qualification or perfection of the processes used to form organisms. In other words, clean animals (as referred to in Leviticus 11) are like the perfection or the highest level of development of the processes or program of life, which formed the living things. Clean animals are like the masterpiece, the perfection, or the highest level of achievement of God's creative work according to the kinds of animals. By kinds of animals, I meant groups or families of animals like fish, birds, creeping things, reptiles, mammals, insects, etc. Among all of those kinds of animals, reptiles are the

CHAPTER 28: ENCRYPTED CODE BEHIND CLEAN AND UNCLEAN ANIMALS

only ones that contain not even a single clean species. I came to realize that the uncleanness of all the species of reptiles is due to them being formed by an incomplete or least developed process. Hence, some reptiles do not even have legs or arms like mammals do. In other words, the clusters of matter used to form reptiles did not reach the highest level of biological turbulence to allow the program of life to form a reptile that reached the highest level of development worthy to be declared cleaned. I also discovered that this incompleteness of the process that formed the organisms was also found with other types of organisms besides reptiles. For instance, although the precursor of some birds, insects, mammals, and fish reached the highest level of perfection during their formation, that of others did not. Hence some birds, insects, mammals, and fish are clean, while others are not. The criteria by which animals were qualified as clean or not varies according to their class. Below, I will review some of those criteria according to the examples provided in the Bible.

For instance, the presence of fins and scales on a fish as a symbol of its perfection or cleanness. The dictionary defined scales are thin horny or bony plates protecting the skin of some fish and reptiles, usually overlapping one another, while fins are flattened appendages on many parts of the body of some aquatic animals, used for propelling, steering, and balancing. I found out that it could have taken an extra level of development or perfection of the precursors of fish before they formed fins and scales. In other words, fish that have fins and scales could be more advanced in their development than those that do not have them. Putting this differently is to say that fish lacking fins and scales are less "mature" or less perfected than those which have them. Hence, wanting His people to eat the best things and to be as clean as He is, God forbade the believers from eating any fish that lacks fins or scales (Leviticus 11:9-12). If fins and scales were cheaply made and/or were not advanced organs that advantage the fish having them, God could not have used them to classify the clean and the unclean fish.

The hoof is the horny part (hard and rough resembling a horn) of the foot of ungulate animals, while the cud is a partly digested food returned from the stomach of ruminants to the mouth for additional chewing. The separation and complete division of the hoof, and the ability of an animal to chew the cud are also symbols of perfection of some animals. Hence, among the land animals, God allowed the Israelites to eat only such animals (Leviticus 11:1-8). Therefore, animals like the coney, camel, and others that chew the cud but do not have a divided hoof are not perfect and are not good for human consumption, because, as I mentioned above, their precursors could not have reached the highest level of formation. Likewise, according to the Bible, the pig is not a clean animal, for although its hoof is separated and completely divided, that omnivorous (meaning animals which eat everything) does not chew the cut.

In other words, having a divided and completely separated hoof and also the ability to chew the cud is not by chance, but the product of a certain level of perfection in the formation of the first kinds of some animals. I later realized that the ability to chew the cut indirectly refers to a certain kind of digestive system found only with ruminants (meaning animals that chew the cud like cows). As I studied the digestive system of the animals who chew the cud, I realized that it is more advanced and more compartmentalized or divided into many sacs, which activities allow grasses eaten by these animals to be properly processed so that the digestion and the intake of the nutrients contained in them can be perfected. In other words, as I dug into the matter, I found out that animals that chew the cud have a specific type of stomach with many compartments different from what all other types of animals have. Indeed, while the stomach of ruminants consists of 4 chambers (i.e. rumen, reticulum, omasum, and abomasum), that of a monogastric consists of a single stomach. Like all ruminants, cows swallow their food without chewing it much at first; then, after regurgitating the cud, they chew it well before swallowing it. More precisely, after ruminants eat grasses, the feed goes into the rumen and reticulum, where symbiotic bacteria contribute to breaking them down. It is believed that the reticulum sort whatever ruminants have eaten and send the feed back to the ruminants' mouth, where it is chewed more. Hence, ruminants spend hours regurgitating, chewing, and rechewing whatever they have eaten in the field in an effort to better break down any hard things including fibers. Afterwards, they swallow again the rechewed feed. In the process, the omasum (a section of the stomach) is believed to remove water from the feed and more digestion continues in the abomasum (another section of the stomach) before the food is sent to the intestines.

Also, animals that chew the cud usually eat only grass and do not eat everything like the omnivores (most of which do not chew the cud). All clean mammals are herbivores. It is interesting that even scientific evidences today suggest that it is better for human beings to eat vegetables and plant-based foods rather than meat. Those who have stuck to plant-based diets have shown healthier signs of life including better control of weight. In fact, in the beginning, man was not supposed to eat meat!

During the formation of the stomach of ruminants, extra processes must have been followed before the precursor of the stomach could split-gather into many compartments allowing the digestive tract to have features that make the cud chewable. In my research I also noticed that, the digestive system of the animals that do not chew the cud (e.g. lion and other carnivores) is usually very short, which also causes the food eaten by most carnivores to quickly pass through the digestive tract without being efficiently digested. The time the grasses eaten by ruminants spend in the digestive system can also help to better extract nutrients from them. In other words, the long multi-stomach digestive system of ruminants also allows them to better remove toxins from their food before it is absorbed by their bodies. The length of the digestive tract of most ruminants is 6-12 times the length of their bodies, therefore giving them time to process and eliminate poisons or toxins from their bodies (Tessler, 1996). According to the previous author, the digestive system of pigs for instance is very short and simple, and does not allow them to detoxify

CHAPTER 28: ENCRYPTED CODE BEHIND CLEAN AND UNCLEAN ANIMALS

their food before it reaches their flesh. On top of that, pigs like dirt, mud, and anything messy. Even scientific studies showed that pigs and other unclean animals have significantly higher toxicity levels than clean animals like cows (Russell, 1996). The eating of unclean meat like pigs has significantly changed the blood chemistry of the subjects (Fallon, 2001). It is not surprising that, until today, observant Jews (to whom the dietary law was initially given to) do not eat pork and shellfish.

Furthermore, I noticed that, during the formation of organisms, extra processes were needed so that the precursor of the hoof could be separated and completely divided. In other words, the animals which hoof is separated and completely divided are more advanced than those which hoof is not separated and not completely divided. The completeness of the division of the hoof and its separation are therefore encrypted symbols of the advancement of the process of the formation of some animals. In the end, animals that have a separated and completely divided hoof are more fully formed than those which do not. Hence, the cleanness and uncleanness classification of animals by the Bible should not be taken just as a ritual or religious thing.

Several animals in the air are imperfect and some examples given in the Bible are eagles, vultures, ostriches, hawks, owls, cormorants, pelicans, herons, and bats. Despite their size, strength, speed, and other abilities, these animals are unclean probably because the cluster of matter used to form them had not reached the highest level of molding. Some of these animals eat the carcasses of dead animals. If God can ask His followers not to even get close to dead animals, He cannot approve any animal that eats from the carcass of a dead animal to be classified as clean.

Among the "winged swarming animals" having 4 feet, only those that have jointed leg above their feet, enabling them to jump, are the best and worthy to be eaten (Leviticus 11:20-25). Examples of these include locusts, grasshoppers, katydids, and crickets. All other animals going on 4 feet are detestable. Any animal having 4 feet, and which goes on its paws is also unclean. This means that the ability of insects to jump and to have jointed legs above their feet is a level of perfection. The ability to jump and not just go on feet or paws must be associated with other internal advancements that must have required additional processes. Hence the animals that can jump have an advantage over those that cannot, but which just go on their paws or feet.

Among the small animals that swarm on the ground, the following are unclean: mouse, lizards, gecko, weasel, crocodile, chameleon (Leviticus 11:29-33). Any animal that swarms on the ground or moves on its stomach, or goes on all four, or has many legs are not to be eaten (Leviticus 11:41-44), meaning their formation did not reach the perfection level worthy to make them clean and edible.

As I pondered on the dietary laws revealed in the Bible, it sounded to me that, the law of the clean and unclean animals is not just a religious regulation, but an encrypted code used to form animals and to show how the level of the completeness of the execution of the program of life ended up deeming some animals clean and others unclean regardless of their size and kind. I understand that some people may still not agree that some animals are clean while others are unclean; however, I hope they will not argue with that, with all their love for dirt, mud, and filthiness, pigs are not the cleanest animal on Earth. One can wonder why one of the dirtiest animals on Earth ended up being one of the most supplied meats on the shelves of most food stores. Likewise, despite their strength and power, with love for dead and filthy animals, eagles and vultures are not the cleanest animals on Earth, yet they are highly appreciated by some people and even were made the national bird of one of the most advanced countries in the world. On another note, I hope you would agree that crocodiles are not the friendliest animals on Earth. Similarly, I hope you will agree that snakes and other kinds of reptiles filled with poisons are not the kindest or most loving animals on Earth although some people perceive that they are great pets and keep them in cages in their home. To me, it would be like putting a shark in your swimming pool and still jumping in your swimming pool on a hot day and ignoring the fact that the shark has its own natural instinct to want to eat you, it does not work well that way. I could go on and on and explain how the word "clean" and "unclean" animals used in the Bible is not just a simple religious classification, but a deep biological mysterious revelation that people have unfortunately thrown out the window of their car like the trash of their half-cooked fast food (made with pork) sandwich bought at a restaurant. To learn more about my decoding of the significance of the classification of animals into clean and unclean, please refer to the chapter I wrote on that topic later in this book.

To wrap things up, I would say that, according to the Bible, clean and edible mammals must have a split hoof and must chew the cud. Cattle meet those requirements. Although they are ruminants, rabbits are not clean because they do not have split hooves. To be eligible as clean and be eatable, a fish must have fins and scales. Examples of qualified fish include salmon, tuna, bass, perch, snapper, sardines, trout, etc. Because they do not have both fins and scales, scavengers are not clean. Example of scavengers are catfish, shark, and shellfish (shrimp, lobster, oyster, and clam) that many people across the globe eat delightfully despite evidences supporting their "filthiness" regardless of their beauty or the sweetness of their juice. Among birds, chicken and turkey are not listed as unclean, hence they are edible. Among birds, scavengers are also prohibited (vultures and buzzards). Many ecologists and dietarians will agree with me that scavengers are not fit as human food but fit to clean the environment from wastes including other dead organisms. I have seen many unclean animals eating garbage and sewage, but I have not seen clean animals enjoying the same yet. Among insects, only locusts, crickets, and grasshoppers are listed as clean. Some of these insects are even richer in protein and poorer in cholesterol than the meat of some cattle. Yet, not all insects are good. For example, please don't tell me that cockroaches are the best meat in the world. In fact, if you would not mind, I will finish this chapter by telling you a few stories I had with cockroaches, rats, snakes, and other wild animals many years ago.

Indeed, when I was a second-year university student, I used to live in a one bed

CHAPTER 28: ENCRYPTED CODE BEHIND CLEAN AND UNCLEAN ANIMALS

room (condo). Because I did not come from a wealthy family and my stipends given to me by a national fellowship was not enough, I could not even afford having electricity in my house. I just used a lantern for my studies which light up my room. When the kerosene finished in the lantern, I could not see until I could afford to buy more. One day, I cooked food and I had some left-overs. Early in the morning, just before I left for school around 6 AM, I took some of the leftover food and was eating. It was a sauce with a few vegetables and some truite. As I was eating it, my hand went over something that felt like some truite, a dry tiny fish I used to buy and eat in those days. Because I did not have light and was operating in the darkness, I was happy with what I felt was in the food, for I thought it was a fish in my food. I grabbed it, in my hand, mixed it with the vegetables (leaves) in the food, and put it in my mouth. Then, I chewed it forcibly, it became juicy, and I felt a cracking noise as breaking something harder than what I had cooked the days before. But, because fish can be bony, I did not bother checking it, and continuing to eat it until the thing started smelling like something I had never smelt before. Yuck! I had to expel it out just to realize it was a cockroach that was competing with the sporadic fish I managed to put in my food. Did I finish the food or throw it away after that experience? Please do not ask me to tell you. I washed my mouth (at one point), not with toothpaste or mouthwash, for I never used those fancy and expensive things until I moved to the US, where I could no longer find nice pieces of wood I could cut, dry, and then use as my personal, customized toothbrush! In Africa, I used to go to the bush just to seek and cut some tree branches, wash them, and dry them, just to use as teeth cleaning materials. So, is cockroach a good food? If you have never eaten it, please hear it from me, you don't have to ever try tasting it! It is useless! Thank God if you have a light in your house that you can switch off and on as you wish, for many people across the globe do not have that privilege. If you have a refrigerator and microwave, be more grateful, for countless people do not have those household appliances and some have never had a chance to know what putting meat in or getting cold water from a refrigerator means. Until I was about 30 years, the few times my family could buy and cook a significant amount of meat to the point of having left-overs (which occurred only on Christmas and New Year), we always kept the rest just in a bowl or pot, which we just covered. If we were far away from home like at the farm, we just covered the meat with some palm branches. If the meat had to last for days, my mother used to warm it over fire wood every day to ensure no bacteria developed in it. Yet, after a few days, we can sometimes smell organisms growing in the food, sometimes including cooked sauces that were not warmed up early in the morning. Do we get rid of those foods that can occasionally smell? Please, don't ask me that question! My belly in those days is my witness! If you were to ask me in those days, I would probably tell you that all bacteria are not bad and if bacteria could be in my food, it better be a good one, else that was the day that bacteria must die in my mouth after I cooked it, provided I had enough firewood to do so, else, in those days, it must die alive under the crunching of my teeth! When I was growing up, my parents could not afford

TURBULENT ORIGIN OF LIFE

buying much clean meat for the family and most of our meat was pork that we raised and fed using our own poop and leftovers, and grasses. With my siblings, we used to hunt to kill rats, mice, rabbits, caiman, cane rats (all of which are unclean) and we ate them, for I did not know or have better in those days. I remember cane rats (also called grasscutters) were so expensive because everything in them is edible. In other words, when we managed to catch cane rats, after removing their hairs that we burned with fire (and enjoyed the smell of the fire), nothing inside those cane rats was thrown away. Even the intestines, we ate them for they tasted good. A few years ago, my children were asking me to tell them some stories from when I was growing up, and I told them some of the things I mentioned above, but they thought I was joking, while I was telling the truth! Sometimes when I see them living an American lifestyle, eating chickens without properly cleaning the bones, I felt like we were wasting food, for up to about my 30 years of age, I could not eat a chicken without breaking the bones to find what I could squeeze out of the marrow. Sometimes, if breaking the bones to the marrow was too hard, we would take a hammer to break it and put it over a fire to allow the heat to release every iota of lipid and anything else still hiding somewhere around or inside the bones! The very moment I was typing this paragraph on February 2, 2022, I just went to the kitchen to see what my wife was fixing for me, for yesterday she did a chicken soup, which I liked except that, she cooked the entire chicken without splitting it into pieces for the chicken was frozen when she decided to cook that day, and, instead of thawing it first, she just cooked it as was so she could also be at work on time. I was hoping to eat the leftover of that chicken that day. Because she was taking longer in the kitchen than I was expecting, I went to the kitchen to see what she was going. As I came there, I opened the pot in which was the meat and could not find them. I asked her where the chicken meat is and she said she removed the bones. I could see the pieces of the meat in the sauces, but I could not see the chicken neck which I saw yesterday and which I planned to eat that day. I asked her where the bones were and she said in the trash. I opened the trashcan and to my surprise, a very nice and long chicken neck welcomed me. I was a little bit sad, not because I did not have food to eat, but because my wife still could not remember that I would have liked the neck and even cleaned the bones of that chicken cooked in those collar green vegetables from my garden. Anyway, I picked myself up and did not complain and had to eat the shredded pieces of meat. Likewise, usually, when I eat meat, I really feel sorry for the waste of meat she calls cartilage and bone and throws away because she could not clean it. May be, she was raised not to eat them, while I was raised to eat them. That life in America, where about half of the meat some people cook is thrown into the garbage! My wife and children were shocked when I told them the kind of meat I have eaten in my life: alligator, apes, bats, cane rats, buffalo, caiman, cartilaginous fish, catfish, cats, chicken, crab, crickets, deer, duck, frogs, grasshoppers, hippopotamus, lamb, lion, lizards, lobsters, monkeys, mouse, pigeons, porcupine, rabbits, rats, sheep, shrimp, snails, snakes (boa, python, etc.), squirrels, tilapia, turtle, and many more. Hearing this, my kids said it is gross, but it was not for me, for I

CHAPTER 28: ENCRYPTED CODE BEHIND CLEAN AND UNCLEAN ANIMALS

still bear scars caused by bushes cutting my legs as I was chasing some of those animals for my own survival about 30 years ago! This was some aspects of my life in some countries in the world. Some people may think those experiences are struggles, but as for me, except the cockroach experience, I enjoyed everything in those days, for I did not have better choices. However, since 2011, I changed my diet and started eating according to Leviticus, for in that year, I became a Messianic believer. Since then, I never ate pork, scavengers, or any other animals termed as unclean.

'Science180 Academy' Success Strategy:
SCIENCE180 BOOKS THAT WILL HELP YOU!

I, Nathanael-Israel Israel, broke down my discovery about the formation of the universe into many books so that you, the readers, can pick the ones that correspond to your needs and interests without disappointing you or wasting your precious time. These books come in many versions (e.g. scientific, public, chemical, biological, biblical or prophetic, pseudepigraphic, and a children's version) targeting people according to their expertise, educational background, and interests as briefed below:

1. "TURBULENT ORIGIN OF THE UNIVERSE" (This is the scientific version of my book tailored to scientists and anyone interested in the detailed scientific demonstration of the universe formation). In this book I used the "mother of all turbulences" to scientifically demonstrate the formation of the universe so that scientists can understand and reorient the course of their research, teaching, and publishing, and accept the truth to better live today and forever. Get *"Turbulent Origin of the Universe"* today to begin an incredible journey of accurately decoding the universe and change your life forever! Learn more at Science180.com/scientific

2. "RECONCILING SCIENCE AND CREATION ACCURATELY" (this is the book that I called the "Biblical or prophetic version of my book on the universe-origin, and it targets Christians and anyone interested in knowing the Biblical perspective of the creation of the universe). This important book accurately demonstrates the marvelous creation and formation of the universe by God in six consecutive 24-hour-days, and answers many questions about the universe creation, so that after acknowledging Him (who deserves all the glory now and forever), human beings can choose life and avoid the terrible judgment awaiting the unbelievers in the world to come. Get this thoughtful book now to figure out what happened at the beginning, what is coming up, and why it is time to urgently rethink everything you have been told about the universe-origin so you don't eventually regret! Don't say I did not tell you! Learn more at Science180.com/biblical

3. "TURBULENT ORIGIN OF CHEMICAL PARTICLES" (Called the "chemical version" of my book on the universe-origin, this elegant book targets chemists, biochemists, and anyone interested in chemistry). With *"Turbulent Origin of Chemical Particles"*, the accurate decrypting and understanding of the formation of chemicals has never been profitable and easy. Hence this great book is THE ultimate how-to guide for great people wanting to correctly decode the origin of the chemicals and positively transform their lives. Get this celebrated book today. Learn more at Science180.com/chemical

4. "ORIGIN OF THE SPIRITUAL WORLD" (This book is what I called the pseudepigraphic or hidden version of my books on the universe-origin, and it is meant for believers who want to tap into a higher level of scriptural secrets that most people may not believe). This book draws the attention of the world toward the pseudepigrapha (a collection of hidden and rejected books, yet filled with deep secrets still valuable today) and explaining how, since thousands of years, God has already revealed deep details about the supernatural origin of the universe, but people (including those who believe or claim to believe in Him) have just refused to literally accept God's mysterious story of creation, which can never be understood by just sticking with conventional science. If you believe in God, have some origin-related questions which answers you cannot find anywhere, not even in the Bible, and if you want to tap into historically neglected revelations to answer fundamental universe and life questions, then be sure to get a copy of *"Origin of the Spiritual World"* today. Learn more at Science180.com/pseudepigrapha

5. "FROM SCIENCE TO BIBLE'S CONCLUSIONS" (I called this book the "public version" of *"Turbulent Origin of the Universe"* and it is tailored for the general public, and it is a great summary of the scientific version from a perspective that laypeople will fully understand). In this book, I, Nathanael-Israel Israel, broke down the complicated (scientific, philosophical including religious) data about the origin of the universe in a simple language that the general public can fully understand, and know in order to live happily forever. Quickly grab and read this scientifically verifiable, bestselling book to finally get the accurate, jaw-dropping answer that has been rationally shaking both believers, skeptics, and all freethinkers. Don't wait! Learn more at Science180.com/public

6. "TURBULENT ORIGIN OF LIFE" (This is the biological or life version of *"Turbulent Origin of the Universe"*).

CHAPTER 28: ENCRYPTED CODE BEHIND CLEAN AND UNCLEAN ANIMALS

7. "*HOW BABY UNIVERSE WAS BORN*" (How was the universe formed? Did God really form it like some people believe, or did it come out of some long processes? How can we scientifically prove and break down this difficult mystery in a language that children will fully understand and like?) Get the answers as you read this book that I called the "children version" of *"Turbulent Origin of the Universe"* and life. Accurately explaining the complex formation of the universe and of life to children can be very hard in our modern world, but by getting *"How Baby Universe was Born"*, you will know the proven formula to help children to easily understand their huge universe-origin and life-origin questions with confidence, humor, and joy. Learn more at Science180.com/children

8. "HOW GOD CREATED BABY UNIVERSE". The most difficult part of writing scientific things to children is how to break down complex technical concepts into simple words that they and even anyone who can read and clearly understand (without losing the accurate details and facts). When the topic to address is about the origin of the universe, the task is even more challenging for most people, but not for Nathanael-Israel Israel. As long as you can read, you will find this amazing book extremely helpful to grasp all complicated concepts needed to properly crack the origin of the universe in a language that even children ages 7-12 and anyone who did not go very far in school can fully comprehend.

9. "*SCIENCE180 ACCURATE SCIENTIFIC PROOF OF GOD*" (Whether you are a believer, an unbeliever, a freethinker, administrator, politician, curriculum designer, curriculum specialist, education policymaker, librarian, school board member, parent, researcher, student, teacher, clergy, or a layperson, as long as you are really seeking to scientifically understand the rational proof of the existence of God, *"Science180 Accurate Scientific Proof of God"* is the much-admired book written for great people just like you). As long as you are interested in the first and the only scientific book that talks to anti-creationists, evolutionists, big bang proponents, atheists, and all other freethinkers and rationalists about the universe formation and they bigly beg to know more about God, the creator, that they mistakenly deny; then this book is for you. If you want to have the entire big picture of my discovery of the origin of the universe, life, and chemicals, and to enlighten your life and career, then plan to get all or some of these books that best suit your needs and interests. For more details, visit Science180.com/books. As long as you are really seeking to scientifically understand the rational proof of the existence of God, *"Science180 Accurate Scientific Proof of God"* is the much-admired book written for great people just like you. Grab it today and start reading it. Don't wait any longer! Learn more at

CHAPTER 29

I ASKED SMART SCIENTISTS THE BEST WAY TO SCIENTIFICALLY TEST EVOLUTION–THEY ALL SAID ALMOST THE SAME THING ... EXCEPT THIS SCIENTIST WHO CAME OUT OF NOWHERE AND SHOCKINGLY PROVED SOMETHING YOU CAN'T LEARN AT ANY CHURCH OR PUBLIC SCHOOL

- Do you agree with the National Academy of Sciences that "creationism has no place in any science curriculum at any level"?
- Do you agree with Evolutionists that creationist arguments against evolution don't hold?
- Do you celebrate that, as of 2025, the teaching of the Biblical creation in the US schools is legally ruled unconstitutional and unscientific?
- Must biology be really founded on evolution?
- Will creationist efforts to explain the origin of life really erode quality science education?
- Do you think that secular science is the single method by which humankind can understand nature and the origin of life?
- Is it true that creationists only criticize evolution but never offer a comprehensive scientific theory of creationism?
- Is it true that evolution is a "major unifying concept" in science?
- Can we really prove that, without evolution, the literacy level of students and professionals will be limited?
- Can we deny all religious interpretations of natural phenomena and still decode the origin of life?

CHAPTER 29: EVOLUTIONISM AND OTHER UNBIBLICAL OR ANTI-CREATIONIST ACCOUNTS OF THE ORIGIN OF LIFE

- Is it safe to deny the Biblical explanations of the origin of life but accept only empirical evidences?
- Was it really proven that all scientific domains or disciplines show evidences that evolution has occurred and is still occurring?
- Did anyone really prove that the Earth is millions of years old, that God did not create life, or that life did not appear suddenly, or that all organisms have evolved?
- Are you sad that evolution has infiltrated all areas of biology education?
- Do you wish that all creationist efforts to teach creationism as an alternative to evolution in science classrooms to fail?
- Or do you think that the teaching of creationism in public schools would advance the belief of the existence of a supernatural creator and impermissibly endorse religion?
- Do you really think that the Biblical account of creation cannot be confirmed or denied by using the scientific method, therefore, it must be removed from science curriculum?
- Did you ever ask yourself what if evolution is not the real explanation of the origin of life on Earth?
- Can we explain the formation of the universe through natural processes without evoking evolution?
- Do you want to learn about a comprehensive scientific theory that does not just criticize evolution, but that uses modern science to rationally test whether God created the universe?

So far, I have extensively explained how I think life arose on Earth. However, when the question of the origin of life is raised, while some people point their fingers toward God, the Creator, others think that the answer was already provided by the English naturalist Charles Darwin in his book (*On the Origin of Species*) published in 1859. Because evolutionism has dominated the secular world for many decades, I felt like I cannot finish writing a book on the origin of life without referring to some key points where it contrasts with the views I defended in this book. By the time you will finish reading this chapter, you will understand whether or not evolutionism is a sound theory to explain the origin of life.

When I came to this point in the writing of this book, I was about to write a chapter on the viewpoints of the other religions in the world, but, as I started reviewing the related literature, I realized that, besides Judeo-Christian religions, most (if not all of the) religions in the world support evolutionism one way or another. In other words, although some Christians strongly believe in evolutionism, no world belief (faith or knowledge) system opposes evolutionism as much as Judeo-Christian religions. Nowhere in the Bible is anything mentioned about evolutionism. However, I have come to realize that, although Islam also claims creationism as the origin of the world, many of its adepts support evolutionism to a point that some Muslims and Muslim scholars accept evolutionism as the main process by which life was formed. In other words, the staunchest opponents of evolutionism are found among the Judeo-Christians. Therefore, it would be unthoughtful for me to write a book on the origin of life in the 21st century without properly reviewing creationism and evolutionism. In the precious chapter, I talked about Biblical creationism.

I understand that many people accept the theory of evolutionism while several others deny it. Therefore, instead of reviewing everything that people believe about the origin of life, I felt like focusing on evolutionism and creationism would suffice to handle most of the religious and secular views about life. Hence, after devoting the previous chapter to creationism, I will focus the current chapter on evolutionary views. I will not review every aspect of evolutionism, but I will critically focus my review on 2 main things: the assumption of a common ancestry of all forms of life and adaptive abilities.

29.1. Errors in the concept of common ancestry postulated by evolutionism

Published on November 24, 1859, the 502 pages of the book of Charles Robert Darwin (1809-1882) untitled *"On the Origin of Species by Means of Natural Selection, or the Preservation of Favoured Races in the Struggle for Life"* has been very influential among those who believe in evolutionary biology. The main theme of that book is that all forms of life arose from a so-called common descent over a set of processes formalized as evolutionism.

Some people think that life arose from chemical compounds through a gradual process termed abiogenesis, and according to which simple chemical compounds were allegedly combined to yield complex forms of living things. In the name of such processes, some people assumed that life appeared billions of years ago. The proponents of that theory of occurrence of life postulated that life appeared more than 200 million years after the formation of the Earth.

Because living organisms have some fundamental molecular mechanisms in common, some people think that they are descendants of a common ancestor. For instance, about a century ago, living organisms had been classified into just 2 groups: prokaryotes and eukaryotes. Because eukaryotes seem more "advanced" or more complex than prokaryotes, some people have postulated that eukaryotes descended or evolved from prokaryotes. Yet, both were formed in the same environment, but shaped by different turbulences.

At the molecular level, problems still exist concerning the origin of some biomolecules. For instance, some people thought that genes were formed before proteins. But because genes are needed to form proteins, and vice-versa, people who advocated for gene first theories or those who supported protein first theories found themselves in a debate similar to that of those who argue the origin of the egg and chicken. To avoid this issue, some human beings accept that genes and proteins arose independently. This same problem and logic applied to the origin of the ribonucleic acid (RNA) and deoxyribonucleic acid (DNA). Yet, like I demonstrated earlier in this book, at the beginning, the RNA did not descend from DNA, but both were formed.

CHAPTER 29: EVOLUTIONISM AND OTHER UNBIBLICAL OR ANTI-CREATIONIST ACCOUNTS OF THE ORIGIN OF LIFE

Unlike what most scientific theories have postulated and that Encyclopedia Britannica reported, I do not think that "life is a material process that arose from a nonliving material system spontaneously" (Sagan, 2021). I showed that life is a spiritual entity that can be manifested in and limited by the physical matter that can host it. Just as people have thought for generations that life can arise spontaneously from nonliving matter before being proven wrong in the 18th century when "advancement" was made in the field of life sciences, so also people will realize that no matter how long chemicals can be mixed or left to themselves or to the "best" possible environment, they can never produce life by themselves.

Although not typically defined as organisms, viruses have genes that code for energy metabolism and protein synthesis. This discovery made some people to wonder whether viruses are organisms or whether they once were able to metabolize. Some people even postulated that viruses and their hosts could have evolved together. It is very sad how some people think about the origin of organisms or of things in alignment with the degree of similarities in physical appearances.

According to the so-called timeline of the evolutionary history of life, some people think that all organisms have a universal common ancestor, something I don't believe in. For instance, some theories defined the so-called "*last universal common ancestor (LUCA) as the most recent organism from which all organisms currently living on Earth descend*" (Theobald, 2010). Some people support the concept of common ancestor because all organisms have nucleic acids and use the same twenty amino acids as the building blocks for their proteins. Again, the similarities in the translation of RNA into proteins by all organisms have caused some people to think of the existence of a common ancestry of all forms of life. Although I know that the concept of common ancestor of all organisms is a mistake made by some people trying to explain life, I have showed that all organisms are a product of the biological turbulence that acted on some water and soil on earth.

Usually traceable at least back to the days of Darwin (in the 19th century) but believed by many long beforehand, the concept of natural selection that most evolutionists espoused is also grounded in the assumption that all organisms do not survive to leave descendants with identical chance of survival. Hence, evolutionists perceived organisms that are alive today as having survived some challenges, threats, or struggles that presumably killed others. This assumption implies that current organisms were not formed as they are today, or that no perfection was imprinted into the makeup of their so-called evolutionary precursors, else they could have all existed and not failed the so-called survival test or struggle. However, based on my understanding, organisms were neither produced by evolutionism nor did they share the same ancestor. The notion of ancestor in itself does not stand well with what happened. My investigation of turbulent origin of life taught me that the concept of natural selection is a mistake.

On November 30, 2021, as I was reflecting on the origin of life with the perspective of turbulence, I found other errors in the concept of a common ancestor. Indeed, if the concept of a common ancestor based on similarities and

TURBULENT ORIGIN OF LIFE

differences between species could be applied to the origin of the celestial bodies, some celestial bodies located many light years away could be wrongly found to be the ancestor of others located lights years away, which is incorrect. For on the scale of the universe, many galaxies and the celestial bodies (e.g. stars, planets, asteroids, and satellites) that they contain could have significant similarities, which can easily cause some linear thinkers to believe they descended from one another and then migrated. Hence, some people think that collisions occurred during the migration of celestial bodies, something that never happened, and because the distance between the celestial bodies is huge, such a linear explanation of the formation of the universe made some people to mistakenly think that the universe could have been formed after billions of years of processes. Likewise, because organisms on Earth are close to one another, those who believe in evolutionism tend to neglect how their argument is biased and could also lead to ridiculous implications if applied to other things in the universe which also had a beginning.

The genetic code has been used to explain how a certain sequence of DNA can be transcribed into a certain sequence of RNA, which can be translated into a certain sequence of protein. But it is not sufficient to explain why, despite the similarities of genes across species and even within the same species, organisms are differently wired in their growth, size, functions, abilities, timing of many events in life (e.g. birth of progeny, "ambitions", life cycle, lifestyle, death), etc. For beyond or beneath the genetic code lies other codes, which science will fail to understand if it cannot look for truth beyond the physical realm.

In other words, just because all the main amino acids found in organisms have been narrowed down to 23 that one can assume that they came from a common ancestor. Although scientists have discovered thousands of genes "common" to all animals, these organisms did not come from a common ancestor, unless by common ancestor you mean God the Creator of all things and from whom all creatures emanated. The similarities of some of the processes involved in the formation of organisms explains why for instance the genetic code (which explain how DNA translated into proteins) is said to be almost identical for all organisms. The fact that different nucleotides can code for the same amino acid leading to the same protein gives also a clue that things can look alike yet they can have different backgrounds. It also implies how the forces that control life are above just chemicals. Similarly, all amino acids found in proteins are left-handed because they were formed using similar mechanisms synchronized with other laws behind chemicals in the world. The so-called phylogenetic trees, phylogeny, or genealogic trees of species built on similarities and differences of morphological appearances or even of molecular or genetical traits are misleading in the efforts to explain the origin of life.

Recalling the formation of the celestial bodies, although there are many planetary systems for instance, all of them consist of a planet surrounded by satellites of various sizes and characteristics. This is because the processes that formed planetary systems are almost the same but the differences among these bodies can be related to the reigning conditions of their formation. Likewise, the mechanisms that support life had many similarities because the processes of their formation have certain

CHAPTER 29: EVOLUTIONISM AND OTHER UNBIBLICAL OR ANTI-CREATIONIST ACCOUNTS OF THE ORIGIN OF LIFE

things in common although all forms of life were not born at the same time. Considering what I learned from the code of the formation of nonliving matters, I felt like the amount of matter, the energy, the size, the position, the environment, and all others factors (turbulence and mission) that affected and surrounded the formation of life explain the calibration of the similarities and differences among all organisms. The universality of the codes (e.g. genetic, turbulence, and mission, etc.) and the functioning of all organisms on Earth is in favor of a common set of processes of formation of life calibrated differently, but not in favor of a universal common descent (besides God the Creator) as some evolutionists tend to claim.

The notion of transmutation of species (according to which some evolutionists think that some species were progressively transformed or mutated into others) is like saying that the Sun became the Earth, and vice versa or that the Moon is able to be transformed into a galaxy such as the Milky Way Galaxy. That has never happened and will never happen either. No matter the number of years (or even billions of years) given to the Moon or its precursor, it can never become a galaxy. Similarly, no passing of time can allow a star like the Sun to become the Moon, yet everything in the Solar System came from a common precursor like I explained. The understanding of the processes which shaped the celestial bodies and also the organisms will help people know that although things and beings in nature share some initial matters in common, they were not formed by an evolution process, but by a very fast mechanism, some which I explained by the turbulent split-gathering processes that I spearheaded.

All I said in this book contributed to making me know that, evolutionary biology will never be able to produce the real history of life (on Earth and outside of the Earth) if its proponents continue to focus on a common descent, natural selection, and speciation. No additional number of breakthroughs in biology, biogeography, ecology, geology, genetics, paleontology, physics, systematics, and in any other discipline related to or feeding evolutionary biology will suffice to prove the origin and fate of life besides some of the turbulence processes I spearheaded in this book and other books I wrote on the origin of the universe (see www.Israel120.com/books for more details).

Another Book by Nathanael-Israel Israel:
FROM SCIENCE TO BIBLE'S CONCLUSIONS

THE # 1 UNIVERSE-ORIGIN MASTERPIECE OF ALL TIME … AND THE MOST ACCURATE SCIENTIFIC FORMULA THAT STOOD AND WILL STAND THE TEST OF TIME AND OF MATHEMATICS

The real reason scientists have been struggling to accurately understand the universe-formation is because they have spent centuries collecting expensive, complicated, and massive amounts of data, but learned very little, if not nothing, about how to unconventionally step back to properly analyze it to decode the universe. Consequently, people learned to collect all kinds of data everywhere to build models and imaginary concepts that betray their discernment, but they never learned to unlearn wrong theories, nor learned how to stop trashing great raw data hidden in theories they dislike or misunderstand, never knew where to find and how to properly combine the fundamental variables without which it is impossible to ever clear the way so their data can properly work for and precisely lead them to the real origin of the universe. How can people abandon the dangerous theories they think are correct because they don't know any better ones?

Lucky you, that is where Dr. Nathanael-Israel Israel, the founder of Science180 (Science180.com) came in to properly reanalyze and put under control these costly, underrated data to provide the accurate and simple solution people have been looking for throughout the ages, but that they have ignored.

In *"From Science to Bible's Conclusions"*, you will:

- Get a world class explanation of the 4 fundamental variables without which it is unquestionably impossible to ever decode the universe-formation scientifically
- Save time and money, and enjoy a life filled with the wonderful peace that the accurate understanding of the universe-origin can create
- Discover the errors in the scientific theories and religious belief systems about the universe-formation that are putting you at risk, and learn how to take control over cosmological threats lurking at the edge of your rational mind, faith, disbelief, or doubt
- Unlock the accurate scientific formula to rationally test the existence of God in a historic way that uncompromisingly satisfies both believers and skeptics (*Science180.com*/public)
- Get all you need to become a knowledgeable person who will never again need anybody else to explain to you the origin of the universe, for, you will fully understand and articulate it yourself and rationally know whether science is really at war with religion

CHAPTER 29: EVOLUTIONISM AND OTHER UNBIBLICAL OR ANTI-CREATIONIST ACCOUNTS OF THE ORIGIN OF LIFE

- Receive deep insights that even those who went to university for years were not able to decrypt by themselves, so you can equip yourself to eliminate all forms of scientific and religious universe-origin prejudices
- Discover whether the scientific data finally confirms that the formation of the Earth was completed on the 3rd day, while that of the Moon and the Sun was on the 4th day of creation like the Bible says, or whether the data proves that it took billions of years to progressively form the universe
- Understand the celebrated scientific formula that rationally puts to rest all debates about the relationship between science, faith, and all theories about the universe-origin so you can properly develop yourself, expand your network, and shape your future

Quickly grab and read this scientifically verifiable, bestselling book to finally get the accurate, jaw-dropping answer that has been rationally shaking both believers, skeptics, and all freethinkers. Don't wait!

Dr. Nathanael-Israel Israel has had the honor to be acknowledged as the #1 universe-origin, life-origin, and chemicals-origin expert. He is the author of *"Turbulent Origin of the Universe"*, *"Reconciling Science and Creation Accurately"*, *"Turbulent Origin of Chemical Particles"*, *"Turbulent Origin of Life"*, *"How Baby Universe Was Born"*, *"Science180 Accurate Scientific Proof of God"*. Visit Israel120.com to learn more about this world's most trusted expert that helps scientists and laypeople to properly decode the origin and formation of the universe, life, and chemicals so people can live more effectively nonstop.

29.2. Most adaptive abilities were born with organisms, but expressed and enforced during their lifespan

Unlike what some people think, adaptation was not a stage reached by organisms during a "lengthy" process of their formation, but instead it was a set of abilities given to them or that can be switched on and off when needed to allow them to adjust to changing environmental and internal conditions. In other words, organisms were formed with the ability to adapt to their environment, and those that could not always could have died or migrated to other tolerant environments, and if they could ever meet conditions that they could not adapt to or move away from, they could have died as well. Other animals could have lived in environments that had to change (including genetically) before they could express some new behaviors to adapt to new conditions. Yet, the benefits that adaptation can give to an organism to live longer even in rough conditions in which others could die are not a reason to think that any organism was formed after a long evolution process during which so-called weaker organisms could have died, while stronger ones survived the so-called struggle. After paying careful attention to the matter, it appeared to me that the real struggle in life is that of human beings who are struggling not only to understand how life arose but also how to properly improve their lives today and forever.

Some organisms could well have the genes and strategies required to adapt to harsh conditions but failed to express them due to some unknown circumstances beyond their ability or ecological niche. In other words, adaptation is not just related to organisms. Hence, it can be partially unfair to pair adaptation with only the survival ability of the organisms. If adaptation abilities had to be acquired after organisms were formed, they could not have survived the harsh conditions during which these abilities would have been formed. For instance, abilities of amphibians (e.g. frogs) to cope with land and water were not developed after they were formed. Else, the first time a frog would have entered the water (with no ability to swim), it could have drowned. Likewise, a soldier who has never learned how to fight and how to survive in a difficult warzone will probably die if sent to such a zone with the hope that he will adapt to the deadly environment without proper preparation. Saying that adaptation was a set of skills that organisms progressively acquired during the process of their so-called evolutionist struggle for survival is like maintaining that human beings who have never swum could jump into the ocean and hope or expect to adapt to the waves and learn how to swim and even how to invent a boat to live on in that ocean. What a dream that cannot happen no matter the number of years (even after billions of years) that can be devoted to that blind test (if the one doing that test can be alive long enough)!

Even long after organisms were formed, the environmental stresses that they have to cope with are not always friendly and if the organisms did not have abilities in them to "fight" them (from Day 1 of their existence), they must have died a long time ago and most of them could not have even ever existed. Without the foundational abilities for organisms to adjust to the stress of their surrounding environments, no organism could endure long enough to get used to such changes. The ability to adapt can be affected by the age and experiences of the organisms of course. For instance, children cannot "adapt" to some conditions as their parents. Likewise, some survival skills require education and training. But some advanced ways of life and education cannot be expected from wild animals beyond the limit of their instinct, mission, and calibration. At the same time, some stresses that organisms must respond to today were not present long ago, meaning that some genes that were not much (or never) used before could be used later by some organisms when their environments became more demanding. Although some organisms can develop strategies to improve their lives and avoid harsh environmental stress today, that ability was given to them in some form since the days they were formed according to their kinds. For all organisms do not have the same "mindset" and ability to cope with their environment the same way. Even in the same environment, organisms behave differently according to their types. In

CHAPTER 29: EVOLUTIONISM AND OTHER UNBIBLICAL OR ANTI-CREATIONIST ACCOUNTS OF THE ORIGIN OF LIFE

contrast, even in different and very distant environments, some organisms of the same kind behave almost the same way. For instance, everything being the same, cattle in Africa do not behave much differently than those in America. Some may be cold and others warm indeed, but the intrinsic behaviors of animals (of the same kind) seem the same as that of a program they tolerate, which is not something they developed themselves just as human beings can invent new ways to cope with some challenges. In other words, the similarities of responses of wild animals of the same kinds to environmental stresses regardless of their geography stress suggests that the foundation of those responses or strategies were not developed by the organisms themselves, but were imprinted in them, hence expressed almost the same way everywhere. Else, how did they communicate those strategies with one another even when they are located in very remote locations?

Organisms were formed perfect and able to adapt to live longer according to their lifespan. Based on my understanding of life, every species was originally formed in the environment that suited it, and those who landed in unsuitable environments must adapt, else they died or will die. The concept according to which species were born in unsuitable environments and had to struggle and fight for survival does not consider the fact that, just as the characteristics of chemical particles and celestial bodies were framed in agreement with the constraints and opportunities of the environment they were formed, so living things were initially formed in environments matching their characteristics so that life does not cease right after birth just because organisms were born in completely unsuitable territories. In other words, the environment was first made to fit the organisms according to the range of their ecological niche. This matching of life and the environment hosting it can explain why organisms adapt and/or move away (migrate) when environmental conditions are detrimental. Else, they will suffer and even die. For all organisms were formed perfect and were endowed with the ability and strategies to adapt to environmental changes to some extent by activating some internal response programs (involving but not limited to genetic responses) to keep living even if sometimes they have to momentarily change their lifestyle (including hibernating) until their environmental conditions are again very favorable. The continual change of the environmental conditions and the adjustments that organisms must make are part of the key encrypted codes that make life dynamic, less monotone, and which contribute to helping organisms enjoy the variations and options they have during their lifespan; else, life can be boring and less fulfilling.

29.3. Law of species belonging

All of the processes and circumstances surrounding the birth and life of organisms have set them to behave within certain limits according to their types. I have realized that organisms were formed to belong to a group or clusters of bodies like them even if they are mixed with others. According to the law of belonging, which came into my mind in 2013, each living organism knows how to belong to its species and act like the other members of that species. Hence, wild animals or plants of the same species behave the same. Chemical differences and variations of internal characteristics played a role in the difference among organisms. According to that same law, nonliving things such as matter know how to behave like the other matter of the same kind. For instance, chemical elements of the same kind behave the same as long as they are put in the same conditions and their internal composition is the same.

However, when it comes to human beings, they do not generally act the same. Because of their free will and the immensity and diversity of their imagination and goals, human beings act according to what they want, provided nobody is blocking them. Sometimes, they hurt each other and try to kill each other for diverse conflicting reasons. Wild animals of the same kind do not usually do this kind of wicked thing to each other. Some bulls can sometimes fight, but I am not aware of any wild animal that lays a snare for other individuals of that species just to catch, trap, hurt, or kill them. Nevertheless, human beings do that! Some human beings even eat human meat.

Birds do not act like fish and vice versa. No mammal can decide to fly. However, human beings know how to take a drug or do something to themselves or to others to alter their state of mind so they can act like another species, even until reaching a certain level of foolishness. Human beings know how to elaborate laws that can catch their political opponents and then put them into jail or even kill them. Human beings know how to collaborate with evils spirit to travel in the air just to go hurt others. Some use austral projections or trips over the broom to travel in the air.

The impact of chemical elements variation on the lifestyle of organisms plays a role in the expression of the law of belonging. Life is not about chemical composition only of course, but it depends a lot on it. For example, the difference between the behavior of some animals across the species can be boiled down to just a difference in the chemical composition of some of their organs (e.g. brain), and how the turbulence that formed them wired their entire being to function as a whole. For example, what causes big carnivores (e.g. lions, tigers, leopards, alligators, and crocodiles) to behave ferociously as they do, while some herbivores (e.g. goats, sheep, deer, and cows) behave peacefully and docilly can be a matter of how they are wired. When it comes to human beings, some behaviors can be explained by chemical imbalances. For instance, medical analyses have shown that some human beings behave as insane just because their thyroidal hormones have changed significantly as compared to their normal levels. In such a state, because some of their hormones are not balance with the physiology of their bodies, some human beings behaved like "wild" animals and remove their clothes, walking in public naked as if that is normal. Subsequently, some people who are insane or foolish, become normal soon after taking some pills that remove their chemical imbalance. Hence, the behaviors of some people called fools can surprise those who are considered normal, while for the so-called fools, their behaviors are normal in their state of mind and they think it is those who are normal (or who are called normal) who are foolish. If some people called normal can enter into the life or world of the so-called fools or insane, and then manage to come back into a so-called normal life,

CHAPTER 29: EVOLUTIONISM AND OTHER UNBIBLICAL OR ANTI-CREATIONIST ACCOUNTS OF THE ORIGIN OF LIFE

some people could better understand and appreciate life and the boundaries set to keep each species within its ecological niche and mission. The variation of some endocrinal biomolecules can be literally detrimental to some organisms, to the point of making some human beings to behave like wild animals. Likewise, the variation in the shape, organization, size, volume, energy make up, and mission of organisms and their constituents can affect how they behave. In the end, life is not just a matter of chemistry, but also about spiritual things that mere science cannot apprehend.

29.4. Why the birthdate of organisms cannot be determined using their size or the level of their complexity?

On September 12, 2021, I was tempted to think that the biological turbulences that shaped the organisms could have occurred according to their ranking and intensity. In other words, I almost mistakenly thought that smaller or less complex organisms could have been formed before more advanced ones. But, on February 8, 2022, as I was wrapping up the writing of this segment, I felt like it may be wrong to think that it took less time to form apparently "inferior" living organisms than to form so-called "superior" organisms. I put the term "inferior" and "superior" in quotation marks because I do not feel very comfortable considering any organism inferior. For instance, although plants can be considered "inferior" to humans being (who are more advanced), but as of today, human beings cannot live without them. I do not think I need to go into any details concerning the fact that plants are the main producers of oxygen on Earth and without them, human beings and wild animals will not have the oxygen or the food they need to support themselves. Therefore, it is understandable why plants were formed before animals. Because they are more complex than any other organism, human beings were the last to be formed. However, another key fact that I deduced from my turbulence insight is that, the stage of development of the biological turbulences that shaped some higher organisms was different from that which shaped so-called inferior organisms. If the so-called superior organisms were formed first, the so-called inferior could have ended up having more complex structures in them (than they do as of today) which could have been left in the environment of their formation. For instance, imagine you want to build a house with clay, another one with sand, and another one with concrete, but all of them need to be formed using the same tools, the same floor plan, the same design, and the same container or same equipment. If the house made up of sand is formed after that in concrete, it is more likely that the one made out of sand will end up having some pieces of concrete left in the soil after the construction of the concrete house. But if the house made up of sand is built before the house in concrete, no major contamination will occur to any of them for sand is usually used in concrete. Another example is that if you want to use the same bowl to make two kinds of salad (one seasoned with just salted water and another one with salted oil) without rinsing that bowl between the 2 types of dishes, you need to be careful which one you make first. If the salad dish with salted oil is done first, the salad with salted water will undoubtedly have oil in it. For the oil can stick inside

TURBULENT ORIGIN OF LIFE

the bowl and be passed on to the next dish made in that same unwashed bowl. But if the salad with salted water is done first, and then the oily one follows, the later will be less impacted. The same logic can explain why fish were formed before land animals and why insects and reptiles were formed before human beings. Likewise, plants were formed before the other forms of life. In other words, some organisms that contain less organelles or biomolecules may have finished their formation before some more complex ones. Although this logic may sound well for some organisms, alone, it cannot explain the timeline of the formation of all forms of life. For some so-called less-complex organisms could have been formed last in some cases according to the time that their precursors were split-gathered by the biological turbulence that birthed them. For instance, in the Solar System, it will be wrong to try to deduce the birth date of the planets, asteroids, and satellites by focusing only on their size. Like I explained in *"Turbulent Origin of the Universe"*, many asteroids in the Solar System were formed before the planets, and many satellites were also formed before their primary planets. I showed that all of the planets inward to Jupiter (i.e. Mercury, Venus, Earth, and Mars) were formed before Jupiter. Although those inner planets are smaller than Jupiter, other planets located outward of Jupiter (e.g. Saturn, Uranus, Neptune, and Pluto all of which are much smaller than Jupiter) were formed after Jupiter. For instance, Pluto is even smaller than Earth, yet formed long after the Earth and Jupiter. Many asteroids located far beyond Pluto were formed after the latter. In other words, if people would base the understanding of complexity on size only, they will think that (because of its small size) Pluto was formed before Jupiter (the largest planet in the Solar System, yet it was the opposite. The same mistake could have caused some people to think that, because of their small size, the outermost asteroids were formed before the planets in the Solar System. Recalling the things that I discovered in cosmology, if I can use some of the evolutionist linear logic to explain the nomenclature of the celestial bodies, I may end up not considering all of the planets in the Solar System as planets. For instance, the organization of the giant planets will not even allow some people to name them as planets as the terrestrial planets, which are clearly different and some of which do not even have satellites.

I showed that the birthdate of the celestial bodies depended on the position of the fluids of their precursor in the stack of fluids of the precursor of their mother and the timing of their split from the remainder of that stack of fluids. Likewise, as a biological turbulence was forming living bodies, the position of a cluster of matter in the stack of materials (e.g. soil and water) used to form a group of organisms can affect the timeline of the formation of a body in that cluster of matter. Some smaller bodies may have come from clusters of matter that were split-gathered last, others could have been formed from the intermittence of the split-gathering of the precursors of bodies, while some larger organisms may have been formed more quickly than some small ones. Hence, it can be biased to base the date of birth of the organisms on their size only.

All that I said above contributes to emphasizing why it is dangerous to always try to find an evolutionary parent for all organisms. I understand that, so they can

CHAPTER 29: EVOLUTIONISM AND OTHER UNBIBLICAL OR ANTI-CREATIONIST ACCOUNTS OF THE ORIGIN OF LIFE

validate their research, some scientists want to constantly assign each organism to a specific group including species. However, these efforts should not serve as excuses to always try to evolutionary link every species or organism to a remote descendant. In other words, although I have no problem with scientific efforts to name organisms according to various objectives going beyond scientific research, I am really shocked and concerned by all attempts aiming at always seeking for an evolutionary parent or descendant for every organism as if everything has really evolved after billions of years of processes claimed by evolutionists. The linear effort of inventing a theory of origin of everything in the universe has complicated the path that could be taken to properly understand the world, its functioning, and ultimately its fate and origin.

In *"Turbulent Origin of the Universe"*, using distance between bodies, speed, and size, I was able to calculate the exact date that the celestial bodies were formed. But here, in this book on the origin of life, I chose not to engage in any mathematics aiming at calculating their birthdate. Because some people missed the story of the origin of the universe and life, they also denied the story of how life is supposed to be lived and how it will eventually end on this Earth, which some people mistakenly think is here to stay. Due to the foundational mistake of evolutionism, I am not surprised that, among other things, Darwin thought that "*women are inferior*" and that "*blacks are a lower race*" (Encyclopedia Britannica, 2021). I bet most people did not know he incorrectly thought that. Human beings were created by God for unique purposes, and if we can focus on our mission and purpose, we don't need to be asking who is superior to who, and which organization is higher than which! We are all different and unique.

29.5. Struggle to classify living organisms into distinct and nonoverlapping categories (e.g. species)

The problem that life science experts have to concisely and comprehensively define life is like the problem that systematists have to classify organisms into distinct categories that do not overlap. Hence, the classification of living things has been changing over the centuries and certain terms that were accepted are now rejected as new evidences are found, while others rejected in the past are now embraced for similar reasons including previously overlooked characterizations. For life cannot be defined only by the physical similarities and differences between the anatomy and physiology of organisms but also by characteristics beyond what can be seen or perceived physically. Unaware of this reality, some people use the similarities between living organisms to postulate things about their origin as if they really descended from one another according to the level of their simplicity or complexity. The same mentality caused some people to base their life-origin theories on similarities and differences between DNA, RNA, and protein sequences and functions of organisms. It can be ok for scientists to try to classify living things in order to facilitate their labelling and study, but it is not ok to use only such a classification to postulate the origin of life.

TURBULENT ORIGIN OF LIFE

For reasons similar to why it is difficult to properly define life and the forms of life into distinct groups, it is also very difficult for biologists and taxonomists to classify all organisms into species which could be properly distinct from one another without similarities or overlaps. The presumed hierarchical processes that some systematists and taxonomists have been using to classify organisms into species do not consider the real processes that ruled during the formation of life, and which are responsible for the similarities and differences among them, instead of a descendance or (macro)evolutionist processes. The definitions of species assume that organisms obtained their genes from their parent(s) just as "daughter" organisms would. Following this stream of thought, some people alleged that an "evolutionary process" so-called speciation had existed and by which species were born. In reality, as I have explained, no organism got most of their genes from another one as part of the process of their formation. It is true that, after the formation of the organisms, some genes were transferred across them. I will revisit this concept in the chapter on evolutionism.

Now, I will say a few words about organism species. Indeed, just as a cell is the structural and functional unit of living organisms, on the scale of the biosphere, a species is a "basic" unit of biodiversity. Species are clearly distinct from one another, but similarities exist between them. While since the work of Charles Darwin in 1859, "On the Origin of Species", many people have quickly accepted that the species across the world arose through a natural selection, many top scientists also do not believe in that theory. For them, the evidences do not militate in favor of that assumption. For instance, unlike what the evolutionists may say, the anti-evolutionists advocate that the variations among species have not evolved or arisen from mutations and recombination, but that most of them were present since the moment the organisms were formed. Some variations arose due to the environmental constraints of course, but anti-evolutionists support that these variations are not enough to supersede or override the massive variations among the species since the Day 1 of their formation. Likewise, the response of some species to environmental conditions have affected the way their genome responded and expressed some genes, which, in the end, made those organisms different from others even within individuals of the same species. In the same manner, some individuals may look alike, yet their genetic makeup put them in different species. Although some species may have affected the gene expression of others, the similarities of the genetics of groups of individuals or so-called species cannot be explained by the transfer of genes between them or their precursors. This can serve as a reminder of how certain things that people are exposed to can affect the health or their offspring. Such things like agent orange, which was used in war (e.g. the Vietnam war) is said to affect those exposed and their offspring. Some toxic things in the environment today can affect unborn babies and cause birth defects. We may not see how those things in the environment affect other organisms like they do to humans, but that could be because the weakness of those organisms may not survive their predators or environment. However, humans have developed sophisticated equipment to aid some of those affected by such agents.

CHAPTER 29: EVOLUTIONISM AND OTHER UNBIBLICAL OR ANTI-CREATIONIST ACCOUNTS OF THE ORIGIN OF LIFE

The assumption of living organisms having descended from others worsened the comprehension that the scientific community has about life. Hence, every time some scientists want to talk about a species or a new organism, they usually think of or try to point at which existing or extinct organism(s) could have fathered it. This linear way of thinking led to even a linear and branchial classification of organisms, which a few centuries ago was based on morphology, but which today, due to the so-called "advancement" in genetics is much more based on genetic characteristics. And because genetic traits do not always translate into phenotypic characteristics, what is physically seen is not always an expression of the genetics of the organisms. Hence, even with the scientific "progress" in life science disciplines, no consensus exists yet about how to define life or the species which most scientists want to consider as the unit of biodiversity. In other words, because the species did not descend from one another, all efforts based on morphology, physiology, genetics, reproducibility, and all other characteristics used to define them will not allow to have a clean and unquestionable linear or hierarchical explanation of living things based on evolution. For no ancestor of living organism evolved to become any organism. Likewise, no star evolved to become a planet and vice versa, but yet both stars and planets have their unique pathway of formation.

I cannot finish this segment without talking a little bit about the intraspecific and interspecific diversity with the perspective of turbulence. Indeed, variations exist between species at many levels. For instance, even between individuals of the same species, significant variations can be found. For, the laws surrounding the formation and reproduction of living things did not and is not and will not allow the same kind of biological turbulence to be replicated exactly the same across all kinds of environments regardless of the scale. Some of the laws surrounding the recombination of DNA did not and still do not allow exact same progeny or offspring to be formed from the same parents. The differences between the organisms may not be easily perceptible, but they exist and are real.

Just as how intraspecific diversity relates to the variation (e.g. genetics) within a single species, so also within a type of celestial bodies (e.g. planets), the intra turbodiversity can relate to the variation among all the planets. In other words, just as the ensemble of main traits they have in common can be used to group organism into species, the traits that celestial bodies have in common have been used to group them into types of celestial bodies. For instance, because of some characteristics (e.g. density, size, and atmospheric composition) they share in spite of their differences, some planets are called terrestrial, gas giants, or ice giants, etc. This is because the conditions (e.g. size, position, composition, movement, and speed) of the precursors of these bodies were different and the processes that shaped them led to the formation of different bodies. In other words, it was not that the turbulence that the precursors of the giant gas (planets) went through was that different from that of the giant ice (planets) or the terrestrial planets, but the fact is that, if the initial conditions of the precursors were different, the bodies formed could have been different, even with the same kind of turbulence.

TURBULENT ORIGIN OF LIFE

The same thing occurred on the precursors of microscopic particles. Hence, just as several types of atoms were formed although they are not all the same, there are also several types of planets, asteroids, stars, and satellites yet, they are not the same. In other words, a species can be for living things what a type of celestial body or type of microscopic particle is for nonliving things.

Just as individuals of one species are different from individuals of another species, so celestial bodies of one type are different from celestial bodies of another type. For instance, even within the same stellar system, planets are different from satellites and asteroids.

I will close this portion with a few words about gene exchange between species. In fact, as a specialist in microbiology, plant biology, and genetics, I cannot rule out that, since the formation of living things, some genes could have been leaked or transferred between organisms of the same and/or different species through processes including hybridization, viral gene transfer (e.g. infection), bacterial gene transfer, etc. However, these gene transfers cannot explain the so-called "horizontal gene transfer" that evolutionists use to support the exchanges of genes between species. Organisms were fully formed before some environmental conditions could have caused them to get or lose some genetic materials or to get them repressed and unable to express and show certain phenotypes. In other words, gene transfer was not what allowed some organisms to evolve and become like others or different from others. For instance, viruses and bacteria can transfer genes between one another and between organisms of other species of course, but those kinds of gene transfers did not lead to the so-called evolution of these organisms. The same way it is hard to classify organisms into distinct and nonoverlapping groups, it is also hard to categorize all types of turbulence into nonoverlapping processes. For living things were made to live together and the processes of their formation have many overlaps.

CHAPTER 29: EVOLUTIONISM AND OTHER UNBIBLICAL OR ANTI-CREATIONIST ACCOUNTS OF THE ORIGIN OF LIFE

Another Book by Nathanael-Israel Israel:
SCIENCE180 ACCURATE SCIENTIFIC PROOF OF GOD

THE FIRST AND THE ONLY SCIENTIFIC BOOK THAT TALKS TO ANTI-CREATIONISTS, EVOLUTIONISTS, BIG BANG PROPONENTS, ATHEISTS, AND ALL OTHER FREETHINKERS AND RATIONALISTS ABOUT THE UNIVERSE FORMATION AND THEY BEG TO KNOW MORE ABOUT GOD, THE CREATOR, THAT THEY DENY.

As you read this historic book, you will:

- Scientifically know what is the one clear sign you should always pay attention to in your efforts to decipher the primary cause and the key drivers of the fundamental processes responsible for the universe-formation
- Discover the only way to scientifically know if God exist, and if so, which of the thousands of beings worshipped across the globe is the true God
- Accurately answer the most critical universe-origin and life-origin questions so you can stop standing in tension with consequential question marks including those related to religion and reason or the so-called war between science and the Bible
- Discover the errors in the scientific and religious theories about the universe-origin and life-origin that are putting you at a high risk you will never recover from if you don't quickly and confidently learn how to rationally take control over threats lurking at the edge of your efforts to understand the universe and life today
- Challenge the cosmological status quo and embrace the real change that will disrupt the hidden cages that may be holding you and that you ignore
- Definitively answer all your doubts about the source or author of the universe and life … (learn more at Science180.com/godproof)
- Satisfy your burning desire for freedom from beliefs and scientific theories about the universe-origin and life-origin that suffocate you and bind your mind, faith, unbelief, heart, and education
- Scientifically set on fire all false theories or dogmas about the existence of God, the Creator, that are enslaving humankind

Whether you are a believer, unbeliever, freethinker, administrator, politician, curriculum designer, curriculum specialist, education policymaker, teacher, librarian, school board member, researcher, parent, student, clergy, or a layperson, as long as you are really seeking to scientifically understand the rational proof of the existence of God, "*Science180 Accurate Scientific Proof of God*" is the much-admired book written for great people just like you! Grab your copy today and start reading it! Don't wait any longer!

Dr. Nathanael-Israel Israel is a Beninese-American scientist, entrepreneur, and international consultant, who shows people of all ages and educational backgrounds how to scientifically decode the formation of the universe and of life, and who is acknowledged as the creator of the Chemicals Turbulent Origin Formula™, the inventor of the Life Turbulent Origin Formula™, and the discoverer of the Universe Creation Formula™. He is the Founder of Science180 Academy, which is trailblazing the reconciliation between science and the creation.

CHAPTER 30

TO BE SCIENTIFICALLY 100% SURE ABOUT WHETHER GOD CREATED LIFE OR EVOLUTION PRODUCED IT, PAY ATTENTION TO "SCIENCE180 MODEL OF THE ORIGIN OF LIFE"

At this point of the reading, you have demonstrated that you are really interested in knowing the true origin of life based on the perspective of the first person in history to crack the code of the formation of the universe. Indeed, as I was working on my books on the origin of the universe (see www.Israel120.com/books), and reflecting along the way on the origin of life, it amazingly appeared to me that the processes that shaped the celestial bodies and chemical particles have some similarities with those that shaped the forms of living things on Earth. If you have read this book from the beginning until this point, you would have understood that strong similarities exist between the process that birthed the nonliving things in the universe and that which birthed the organisms. It is not surprising that living organisms are affected by the characteristics of the nonliving matters they are made of. By comparing the shape, organization, and other characteristics of the living and nonliving things in nature, I understood the mark or footprint of turbulence, a phenomenon investigated by physicists for centuries, but in which little progress was made for the last five centuries since scientists started studying it. Using what I learned from the turbulence that molded the celestial bodies and chemical particles, I was able to crack the code of what I called "biological turbulence", a process by which life was formed using raw materials in the soil, water, and air. In other words, life was formed after the formation of the environment that hosts it.

Highlighting the groundbreaking discovery of the true origin of life, this book is framed after the discovery of the code of the formation of the universe. In it, I reviewed how the secular world defines life, and I showed how the ignorance or negligence of the spiritual aspects of life has prevented scientists from delving deep into the meaning and origin of living things. Although I also explored the origin of spiritual beings like angels and demons (that many people believe in), I focused my demonstration on the scientifically accepted forms of life: plants, animals, fungi, protists, archaea, and, bacteria. Although not considered as living by some references, viruses were also a topic of my investigation.

I showed that part of the difficulty that scientists have in defining life is that some functions present in some organisms today are found with certain clusters of matter or clusters of chemicals not technically labeled as living things. I showed that the difficulty that life science experts have been having to come up with a single inclusive definition of life is almost like the difficulty that physicists have been having to use a single definition to label all types of celestial bodies and microscopic matter in the universe. I demonstrated that the plethora of definitions of life is an encrypted message about how little the scientific community knows about the origin of life and of the features or proprieties of living entities. The difficulty scientists have to fully appreciate life and split its components into distinct categories also affected the inability of the scientific community to fully grasp all aspects of the investigation of life without having overlapping disciplines. Based on my discoveries on turbulence, I realized that, if people had understood the law of split-gathering and of intermittence that I spearheaded in my books on the origin of the universe, they would have already realized that, the way the precursors of living things were split-gathered also allowed the formation of various living entities having different constituents on various scales (e.g. macromolecules, organelles, cells, tissues, organs, apparatuses, and organisms).

Although chemical elements are found in all forms of life present on Earth, the formation of life was more complex than a mere association between chemicals. To allow the readers to understand where I am coming from with the use of certain concepts I learned from my insight on the formation of the universe, I spent a lot of time reviewing key aspects of the formation of celestial bodies which I also noticed with the formation of living things. During the formation of a system of celestial bodies for instance, I showed that a mother precursor split-gathered to usually form the precursor of a primary body and the precursor of secondary bodies. I showed that in the process, the precursor of the secondary bodies escaped the precursor of their primary bodies and, as it flowed, it was organized into fluid layers, and branched out the precursor of each secondary body as time passed by until the bottom fluid layers were reached. Interestingly, when I studied living things and what is known about development biology, I understood that, during the formation of life, precursors of life were also organized as fluid layers, which flowed and split according to a branching pattern similar to how some trees branch out. In other words, when I studied the organization of bodies formed in turbulence, I sensed features looking like a trunk, branches, roots, and leaves of a tree. To better express the similarities that I found, I coined the term turbulent trees, turbulent branches, turbulent leaves, turbulent neck, turbulent head, turbulent belly, turbulent spine, turbulent arms, and turbulent tail. For instance, the precursor-tree of a system of

CHAPTER 30: CONCLUSION–SCIENCE180 MODEL OF THE ORIGIN OF LIFE

bodies is a graphical description of what the fluid layers of the precursor of the system could have looked like during the formation of all of its daughter bodies, with the root of that tree positioned as the precursor of the primary body of that system, the trunk of the tree being the stack of fluid layers of all the secondary bodies orbiting, moving, or placed around the primary body before they branched out, and the branches of that tree are all the systems of bodies that emerged from that trunk. Just as all parts of a tree are connected together from the roots all the way to the top of the trunk or main stalk passing by all the branches (primary and secondary), with the leaves attached to them, it is also so when I talk about the precursor-tree of the celestial bodies, I maintain an apparent connection between the components or parts of that tree, although in reality, the fluid layers of the precursors of the systems of bodies were not connected altogether anymore once their split gathering started. For instance, on the scale of the Solar System (which is a stellar system), the precursor-tree consists of a tree which roots are the precursor of the Sun, the trunk being the precursor of all the bodies orbiting the Sun, the systems of primary branches connected to the trunk being the precursor the planetary systems or asteroid systems, while secondary branches are like the satellites of the planets and asteroids. Leaves are like some of the smallest levels of organization of matter on the scale of these celestial bodies, meaning they are like satellites and asteroids which have no satellite. I showed that the intensity of the turbulence in the trunk, which is partially transferred to the turbulence in the branches gave rise to many orders of branching such as primary branches, secondary branches, tertiary branches and so on and so forth until no branching could occur no more downstream of the flow in its veins. Hence, leaves can be found even attached to the main trunk or attached to a branch. In other words, if all of the branches and the trunk of a tree or plant can be removed without displacing the leaves, what would be left would be a system of leaves which organization may look weird, but which a careful study of the positions and distribution of those leaves could allow to pinpoint the potential positions of where the trunk and branches could have been. To put it another way, by carefully looking at the organization of leaves with respect to one another and with respect to the different clumps or clusters of leaves, it is possible to describe the position of the branches and trunk and even describe how the entire tree could have grown from its roots all the way to the top of the tree. On the scale of the entire universe, all of the celestial bodies formed a kind of tree that I called the "Turbulent Tree of the Universe", or the "Tree of the Universe" in short. I also understood that what has prevented people from discerning these patterns was that they did not know the history of the formation of those bodies and also, unlike real tree parts (roots, trunk, branches, leaves, and inflorescence), which are all connected, in the case of the celestial bodies, the connection between the bodies were broken during the split-gathering of their precursors, leaving behind just systems of bodies which seem independent, yet some of them descended from common precursors, which all together can be traced back to the original matter in the universe, the first matter.

TURBULENT ORIGIN OF LIFE

Some patterns pointing at a biological turbulence in living things can be seen among animal markings, animal segmentation, arrangement of leaves on the stems of plants, and many other anatomical and morphological characteristics of organisms. I showed that the size of living things was defined by the types of biological turbulence that birthed them. The organization of the root system, shoot system, and their components (e.g. leaves, fruits, branches, and roots) bear marks of turbulence. The precursor of the head of an animal could have been the precursor of the primary body, while the precursor of the rest of body was the precursor of the secondary bodies, while arms and legs are like branches emerging or branching of the main bodies just as satellites emerged from primary planets. The huge size and the position of the belly in most animals are shaped after the turbulence that took place during their formation and which also respects the turbulence that their environment can bring on them during their lifespan.

I showed that during the split gathering of the precursor of organisms, the head is like the primary body, while the torso and the rest of the body downward of the neck is a cluster of secondary bodies. Each of these systems of secondary bodies are organized into a system of primary bodies and secondary bodies, which at their turn also are organized into other primary and secondary bodies and so on and so forth until the smallest level of subdivision. I invented the term "turbulent neck" to label the features or the distance located between a primary body (let's say the initial position of a primary body) and its innermost secondary bodies. It can also be viewed as structures formed along the distance separating a primary body from its innermost secondary body. Although a neck is usually perceived while dealing with an entire organism as a whole, many body parts or components of living organisms have features which, according to their scale, are similar to a neck. For instance, the esophagus of the digestive system is like a neck connecting the mouth with the stomach. Likewise, the upper part of the respiratory tract is like a neck connecting the nose with the lungs. On a tree for instance, a turbulence neck can be the distance separating the soil from the first branch. On the scale of a primary branch, a turbulent neck can be the portion of the primary branch separating the innermost secondary branches from the node where the primary branch connects to the trunk. In general, the largest parts of organisms are found downstream or downward or after the neck, the position of the head being upstream. The turbulent spine of plants is the trunk, or the main stalk. Together with the system of branches, the trunk of plants serves as the "skeleton" or the "backbone" holding the plants, and without which it can collapse or crumble. Moreover, arms and feet of animals are secondary bodies that emerged as outgrowths or like a "shoot" or a "bud" from a developing thorax and/or abdomen. In other words, the split-gathering of the precursor of the thorax and abdomen yielded the arms and the feet. Arms and legs are like secondary bodies emerging from the thorax and abdomen and at their turn, fingers and toes are like tertiary parts emerging from the precursor of the arm and leg. Unlike human beings, most animals have a tail, which is usually the outer part of their body. Human beings lack a tail, not because they lost it a long time ago as some people think, but because no human being ever had one.

CHAPTER 30: CONCLUSION–SCIENCE180 MODEL OF THE ORIGIN OF LIFE

I showed that, on the scale of a leaf, the primary vein is like the primary body, the secondary veins are like the secondary bodies, the apex is like the outermost body on a leaf, everything inside the leaf or lamina except the veins is for a leaf almost what rings are for a system of satellites. The variation of the turbulence that took place when plants were being formed can explain the variation of the plant venation. In other words, the variation of the venation pattern of leaves results from how the precursors of the veins split-gathered during the formation of the leaves. I showed that, when animals are dissected in half through the middle of their abdomen and thorax along the direction going from the head to the tail, they can form a leaf-like shape. The shape of most fish looks like an elliptical leaf. The apex of a leaf is like the tail of an animal, while the petiole is like the neck. If most animals can lay on their back or belly, then open their arms and legs wide, the shape they will give is like that of a leaf which main vein is the torso, secondary veins are the arms and legs, while the tail of that animal would be like the apex or tip of the leaf. I showed that the similarities between the shapes of plants and animals are due to the fact that they are all products of a biological turbulence, which at its turn is related to the turbulence that birthed the celestial bodies. Some biological processes today are witnesses of some of the processes used to form the first organisms in the beginning. If a stem cell is able to become any type of cell, we cannot rule out that all types of organisms could have been made using similar processes or similar initial clusters of matter but framed or organized differently.

The mixing of small organisms with large ones is not something that the organisms chose to do, but it is an organization resulting from how organisms were distributed since the day their first specimens or "ancestors" were formed. The migration and the ecological niche of some animals also played a role in clustering some animals according to their environments of course, but still, those ecological realities are not sufficient to explain the underlying rules forcing animals to live in mixture with one another, even if it is a "pocket" of population of one form of life mixed with a "pocket" of the population of another.

As I was describing the formation of life, I realized that, different abilities were given to each form of life. I coined the terms universal pool of qualities, or universal pool of abilities to express all of the qualities found in all organisms. I also defined the universal holistic being as a single organism that is the amalgamation of all of the universal pools of qualities, abilities, traits, functions, features, and characteristics (known and unknown) found with living beings. Then, I showed that, in the beginning of the formation of life, a turbulent program of life was run through the medium where life was formed. In other words, the turbulent program of life is a program that formed the matrix or combination of all possible functions, features, and characteristics that living organisms can express. The turbulent program of life acted by means of a biological split-gathering of matters into organisms. The biological split-gathering is a set of processes by which the precursors of organisms were split and collected together as they were being formed for the first time. Just as a series of split-gatherings of fluid layers of the precursors of the celestial bodies,

TURBULENT ORIGIN OF LIFE

according to their turns, led to the organization of the celestial bodies in various organized systems of bodies, so the way the original biological clusters of matters were split-gathered (as can be seen during developmental processes today) explains how each organism reached its current body plan and characteristics, which, as of today, is reproducible via procreation and others means of germ line "perpetuation".

Any cluster of matter on which the turbulent life program acted became alive and started performing living functions according to its size and the abilities conferred onto it. In other words, the turbulent program of life was a mysterious biological program which, at the beginning of the formation of life, ran through matters to cluster them and impart onto them different shapes and sets of some of the abilities and qualities. The variation of the impact of the turbulent program of life explains the diversity of the forms of life which resulted from its actions. The similarities and dissemblance of the types of living things formed explain why some people ended up classifying organisms into different groups to which various names were given.

When I reached this level of my investigation, I defined a living organism, living thing, or a living being as an entity (physical, spiritual, and of unknown natures) made of an arrangement or a set of the characteristics of the universal holistic being under the influence of the turbulent program of life. In other words, a living thing, or an organism (spiritual or physical) is an expression of the turbulent program of life run through a certain kind of matter serving as a precursor of life. Just as I demonstrated that the same initial matter in the universe was split-gathered to produce various celestial bodies in the universe under the influence of the mother of all turbulences, so also various substrates (e.g. water and soil) were split-gathered under the influence of the biological turbulence and the turbulent program of life to produce the plethora of organisms in the universe.

The split-gathering of the universal pool of abilities in living things explain why they have specific characteristics and no organism has all of these attributes. For instance, some wild animals can sense things that most human beings cannot. Some animals can see very far, while others can smell things located very far from them. Some animals can perceive and detect many remote things including radiation. Others still can see even in the darkness, others can use echolocation to measure distance. Some wild animals work very hard, some are very relaxed. Some animals are clean, while others are very strong. Some organisms can live in water only, but cannot live in the air, and vice versa. Others can live only on the land, but not in the air or water. Some can live in both water and land, but they cannot fly, but they can jump. Some can fly and walk, while others can only swim. Some animals are generally poisonous, yet they are wise. In general, the attitude, characteristics, abilities, and all other things that can be said about living and nonliving things are for a purpose. Ecologists also testified about how the hierarchy and interactions between organisms and their environment are highly aligned with nature to such a point that it would be thoughtless or unthoughtful to deny that nature and its inhabitants have a purpose. Although some people have no clue about what exists beyond the Earth, or even beyond their city or country, we can all agree that without

CHAPTER 30: CONCLUSION–SCIENCE180 MODEL OF THE ORIGIN OF LIFE

a certain number of plants, animals, and all other forms of life on Earth, human beings could not have accomplished what they did.

Sometimes, because they fail to understand the mission of organisms around them, some people mistakenly think that some organisms are not good enough and that some must have evolved from others (to become better) and organisms they think are less-complex or less-advanced are wrongly viewed as inferior, or things that are still evolving and which have not fully attained their ultimate maturity yet. Without seeing life through the glance of a system of properly-scaled missions of various entities, it will be impossible to fully understand the origin and fate of life and of the whole universe. My discovery on the origin of the celestial bodies helped me to recognize the importance of scaling and the environment of formation of things on their diversity, forms, functions, and other characteristics. For instance, differences (even small variations) in the biological turbulence and the size of the matter used to form the bodies had a big impact on their anatomy and physiology. In other words, organisms (e.g. plants, animals, etc.) are what they are (morphologically, physiologically, spiritually, etc.) because, under the influence of the biological turbulence that shaped them, the turbulent program of life was aligned and scaled according to the characteristics of their mission and environment, which, at its turn, was framed according to the laws of the mother of all turbulences, which was at the base of everything formed in nature.

These observations led me to coin the term the calibration and mission code of life. For beyond, beneath, and even before the genetic, epigenetic, and all other known biological codes, there was and there is still a mysterious turbulent code and mission code that allowed the turbulent program of life to form and reproduce life. For instance, the processes that aligned the lifestyle of organisms with their environments are part of what I termed the "turbulent law of calibration". In other words, body parts were calibrated according to their missions in the environment they were made. By calibration of living components, I meant a set of laws that defined the size, morphology, physiology, and many other properties of organisms and their living components of body parts according to their mission and the conditions (including environmental ones) in which they were formed. This calibration is not just about the sizing of the dimensions, surface, volume, energy, abilities of the organisms, but about everything pertaining to the organisms. In my book *"Turbulent Origin of Chemical Particles"*, I explained how the chemical elements that constitute living organisms were not selected by chance. The law of calibration caused the size of the organisms to depend on the turbulence of their precursors. In other words, the size of the precursors of the organisms affected the expression of the turbulent program of life. Likewise, the characteristics of the components of the living organisms were affected by the size of the precursors of the organism. For example, the size of the organs of animals depends on the type of turbulence that the matter used to build them went through.

As I explained the generic turbulent process of the formation of organisms, I showed that, just as a computer program can be written, launched, run by itself, and execute all of its codes, so also the biological systems and beings in the universe have been running under the influence of turbulent programs coded during their formation. Some of these programs have been maintained and/or functioning under programs of reflexes, while others are influenced by the will of humans and spirits, but everything is under the control of a primary plan. Just as cells are believed to carry hereditary information in their DNA, so also, they carry turbulent information in the way the fluid layers of their precursors are and/or will be stratified during the biological split-gathering leading to their birth and growth. The turbulent program of life followed a law of turbulence according to which different organisms were created in line with the intensity of the turbulence that animated the precursor of their body. Just as some systems of celestial bodies don't have bodies in all turbulence zones, so at the individual levels, all organisms do not have body parts in all turbulence zones. Just as all turbulence zones are not present in all systems of celestial bodies, likewise, all turbulence zones are not present in all organisms. Hence, some organisms don't have body parts in all turbulence zones.

I showed that the lifespan of human beings is organized like a series of turbulent events connected to what I called "familial turbulence". For instance, to some extent, children in a family are like products of layers of fluids born according to their rank. From the baby stage until death, passing by the adolescent stage, adulthood stage, and the end-of-life stage, many behavioral changes can be noticed. I showed that, the trouble of puberty and middle age crisis are activities of the most turbulent zone of a human being's lifespan. When I see humans being more turbulent in their puberty years, a period that gives a lot of headache to parents, than in their late years or old years, I felt like this behavior respects the fact that turbulence is more developed in Zone 3, which is neither the innermost or outermost zone just as the middle age is neither the youngest age nor the oldest one. Life between 50-80 years old is like the transition zones toward older ages beyond 80-120 years or beyond, which is the outermost zone leading to the natural death (of the flesh) for a life long-lived according to current standards.

To explain the reason of the turbulent neck, belly, arms, legs, and tail, I demonstrated that, in general, the head is like the primary body of animals and the neck is related to the distance travelled by the fluid layers of the precursor of the secondary bodies (located below or downward of the head) before it started splitting into larger body parts. On the scale of living things, a certain distance usually separates most primary bodies from their closest or innermost secondary body. This does not mean that the head was completely formed before the other parts, or that the neck contains no body part, but that, just as in the case of the Solar System for instance, some planets were formed before the formation of the Sun was completed, so also the precursor of some upstream or upward organs could have been split and formed before that of the head finished its formation. For plants, the root system is like the primary body and the distance separating the upper part of the root from the lowest branch is associated with the distance travelled by the precursor of the shoot system (aerial part of the plant) before starting to split-gather into different parts. Long after the formation of the branches, due to growth and development,

CHAPTER 30: CONCLUSION–SCIENCE180 MODEL OF THE ORIGIN OF LIFE

the distance separating the branches increases of course, implying that the distance separating parts of a plant is not always that travelled before these parts were formed. Just as how the distances between celestial bodies are increasing as the whole universe is expanding, so also the distance I mentioned above for living things are increasing and are not exactly the same as they were when the first innermost secondary split. I showed that some animals have a long neck because the precursor of their secondary bodies had to travel for a long distance after escaping the precursor of the primary body (e.g. head) before the precursor of the living organisms could be amalgamated to a point that allowed a developed turbulence capable of forming the thorax and the abdomen which, among other things, comprise the belly, which is usually the largest part of most animal's body. In other words, the length of the neck of animals may have been defined by the time it could have taken for the cells that were differentiated into the precursor of the body parts downward of the head to elongate before being split-gathered and differentiated into bodies which size is higher than that of the neck. The longer the cell division must continue and cells must pile up before the branching could occur, the longer could be the neck of the corresponding animals. If the precursor of the neck escaped the precursor of the head very fast, it could travel a longer distance before its turbulence could allow the formation of larger body parts downward of the neck. Likewise, the speed of cell division and elongation of the precursor of the neck can also affect its length. For instance, I showed that the long neck of giraffes and dinosaurs were caused by the processes that formed them, not by any evolutionary process that some people alleged to have occurred after the formation of these animals.

Considering all of the things I have been explaining so far, here is the take home message of the generic processes by which living things were formed on Earth. Indeed, after the Earth was formed, the precursors of many celestial bodies in the universe were still going through the processes that would birth them (see details in *"Turbulent Origin of the Universe"*, one of my books on the origin of the universe). By this time, water and soil were already present on the surface of the Earth, and they were the matters used to form the organisms on Earth—in my book *"Reconciling Science and Creation Accurately"*, I explained why and how some organisms on Earth (angels and demons) were not originally formed here, but somewhere else before being transported on Earth. Chemicals in the soil, waters, and atmosphere on Earth were used to form life on Earth. Altogether, the soil, air, and water that were used to form life are the physical precursors of life on Earth. The precursors of life contain enough oxygen, carbon, hydrogen, nitrogen, calcium, phosphorus, sulfur, potassium, sodium, chlorine, magnesium, silicon, iron, fluorine, zinc, and other key chemical elements abundant in most organisms on Earth. Just as a seed that germinates in an environment is able to selectively uptake the chemicals and nutrients it needs to grow until reaching its full size, so also the turbulent program of life, which acted on the precursors of organisms, knew which chemicals to uptake according to the stage of its development so each organism could have been quickly and properly formed according to the biological turbulence of its precursor.

In the beginning of the formation of life on Earth, a biological turbulence broke out in water and soil on the surface of the Earth. The turbulent program of life started acting upon the turbulent structures formed in the water and the soil. Quickly, the impact of the turbulent program of life on the entire Earth birthed the precursor of the biosphere, which, after going through changes and impartation of abilities became the biosphere. Because of the variation of the environmental conditions on Earth, the turbulent program of life acted on the precursors of the biosphere differently. Therefore, according to the variations or the changes they were going through the precursor of the biosphere was split into different lots, which became the precursors of the ecosystems. Due to their size and the diverse impact of the biological turbulence, the precursors of the ecosystems were split-gathered into precursors of communities, which were also split-gathered into precursors of populations, which were split-gathered into precursors of organisms and all of these events happened quickly. In other words, during the formation of life, the soil, the waters (oceans, rivers, lakes, and all other types of water on Earth) were turbulently moved so that the precursors of organisms and systems or clusters of organisms could form. Because of their position in the global ecological niche, plants were the first life to be fully formed. Most land animals originated from the soil, while most aquatic organisms originated from water.

I showed that the hierarchical organization of living things into cells, organelles, tissues, organs, apparatuses, organisms, populations, communities, ecosystems, and the entire biosphere is the product of the biological split-gathering of the "turbulent program of life" being run over the soil and water on Earth at the beginning of the formation of life. I described the process of the formation of each of these components of the biosphere. I explained the process of the formation of some key macromolecules including the nucleic acids (e.g. DNA and RNA), proteins, and carbohydrates. Then, I detailed how each of the forms of life was formed and what can explain the differences between them. I extensively explained why the similarities between the organisms are not synonymous of their descendance or origin. Because I am the one who pioneered the explanation of the formation of the universe and life through the glance of turbulence, I coined the term turbodiversity or diversity of turbulence to label all forms, entities, varieties, or types of turbulences in the universe. I also coined the term turbalogy to designate the study of the types of turbulence.

I devoted a lot of attention to the creation of life according to the Bible, and I also reviewed some secular theories of life such as evolutionism, which I proved do not properly explain the formation of life. I also showed that the linear way of thinking of some people has prevented them from understanding the creation of life and the universe. I also reviewed evolutionism and other unbiblical or anti-creationist accounts about the origin of life postulated by evolutionism. I also demonstrated how most adaptive abilities were born with organisms, but expressed and enforced during their lifespan. I also revealed the errors in the concept of

CHAPTER 30: CONCLUSION–SCIENCE180 MODEL OF THE ORIGIN OF LIFE

common ancestry. I established how the birthdate of organisms cannot be determined using their size only. In the end, I showed how the Biblical account of creation matches the scientific evidences I presented in my books. Considering how I scientifically demonstrated the perfect match between the scientific evidences on the formation of the universe and the Biblical story of creation of celestial bodies, I cannot doubt the Biblical story of the formation of life. For God is the Creator of life and the designer of the turbulent program of life. Yet, in this book, I did not present the evidences in a religious way, but scientifically. I demonstrated that life has a spiritual component that many people have chosen to ignore. I also talked about extraterrestrial forms of life and how the physical death of human beings is not the cessation of life. For the life of humans is a spirit or a spiritual entity hosted or incarnated in a physical matter. The spirituality of life can also explain why dead people are not really dead but sleeping or translated into another dimension that can be reversed or restored via supernatural processes involving resurrection that some people do not want to hear about. I elaborated on these issues in my books:

- "Origin of the Spiritual World"
- "Reconciling Science and Creation Accurately"
- *"How God Created Baby Universe"*
- "Science180 Accurate Scientific Proof of God"

At one point, I wondered why no scientist before me has explained the origin of the universe as I did. But I realized that, because turbulence was not understood and the so-called experts who were "supposed" to explain it, they did not know how to handle it and they wasted their time laying complex equations (having closure issues) as if all it will take to explain complexity must be to lay complex equations. Another cause is that people did not try to ask and answer hard questions based on the observation of things around us. In other words, people are so familiar with things around them that they have seen since they were born that they never tried to question their origin and meaning. For instance, most people never wonder why trees look as they are or why animals look as they are. Yet, it was such "simple" questions that led me to realize the depth of the turbulence hidden behind biological systems.

As I was looking at some images of animals on January 27, 2022, I felt like biological turbulences are a closed version of the abiotic turbulences, which formed celestial and chemical bodies in the universe. That day, as I was fasting and praying, I felt like I did not need to reveal certain things to human beings in the common era, for my mind was downloading things that were supposed to be known not by some human beings in this world, but by some elected people in the world to come. Therefore, as it was 3:50 PM on that cold day of January 27, 2022, I deeply felt like I needed to refrain from sharing certain mysterious details with some human beings, for they can be better understood or accessed via the optic of faith in God, instead of via scientific demonstrations, which can be biased depending on the level and

TURBULENT ORIGIN OF LIFE

stage of the science involved. In all, I know and must confess that God is the Creator to whom all the glory, honor, and power belong. For any errors in this book, I ask God to forgive me, for my goal is to try to scientifically demonstrate how He created life and everything else in the universe, a task that cannot be scientifically done without human errors, which I know can be found in my books (www.Israel120.com/books). Nevertheless, I know I discovered a universal truth about the turbulent origin of the universe and life that needs to be shared and pondered on. I hope this book blesses you, and opens your eyes to a new and fresh perspective of the origin of life and of the universe we live in, and which we can never fully understand without a revelation from the Creator, whom I know is proud of my historic effort and will forgive any imperfection in my thought that I will keep surrendering to Him until I meet Him face to face one day. If you don't know God and if you are not sure you are ready to meet Him one day, I will encourage you to get a Bible and get accustomed with more revelations about the God of creation asap before it is too late.

Before ending this book, I thank you for sticking with the reading until this point. I know you learned a lot from this book and others I wrote on the origin of the universe. I would like to hear from you if you have any comments, suggestions, or questions. Please feel free to reach me at www.Israel120.com. You can also contact me at the same website for any speaking engagements, partnership, collaboration, business opportunities related to my area of expertise. Even if you will forget everything in this book, remember that, just as the universe, life was formed by God, and He loves you. If you disagree, please checkout *"Science180 Accurate Scientific Proof of God"*. But if you agree, checkout *"Reconciling Science and Creation Accurately"*.

NEXT STEPS OF THE JOURNEY

Get free resources on Science180.com
If you have finished reading this book and would like to learn more about my discoveries and how they can help you, you are at the right place. Indeed, I am really committed to helping you address any questions that you may still have concerning the origin, functioning, and fate of the universe, and how you can partner or collaborate with me for greater results.

To get free resources that will help you understand other aspects of the universe formation not covered in this book, visit Science180.com and my personal website Israel120.com. On those sites, I will be sharing guides and strategies to get the most out of my initiatives. I will also be sharing my favorite references, tips, next-steps readings and other important things in the pipeline that will help you regardless of your field of expertise, interest, and needs.

Subscribe to "Science180 Newsletter": The only accurate universe-origin, life-origin, and chemicals-origin newsletter in the whole world!
Be a part of decoding the universe-origin, life-origin, and chemicals-origin! Get origin-related news, information, discoveries, updates, announcements, news, reviews, articles, educational materials, and opportunities, from a holistic perspective not available anywhere else so you can participate in and enjoy decoding the origin, current state, and fate of the universe and its content. You will also receive priceless tips about how Nathanael-Israel thinks, what are his secrets and initiatives, what he has accomplished, and what he recommends. Without any delay, sign up for Science180 Newsletter today at Science180.com/newsletter. It is free!

Speaking engagement
In addition to writing groundbreaking books and engaging in other business endeavors, Nathanael-Israel Israel is a renowned speaker, who you can invite to speak at your organization.

TURBULENT ORIGIN OF LIFE

Values that Dr. Nathanael-Israel Israel can add to your life include:
- Rare expertise and tips that will increase your abilities
- Usefulness that will advance your impact regardless of your field of expertise
- Understanding of the world that will sharpen your perspective
- Critical information that will positively change your life
- Experiences turned into insight that will motivate and guide you
- Irrefutable scientific proofs of the existence of God that will save you time and launch you into a zone of unlimited opportunities
- Unquestionable scientific proofs of how God created the universe
- Accurate demonstration of the historic formula that reconciled science and the Bible
- Enlightenment that will help people including Christians to start using their brain instead of just praying and expecting God to do everything for them

For speaking inquiries, including how you can get Dr. Nathanael-Israel Israel to speak to your organization or at an event, visit Science180.com/speaking for more details.

As the standout scientific authority who accurately decoded the universe, Nathanael-Israel Israel has been helping countless people across the globe to discover and understand the complex origin of the universe without leaving out the challenging questions that people of all ages have been struggling to answer for thousands of years! As the true go-to expert when it comes to the formation of the universe and of life, Nathanael-Israel believes that, regardless of age, background, culture, religion, profession, everyone deserves to understand how the universe and life were formed and how they can leverage on that knowledge to improve lives nonstop. Therefore, his groundbreaking discoveries of the formation of the universe, life, and chemicals have been broken down into books tailored to scientists (including physicists, chemists, biologists, mathematicians), laypeople or general public, believers and freethinkers, philosophers, children, etc., therefore maximizing the benefits to humanity. These historic, internationally-acclaimed origin books (details at www.Israel120.com/books) include:
- "Turbulent Origin of the Universe"
- "Reconciling Science and Creation Accurately"
- "Turbulent Origin of Chemical Particles"
- "From Science to Bible's Conclusions"
- "Turbulent Origin of Life"
- "Origin of the Spiritual World"
- "How Baby Universe Was Born"
- "How God Created Baby Universe"
- "Science180 Accurate Scientific Proof of God"

NEXT STEPS OF THE JOURNEY

When you hire Nathanael-Israel Israel to speak at your organization, you will:
- get specific in-depth knowledge, up-to-the-minute information, ideas, and insights about the universe-origin, life-origin, and chemicals-origin so that you expand your market, cut useless costs, stop wasting time on inadequate projects, and start focusing on the profitable solutions
- get relevant universe-origin stories that are specific to your field of expertise
- learn from a cooperative, flexible, and an easy to work with expert who will respond to your universe formation needs and position you to stay on top of your competitors
- interact with a renowned expert that will not just lecture you, but that will help you sort out your origin-related questions using strategies to tap into deep secrets you ignore
- listen to an experienced expert who discovered outstanding secrets about the origin of all there is
- learn authentic information not from someone who just reads you a PowerPoint, but from the true go-to expert (when it comes to critical cosmological problems) who will share with you both his mistakes and successes that will help you get much closer to the better life you want to live
- revolutionize every origin-related domain with your accurate understanding of the universe-origin
- hear Dr. Nathanael-Israel Israel's personal selection and teaching of key topics that will help you break the code of the universe formation and functioning, and strategically enlighten you, guide you to navigate and filter the massive data collected on the universe and its content so you know how to answer the world's most challenging origin questions, remove any scientific and philosophical cataracts that may be blocking you, and help bring you many steps closer to your best life today and forever
- hear the greatest scientific and philosophic lessons of some top scientists, philosophers, thinkers, and public figures who have realized historic mistakes they made in life (concerning the origin of the universe, life, and chemicals), and that they corrected thanks to the discoveries of Nathanael-Israel Israel, who founded Science180, and who is acknowledged as the scientist that truly decrypted the universe-origin for the first time
- Get world key lessons successful people have learned in life and how people can learn from their experiences to improve lives instead of repeating their mistakes that many people still ignore at their own perils

To book Dr. Nathanael-Israel Israel for a speaking engagement purpose, visit Science180.com/speaking.

How you can make money by joining the affiliate program to sell Nathanael-Israel Israel's books

Greetings,

Do you want to make easy money by selling the #1 universe-origin, life-origin, and chemicals-origin books on your website, newsletter, and by mail? You can start making big money as you help sell Science180 Books including this one on your website and network. Indeed, by now, you know that I operate a website called Science180.com, specialized in helping people across the globe to scientifically decode and understand the formation of the universe, life, and chemicals.

Your contacts, site, blog, forum, podcast, and newsletter may be admired among my target audience. Some of my products and services may be of interest to your audience. My books are the first in history to scientifically demonstrate the match between science and Biblical creation in a way that satisfies both believers and nonbelievers, a historic achievement and discovery that is revolutionizing our view of the origin of the universe, life, and chemicals for the benefit of humankind.

Imagine you have a website where you can talk to people about my books and services and get a great percentage of every purchase they do on my site? Imagine you send a certain link about my books to your friends or network and, when any of your contacts buy a copy of my books, you get a percentage or a certain amount of what they pay on my sites. Imagine you can email your friends and spread the good news about my books and when anyone uses that link to buy my books, I give you something. Well! This is what the affiliate program is about. Apply today or learn more about it at Science180.com/affiliate. Likewise, if you own a website, you can apply for Science180's affiliate program, and I will send you a specific affiliate link that you will place on your website and newsletter, and if people click on it to buy my books, they will be led to my page and after they buy, I will pay you a certain amount, sharing the profit with you instead of just verbally saying thank you.

Would you be interested in reviewing some of my products and services with the aim of becoming an affiliate? We have a wonderful affiliate program and commissions are paid quickly and accurately.

If you are satisfied by the quality of our products and services, I am convinced you will also be impressed by our affiliate program.

I look forward to hearing from you

Nathanael-Israel Israel, PhD

Collaborate or partner with Nathanael-Israel Israel

If you have any lawful idea, initiative, or suggestion for a genuine partnership with Dr. Nathanael-Israel Israel or Science180, please visit Science180.com/partner to inform us.

NEXT STEPS OF THE JOURNEY

How to be trained or mentored by or have a one-on-one consulting with Dr. Nathanael-Israel Israel
Hire Nathanael-Israel Israel to train you or your organization in the best ways to conduct yourself and your organization to align your initiatives with the real understanding of the origin of the universe, of life, and of chemical particles in a way that you will not hear anywhere else. Nathanael-Israel Israel offers training through the program called "Science180 Academy". For training purposes, please visit Science180Academy.com.

Visit Nathanael-Israel Israel's personal website to get for free great resources you won't find anywhere else
To stay in touch with, Dr. Nathanael-Israel Israel, and to get updates directly from him, please visit his website, Israel120.com, and sign up for his popular newsletter at Israel120.com/newsletter for free.

Ask for review
If you are a book reviewer or a professional wanting to review this book or others written by Nathanael-Israel Israel, please contact us at Science180.com/AskForReview

Donate and support Nathanael-Israel Israel's efforts and initiatives
To help humankind accurately understand the real origin of the universe and its content, like I have done in the groundbreaking books I published after 12 years of sacrifice, I need your financial support. Please consider donating to me or to Science180 by visiting Israel120.com/donate or Science180.com/donate.
   Your donation will be used to help me continue doing what I did to birth these books that you enjoyed and that you know will help many people across globe. No amount of money is too small or too big. Whatever you can give, please give.

Quantity discounts: Purchase Science180 books including this one in bulk at a special discount
To purchase Science180 books including this one in bulk at a special discount for sales promotion, corporate gifts, fund-raising, or educational purposes or to create special editions to specifications, visit Science180.com/discount.

Buy a copy of Nathanael-Israel Israel's books for your friends, family, or someone

TURBULENT ORIGIN OF LIFE

If this book has been a blessing to you, and we know it has, please consider getting another copy and giving it to a friend, a family member, or someone you think it may help or challenge. If you want to get many copies, we can even give you a discount; just contact us as we previously explained.

Recommend Nathanael-Israel Israel's books to your organization
Because I know this book has been a blessing to you, I ask that you recommend it and others that I wrote to your organization, class, workplace, church, school, network, or clubs. Recommending this book will help others to tap into the blessing and opportunities that my books will open for them.

Share Nathanael-Israel Israel's groundbreaking discovery with others
To improve more lives, please share the findings of Nathanael-Israel Israel's books with others, for many people out there still do not understand how the universe was formed and sharing your experience of reading this book will help them. If you enjoy Nathanael-Israel Israel's books, please help other people find them by writing a book review on your blog or on online bookstores, or write it and share it with us. Likewise, share and mention this book on your social media platforms (e.g. Facebook, Twitter, YouTube, etc.).

Follow Nathanael-Israel Israel on social media
In our modern world, social medias have become a huge part of how messages spread across the globe today. To ensure more people hear about the good news revealed in my books, I need you to follow me and share my contents on your social medias and in your network. To know the full list of my social media accounts and follow me please visit Science180.com/socialmedia.

Share your feedback, critics, testimony, experience, adventures, story, or comment about this book with me
How has Nathanael-Israel Israel's books and services at Science180 improved your life? I would love to hear from you.

To help me know how I can better help you next and encourage others, I need to know and capture your testimony or critics. Please visit the feedback page, science180.com/feedback, to tell me:
- how this book impacted you or will impact you
- what you like or dislike or disagree with
- what you think, wish, or dream that I need to work on next
- what you wish to see in this book but that was absent
- what shocked you the most

NEXT STEPS OF THE JOURNEY

- what got your heart pumping as you were reading this book
- what you found more insightful or thought-provoking
- what you want to do to be a part of my journey
- how my work changed your life or someone else's life

Message from the publisher of this book
Just like Nathanael-Israel Israel, you can publish your book(s) with us too. To get started and see how we may help you, please visit Science180Publishing.com today.

To contact Nathanael-Israel Israel or Science180
For any suggestions or questions, please visit Science180.com/contact and Nathanael-Israel Israel's personal website: Israel120.com. Feel free to ask me any questions you have about the universe formation, life formation, and chemicals formation.

Another Book by Nathanael-Israel Israel:
ORIGIN OF THE SPIRITUAL WORLD

ONLY ONE ANCIENT BLUEPRINT HAS THE RELIABLE POWER TO HELP YOU TO ACCURATELY DECRYPT THE SPIRITUAL ORIGIN AND HISTORY OF EVERYTHING IN THE UNIVERSE

Countless books talk about the origin of the universe and of life, but this amazing book is the first and the only one that has undeniably explained how the formation of the universe and everything in it was truly revealed in the rejected and hidden scriptures such as the Books of Enoch and others. In *"Origin of the Spiritual World"*, you will:

- Discover deep rejected secrets that have prevented humankind from unearthing the beginning of the universe
- Plainly see the scientific proof (hidden in scriptures) of the formation of the Earth, the Moon, and the Sun in a matter of days, a historic revelation that bizarrely and shockingly matches the scientific data as scientifically proved in *"From Science to Bible's Conclusions"*, a popular book written by Dr. Nathanael-Israel Israel
- Properly use the lost and rejected scriptures to articulate the process by which the universe was formed, and use that insight to improve your understanding of the Bible, innovate in your domain of interest, and improve your life perpetually

TURBULENT ORIGIN OF LIFE

- Empower and align yourself with the historic breakthrough that has done what no other discovery has ever done: accurately unlock and decode mysteries concerning the origin of the cosmos and its content using scientific keys revealed in ancient scriptures that some elites have concealed (*Science180.com*/pseudepigraphic)
- Discover and apprehend the complex formation of the universe and life without leaving out the challenging questions that people of all ages have been struggling to answer for thousands of years, while the answers were hidden
- Find more joy in life through a clear interpretation of old and fresh revelations about the creation of the universe astonishingly backed by modern science, which some people wrongly think opposes the Bible
- Make a difference and blaze new trails for those who depend on your leadership

If you believe in God, have some origin-related questions which answers you cannot find anywhere, not even in the Bible, and if you want to tap into historically neglected revelations to answer fundamental universe and life. questions, then be sure to get a copy of *"Origin of the Spiritual World"* today

---

Dr. Nathanael-Israel Israel happens to be the discoverer of the historic mathematical equations that scientifically demonstrated that the Earth was formed 2.82 days, the Moon 3.32 days, and the Sun 3.69 days after the beginning of the universe, therefore confirming the Biblical account of creation that revealed about 3500 years ago that the formation of the Earth was completed on the 3rd day, while that of the Moon and the Sun was completed on the 4th day of creation. Nathanael-Israel Israel is referred to as the "Undisputable Specialist of all Questions at the Intersection of Science and Biblical Creation". Learn more about this rare scientist at Israel120.com.

REFERENCES

Air & Space Magazine (2021). Falling with the Falcon | Flight Today | Air & Space Magazine. http://www.airspacemag.com/flight-today/falcon.html.

Alberts B, Johnson A, Lewis J, Morgan D, Raff M, Roberts K, Walter P (2015). Molecular Biology of the Cell (6th ed.). Garland Science. p. 2. ISBN 978-0815344322.

Alderwick LJ, Harrison J, Lloyd GS, Birch HL (March 2015). "The Mycobacterial Cell Wall – Peptidoglycan and Arabinogalactan". Cold Spring Harbor Perspectives in Medicine. 5 (8): a021113. doi:10.1101/cshperspect.a021113. PMC 4526729. PMID 25818664.

Alexopoulos CJ, Mims CW, Blackwell M (1996). Introductory Mycology. John Wiley and Sons. ISBN 978-0-471-52229-4.

Anderson, Alyssa M. (2018). "Describing the Undiscovered". Chironomus: Journal of Chironomidae Research (31): 2–3. doi:10.5324/cjcr.v0i31.2887.

Asch R, Simerly C, Ord T, Ord VA, Schatten G (1995). "The stages at which human fertilization arrests: microtubule and chromosome configurations in inseminated oocytes which failed to complete fertilization and development in humans". Hum. Reprod. 10 (7): 1897–1906. doi:10.1093/oxfordjournals.humrep.a136204. PMID 8583008.

Bang C, Schmitz RA (2015). "Archaea associated with human surfaces: not to be underestimated". FEMS Microbiology Reviews. 39 (5): 631–48. doi:10.1093/femsre/fuv010. PMID 25907112.

Benson, R. B. J.; Hunt, G.; Carrano, M.T.; Campione, N.; Mannion, P. (2018). "Cope's rule and the adaptive landscape of dinosaur body size evolution". Palaeontology. 61 (1): 13–48. doi:10.1111/pala.12329.

Benton M. J. (2001). "Biodiversity on land and in the sea". Geological Journal. 36 (3–4): 211–230. doi:10.1002/gj.877.

Bidhendi, Amir J.; Altartouri, Bara; Gosselin, Frédérick P.; Geitmann, Anja (2019). "Mechanical stress initiates and sustains the morphogenesis of wavy leaf epidermal cells". Cell Reports. 28 (5): 1237–1250. doi:10.1016/j.celrep.2019.07.006. PMID 31365867.

Bowen, R. (2006). Gastrointestinal Transit: How Long Does It Take? Colorado State University.
http://www.vivo.colostate.edu/hbooks/pathphys/digestion/basics/transit.html.

Bowman, JL; Drews, GN; Meyerowitz, EM (August 1991). "Expression of the Arabidopsis floral homeotic gene AGAMOUS is restricted to specific cell types late in flower development". Plant Cell. 3 (8): 749–58. doi:10.1105/tpc.3.8.749. JSTOR 3869269. PMC 160042. PMID 1726485.

Buckingham Marcus and Donald O. Clifton (2001). Now, discover your strengths. The Free Press. New York, NY, USA. 260 pages.

Burgin CJ, Colella JP, Kahn PL, Upham NS (2018). "How many species of mammals are there?". Journal of Mammalogy. 99 (1): 1–14. doi:10.1093/jmammal/gyx147.

Carwardine, Mark (2008). Animal Records. New York: Sterling. pp. 11, 43. ISBN 9781402756238.

Casimir CM, Gates PB, Patient RK, Brockes JP (1988). "Evidence for dedifferentiation and metaplasia in amphibian limb regeneration from inheritance of DNA methylation". Development. 104 (4): 657–668. PMID 3268408.

Catry Paulo and Phillips Richard (2004). "Sustained fast travel by a gray-headed albatross (Thalassarche chrysostoma) riding an antarctic storm". The Auk. 121 (4): 1208. doi:10.1642/0004-8038(2004)121[1208:SFTBAG]2.0.CO;2.

Chapman, A.D. (2006). Numbers of living species in Australia and the World. Canberra: Australian Biological Resources Study. ISBN 978-0-642-56850-2.

Cheek, Martin; Nic Lughadha, Eimear; Kirk, Paul; Lindon, Heather; Carretero, Julia; Looney, Brian; et al. (2020). "New scientific discoveries: Plants and fungi". Plants, People, Planet. 2(5): 371–388. doi:10.1002/ppp3.10148.

Clark, David (2010). Germs, Genes, & Civilization: how epidemics shaped who we are today. Upper Saddle River, N.J: FT Press. ISBN 978-0-13-701996-0. OCLC 473120711.

Clements, James F. (2007). The Clements Checklist of Birds of the World (6th ed.). Ithaca: Cornell University Press. ISBN 978-0-8014-4501-9).

Collier L, Balows A, Sussman M (1998). Mahy B, Collier LA (eds.). Topley and Wilson's Microbiology and Microbial Infections. Virology. 1 (Ninth ed.). ISBN 0-340-66316-2.

Cresswell, Julia (2010). The Oxford Dictionary of Word Origins (2nd ed.). New York: Oxford University Press. ISBN 978-0-19-954793-7.

Darwin Charles Robert (1859). On the Origin of Species by Means of Natural Selection, or the Preservation of Favoured Races in the Struggle for Life. First edition. London: John Murray, 502 pages.

REFERENCES

Dayah, M. (1997). Dynamic Periodic Table. Retrieved December 4, 2014, from Ptable: http://www.ptable.com.

Deacon J (2005). Fungal Biology. Cambridge, Massachusetts: Blackwell Publishers. ISBN 978-1-4051-3066-0.

DK (2016). Animal!. Penguin. ISBN 9781465459008.

Eggleton, Paul (2020). "The State of the World's Insects". Annual Review of Environment and Resources. 45 (1): 61–82. doi:10.1146/annurev-environ-012420-050035. ISSN 1543-5938.

Elbein, Asher (2020). "Making Sense of 'One of the Most Baffling Animals That Ever Lived' - Important mysteries have been solved about a reptile with a giraffe-like neck that hunted prey 242 million years ago". The New York Times. Retrieved 14 August 2020.

Encyclopedia Britannica (2021). Side necked turtle. Retrieved December 6, 2021, from www.britannica.com/animal/side-necked-turtle.

Encyclopedia Britannica (2021). Article on "On the Origin of Species". Retrieved on October 17, 2021, from https://www.britannica.com/biography/Charles-Darwin/On-the-Origin-of-Species.

Fallon Sally (2001). Nourishing traditions. The cookbook that challenges politically correct nutrition and the diet dictocrats. New Trends Publishing.

Ferreira L (2014). "Stem Cells: A Brief History and Outlook". Stem Cells: A Brief History and Outlook - Science in the News. WordPress. Retrieved 3 December 2019.

Flint, W.D. (2002). To Find The Biggest Tree. Sequoia Natural History Association, ISBN 1-878441-09-4.

Fischer M., U. K. Franzeck, I. Herrig, U. Costanzo, S. Wen, M. Schiesser, U. Hoffmann and A. Bollinger (1996). "Flow velocity of single lymphatic capillaries in human skin". Am J Physiol Heart Circ Physiology 270 (1): H358–H363. PMID 8769772. Retrieved 2007-11-14.

Gilbert, Scott F (2010). Developmental Biology. 9th ed. Sunderland, MA: Sinauer Associates, 2010: 333-370.

Gilbert, SF. (2013). "Endoderm". Sinauer Associates. Retrieved 14 March 2013.

Giles KL (1971). "Dedifferentiation and Regeneration in Bryophytes: A Selective Review". New Zealand Journal of Botany. 9 (4): 689–94. doi:10.1080/0028825x.1971.10430185. Archived from the original on 2008-12-04. Retrieved 2008-01-01.

Gow, Neil A. R.; Latge, Jean-Paul; Munro, Carol A.; Heitman, Joseph (2017). "The fungal cell wall: Structure, biosynthesis, and function". Microbiology Spectrum. 5 (3). doi:10.1128/microbiolspec.FUNK-0035-2016. hdl:2164/8941. PMID 28513415.

Grabski, Valerie (2009). "Little Penguin – Penguin Project". Penguin Sentinels/University of Washington. Archived from the original on December 16, 2011. Retrieved November 25, 2011.

Green D (2011). Means to an End: Apoptosis and other Cell Death Mechanisms. Cold Spring Harbor, NY: Cold Spring Harbor Laboratory Press. ISBN 978-0-87969-888-1.

Guinness World Records (2014). Guinness Records – Fastest Bird Level Flight". Guinness World Records Limited. Retrieved 12 April 2014, from http://www.guinnessworldrecords.com/world-records/speed/fastest-bird-level-flight.

Guinness World Records (2018). Longest necks. Retrieved December 6, 2021, from https://www.guinnessworldrecords.com/world-records/longest-neck.

Guinness World Records (2021). The fastest diving bird. Visited on 12/2/2021, from https://www.guinnessworldrecords.com/world-records/70929-fastest-bird-diving.

Hassler, Michael (2020). "Total Species Count". World Plants. Synonymic Checklist and Distribution of the World Flora. Retrieved 26 October 2020.

Haughn, George W.; Somerville, Chris R. (1988). "Genetic control of morphogenesis in Arabidopsis". Developmental Genetics. 9 (2): 73–89. doi:10.1002/dvg.1020090202.

Hill, David; Holzwarth, George; Bonin, Keith (2002). "Velocity and Drag Forces on motor-protein-driven Vesicles in Cells". American Physical Society, the 69th Annual Meeting of the Southeastern. abstract. #EA.002. Bibcode:2002APS..SES.EA002H.

Hussain Kanchwala (2021). Why Do Giraffes Have Long Necks? Updated On: 13 Nov 2021, Retrieved on December 6, 2021, from
https://www.scienceabc.com/nature/animals/why-giraffes-have-a-long-neck.html.

ICTV online (2021). Virus Taxonomy: 2020 Release". talk.ictvonline.org. International Committee on Taxonomy of Viruses. Retrieved 21 May 2021.

Israel Nathanael-Israel (2025a). Turbulent Origin of the Universe. Science180, Augusta, USA 683 pages.

Israel Nathanael-Israel (2025b). From Science to Bible's Conclusions. Science180, Augusta, USA 170 pages.

Israel Nathanael-Israel (2025c). Reconciling Science and Creation Accurately. Science180, Augusta, USA 299 pages.

Israel Nathanael-Israel (2025d). Turbulent Origin of Chemical Particles. Science180, Augusta, USA 397 pages.

Nathanael-Israel Israel: Has had the honor to be acknowledged the First Human Being that Scientifically Reconciled Science and Biblical Creation

REFERENCES

Israel Nathanael-Israel (2025e). Turbulent Origin of Life. Science180, Augusta, USA 370 pages.

Israel Nathanael-Israel (2025f). Origin of the Spiritual World. Science180, Augusta, USA 151 pages.

Israel Nathanael-Israel (2025g). How Baby Universe Was Born. Science180, Augusta, USA 130 pages.

Israel Nathanael-Israel (2025h). How God Created Baby Universe. Science180, Augusta, USA 224 pages.

Israel Nathanael-Israel (2025i). Science180 Accurate Scientific Proof of God. Science180, Augusta, USA 214 pages.

IUCN (2010). Numbers of threatened species by major groups of organisms (1996–2010)" (PDF). International Union for Conservation of Nature. 11 March 2010. Archived (PDF) from the original on 21 July 2011. Retrieved 27 April 2011. http://www.iucnredlist.org/documents/summarystatistics/2010_1RL_Stats_Table_1.pdf.

Jessop, Nancy Meyer (1970). Biosphere; a study of life. Prentice-Hall. p. 428.

Jones J. Knox (2011). Mammal. https://www.britannica.com/animal/mammal. Visited on October 17, 2021.

Karam JA (2009). Apoptosis in Carcinogenesis and Chemotherapy. Netherlands: Springer. ISBN 978-1-4020-9597-9.

Krieg N (2005). Bergey's Manual of Systematic Bacteriology. US: Springer. pp. 21–26. ISBN 978-0-387-24143-2.

Kumar, Rani (2008). Textbook of Human Embryology. I.K. International Publishing House. p. 22. ISBN 9788190675710.

Langstroth, Lovell; Langstroth, Libby (2000). Newberry, Todd (ed.). A Living Bay: The Underwater World of Monterey Bay. University of California Press. p. 244. ISBN 978-0-520-22149-9.

Lewis Wexler, Derek H. Bergel, Ivor T. Gabe, Geoffrey S. Makin, and Christopher J. Mills (1968). "Velocity of Blood Flow in Normal Human Venae Cavae". Circulation Research. 23 (3): 349–59. PMID 5676450. Retrieved 2007-11-14.

Linnaeus, Carl (1758). Systema naturae per regna tria naturae: secundum classes, ordines, genera, species, cum characteribus, differentiis, synonymis, locis (in Latin) (10th ed.). Holmiae (Laurentii Salvii). Archived from the original on 10 October 2008. Retrieved 22 September 2008.

Luttermoser, Donald G. (2012a). "ASTR-1020: Astronomy II Course Lecture Notes Section XII" (PDF). East Tennessee State University. Archived from the original (PDF) on 7 July 2017. Retrieved 8 March 2021, from https://faculty.etsu.edu/lutter/courses/astr1020/a1020chap12.pdf.

Luttermoser, Donald G. (2012b). "Physics 2028: Great Ideas in Science: The Exobiology Module" (PDF). East Tennessee State University. Archived from the original (PDF) on 12 April 2016. Retrieved 8 March 2021, from https://faculty.etsu.edu/lutter/courses/phys2028/p2028exobnotes.pdf.

Mahla RS (2016). "Stem cells application in regenerative medicine and disease threpeutics". International Journal of Cell Biology. 2016 (7): 19. doi:10.1155/2016/6940283. PMC 4969512. PMID 27516776.

Mitalipov S, Wolf D (2009). "Totipotency, pluripotency and nuclear reprogramming". Engineering of Stem Cells. Advances in Biochemical Engineering/Biotechnology. 114. pp. 185–199. Bibcode:2009esc..book..185M. doi:10.1007/10_2008_45. ISBN 978-3-540-88805-5. PMC 2752493. PMID 19343304.

Moissl-Eichinger C, Pausan M, Taffner J, Berg G, Bang C, Schmitz RA (2018). "Archaea Are Interactive Components of Complex Microbiomes". Trends in Microbiology. 26 (1): 70–85. doi:10.1016/j.tim.2017.07.004. PMID 28826642.

Mosby's Dictionary of Medicine (2017). Nursing and Health Professions (10th ed.). St. Louis, Missouri: Elsevier. 2017. p. 1281. ISBN 9780323222051.

NCBI (2017). Viral Genome database". ncbi.nlm.nih.gov. Retrieved 15 January 2017.

Parenti Lynne R. (2021). Fish. Encyclopedia Britannica. Visited on October 17, 2021, from https://www.britannica.com/animal/fish.

Paul, G.S. (1988). "The brachiosaur giants of the Morrison and Tendaguru with a description of a new subgenus, Giraffatitan, and a comparison of the world's largest dinosaurs" (PDF). Hunteria. 2 (3).

Penny, P. (2003). Hemodynamic: Blood Velocity (http://www.coheadquarters.com/PennLibr/MyPhysiology/lect5/xpen5.01.htm).

Periodictable.com (2014). Abundance in Humans of the elements. December 4, 2014, https://periodictable.com/Properties/A/HumanAbundance.html

Pester Patrick (2021). The world's fastest animals. https://www.livescience.com/59822-fastest-animals.html.

Pommerville JC (2014). Fundamentals of Microbiology (10th ed.). Boston: Jones and Bartlett. ISBN 978-1-284-03968-9.

Nathanael-Israel Israel: Has had the honor to be acknowledged the First Human Being that Scientifically Reconciled Science and Biblical Creation

REFERENCES

Rafferty, John. (2021a). "Gentoo Penguin". Britannica Online Encyclopedia. Encyclopedia Britannica Inc. Retrieved January 20, 2021, from https://www.britannica.com/animal/gentoo-penguin.

Rafferty, John. (2021b). "Emperor Penguin". Britannica Online Encyclopedia. Britannica Encyclopedia Inc. Retrieved January 20, 2021, from https://www.britannica.com/animal/emperor-penguin.

Raven PH, Evert RF, Eichhorn SE (2005). Biology of Plants (7th ed.). New York: W.H. Freeman and Company Publishers. pp. 504–508. ISBN 978-0-7167-1007-3.

Royal Society Publishing (2021). Courtship dives of Anna's hummingbird offer insights into flight performance limits http://rspb.royalsocietypublishing.org/content/276/1670/3047.

Ruppert, E.E., Fox, R.S., and Barnes, R.D. (2004). "Introduction to Bilateria". Invertebrate Zoology (7th ed.). Brooks/Cole. pp. 217–218. ISBN 978-0-03-025982-1.

Russell, Rex, M.D. (1996). What the Bible says about healthy living. Regal.

Rybicki EP (1990). "The classification of organisms at the edge of life, or problems with virus systematics". South African Journal of Science. 86: 182–86.

Sadhu MK (1989). Plant propagation. New Age International. p. 61. ISBN 978-81-224-0065-6.

Safra, Jacob E. (2003). The New Encyclopædia Britannica, Volume 16. Encyclopædia Britannica. p. 523. ISBN 978-0-85229-961-6.

Sagan Dorion (2021). Life. Visited on October 17, 2021, from https://www.britannica.com/science/life.

Schnabel M, Marlovits S, Eckhoff G, et al. (2002). "Dedifferentiation-associated changes in morphology and gene expression in primary human articular chondrocytes in cell culture". Osteoarthr. Cartil. 10 (1): 62–70. doi:10.1053/joca.2001.0482. PMID 11795984.

ScienceDirect (2019). Apoptosis – an overview. www.sciencedirect.com. ScienceDirect Topics. Retrieved 2019-03-19.

Schöler, Hans R. (2007). "The Potential of Stem Cells: An Inventory". In Nikolaus Knoepffler; Dagmar Schipanski; Stefan Lorenz Sorgner (eds.). Human biotechnology as Social Challenge. Ashgate Publishing. p. 28. ISBN 978-0-7546-5755-2.

Scott, Gilbert (2010). Developmental biology (ninth ed.). USA: Sinauer Associates.

Sharma, N. S. (2005). Continuity And Evolution Of Animals. Mittal Publications. p. 106. ISBN 978-81-8293-018-6.

Smith Precious (2021). World's Slowest Animals: These Creatures Aren't in a Hurry. Visited on 12/2/2021, from https://www.natureworldnews.com/articles/46832/20210724/worlds-slowest-animals-these-creatures-arent-in-a-hurry.htm.

Stephan N.F. Spiekman, James M. Neenan, Nicholas C. Fraser, Vincent Fernandez, Olivier Rieppel, Stefania Nosotti and Torsten M. Scheyer (2021). "Aquatic Habits and Niche Partitioning in the Extraordinarily Long-Necked Triassic Reptile Tanystropheus. 6 August 2020, Current Biology. DOI: 10.1016/j.cub.2020.07.025.

Storer Robert W. (2021). Bird. Encyclopedia Britannica. Visited on October 17, 2021, from https://www.britannica.com/animal/bird-animal.

Taylor MP, Wedel MJ. (2013). Why sauropods had long necks; and why giraffes have short necks. PeerJ1:e36 https://doi.org/10.7717/peerj.36.

Tessler Gordon (1996). The Genesis Diet. Be Well Publications.

Theobald, D.L.I (2010). "A formal test of the theory of universal common ancestry", Nature, 465 (7295): 219–222, Bibcode:2010Natur.465..219T, doi:10.1038/nature09014, PMID 20463738, S2CID 4422345.

Trifonov, Edward N. (2011). "Vocabulary of Definitions of Life Suggests a Definition". Journal of Biomolecuoar Structure and Dynamics. 29(2): 259–266. doi:10.1080/073911011010524992. PMID 21875147.

Vaughan Don (2021). The Fastest Animals on Earth. Encyclopedia Britannica. Visited on 12/2/2021, from https://www.britannica.com/list/the-fastest-animals-on-earth.

Voytek, Mary a. (2021). "About Life Detection". NASA. Archived from the original on 18 March 2021. Retrieved 8 March 2021, from (https://web.archive.org/web/20210318042627/https:/astrobiology.nasa.gov/researc h/life-detection/about.

Waddington Conrad H. (2020). Biological development. Encyclopedia Britannica. Visited on September 8, 2020, from https://www.britannica.com/science/biological-development.

Whitmore, Ian (1999). "Terminologia Anatomica: New terminology for the new anatomist". The Anatomical Record. 257 (2): 50–53. doi:10.1002/(sici)1097-0185(19990415)257:2<50::aid-ar4>3.0.co;2-w. ISSN 1097-0185. PMID 10321431.

Wikipedia (2021a). Life. Visited on October 17, 2021, from https://en.wikipedia.org/wiki/Life.

Wikipedia (2021b). Fastest animals. https://en.wikipedia.org/wiki/Fastest_animals. Visited on 12/2/2021.

REFERENCES

Wikipedia (2022). Penguin. Retrieved March 22, 2022, from https://en.wikipedia.org/wiki/Penguin.

Wildlife Informer (2021). 10 awesome animals with long necks. Visited on 12/6/2021, from www.wildlifeinformer.com/animals-with-long-necks/.

Wood, Gerald (1983). The Guinness Book of Animal Facts and Feats. ISBN 978-0-85112-235-9.

World Atlas (2021). Animals With The Longest Tails. Visited on December 6, 2021, from https://www.worldatlas.com/articles/animals-with-the-longest-tails-in-the-animal-kingdom.html.

World Flora Online (2020). An Online Flora of All Known Plants". World Flora Online. Retrieved 26 October 2020.

Wright Steph (2020). The Slowest Animals In The World. Visited on 12/2/2021, from https://www.worldatlas.com/articles/the-slowest-animals-in-the-world.html.

Zug R. George (2021). Reptile. Encyclopedia Britannica. Visited on October 17, 2021, from https://www.britannica.com/animal/reptile

INDEX

A

Adam 282, 283, 284, 285, 286, 292, 296
Adaptation 138, 319, 320
Africa 20, 55, 56, 57, 58, 63, 65, 297, 307, 321
AIDS 22
Algae 4, 8, 12, 16, 20, 51, 85, 105, 108, 132, 169, 260
American 121, 308, 354, 369
Angels 23, 94, 95, 96, 100, 105, 113, 120, 126, 140, 141, 252, 281, 291, 292, 332, 339
Animals 6, 4, 5, 8, 9, 12, 16, 17, 18, 19, 50, 51, 53, 54, 55, 56, 57, 58, 59, 60, 61, 62, 63, 64, 65, 66, 73, 76, 78, 79, 81, 84, 85, 86, 87, 91, 92, 97, 98, 102, 104, 105, 106, 108, 110, 111, 112, 113, 117, 118, 119, 120, 122, 124, 127, 128, 129, 130, 131, 134,135, 137, 138, 139, 142, 143, 144, 162, 165, 167, 174, 178, 179, 218, 219, 220, 221, 226, 228, 230, 231, 232, 233, 250, 251, 252, 257, 258, 259, 263, 270, 271, 272, 279, 280, 281, 282, 283, 284, 285, 289, 292, 293, 296, 297, 300, 301, 302, 303, 304, 305, 306, 309, 316, 319, 320, 321, 322, 323, 324, 332, 334, 335, 336, 337, 338, 339, 340, 341, 354, 356, 358, 359
Animism 37
Apoptosis 80, 81, 354, 355, 357
Apostle Paul 296, 352, 353, 356
Apparatus ... 7, 101, 123, 124, 143, 153, 156, 157, 158, 163, 165
Archaea 6, 4, 5, 8, 16, 20, 21, 84, 85, 106, 257, 258, 260, 261, 292, 332, 351, 356
Aristotle 5
Arsenic 11
Asteroids 31, 35, 36, 42, 48, 50, 83, 84, 107, 108, 129, 269, 316, 324, 328, 333, 369
ATP 12
Autotrophs 12, 122

B

Baby Universe 355
Bacteria 6, 4, 5, 8, 9, 12, 13, 16, 18, 21, 22, 56, 84, 85, 105, 106, 121, 134, 151, 152, 160, 161, 230, 257, 258, 260, 261, 262, 263, 281, 292, 304, 307, 328, 332
Beninese 369
Bible . 2, 1, 16, 26, 29, 37, 81, 111, 113, 120, 122, 128, 138, 141, 162, 275, 276, 278, 279, 280, 282, 283, 284, 290, 291, 292, 293, 295, 296, 300, 302, 303, 305, 306, 313, 340, 342, 344, 354, 357, 370

INDEX

Biblical account of creation 37, 275, 276, 279, 281, 288, 294, 312, 313, 341, 346
Big Bang 2, 1, 290
Billions of years 97, 289, 290, 314, 316, 317, 320, 325
Biological patterns 49
Biomolecules 8, 150, 153, 154, 157, 193, 209, 210, 211, 261, 263, 314, 323, 324
Birds 221, 222, 227, 322, 352
Bones 11, 12, 51, 64, 79, 101, 111, 120, 128, 164, 219, 221, 228, 232, 271, 272, 308
Books of Adam and Eve 284, 285
Books of Enoch 284, 285
Books on Enoch 284
Branches 180
Buddhism 37

C

Calcium 10, 11, 12, 124, 140, 141
Carbohydrates 8, 19, 147, 148, 157, 340
Carbon ... 9, 16, 21, 111, 123, 124, 141, 147, 148, 149, 151, 164, 260, 339
Carl Linnaeus 15, 16
Celestial bodies ... 5, 6, 7, 15, 25, 26, 28, 30, 31, 34, 35, 36, 37, 41, 42, 43, 45, 46, 49, 50, 54, 56, 61, 62, 66, 73, 83, 84, 87, 89, 91, 94, 105, 107, 109, 114, 116, 119, 120, 121, 123, 124, 126, 127, 128, 129, 130, 131, 132, 133, 136, 138, 139, 140, 142, 143, 144, 145, 151, 152, 158, 160, 161, 162, 176, 179, 180, 181, 182, 191, 193, 198, 211, 215, 219, 228, 230, 252, 257, 258, 268, 269, 270, 276, 278, 281, 282, 283, 290, 300, 316, 317, 321, 324, 325, 327, 328, 331, 332, 333, 335, 336, 337, 338, 339, 341
Cell wall 16, 51

Cells 6, 5, 7, 8, 12, 16, 19, 21, 34, 51, 75, 76, 77, 78, 79, 80, 81, 101, 106, 113, 114, 123, 127, 135, 137, 143, 148, 153, 156, 159, 160, 161, 162, 163, 165, 168, 169, 173, 174, 176, 180, 181, 182, 190, 193, 221, 226, 230, 232, 258, 261, 262, 292, 332, 338, 339, 340, 351, 356
Cellulose 16, 19, 101, 148, 258, 272
Centrosomes 8
Chemical elements 9, 10, 11, 12, 28, 37, 89, 96, 108, 123, 124, 131, 134, 140, 141, 150, 155, 158, 269, 282, 322, 332, 337, 339
Chemical particles ... 5, 7, 9, 15, 25, 28, 32, 35, 50, 76, 86, 89, 107, 121, 123, 126, 130, 133, 140, 145, 252, 269, 298, 321, 331, 347
Chemoautotrophs 12
Chlorine 10, 11, 124, 140, 141
Chloroplasts 8, 16, 19, 20, 148, 158, 258, 259, 260
Christians 2
Chromosomes .. 19, 102, 151, 154, 158
Cilia 8
Clean animals 6, 86, 110, 113, 127, 128, 138, 139, 232, 285, 286, 293, 300, 301, 302, 303, 304, 305, 306, 308, 327, 336
Common ancestor . 269, 272, 314, 315, 316
Confucianism 37
Covid-19 22
Creation 354
Creationism 6, 312, 313, 314
Creator 3, 94, 102, 272, 276, 279, 284, 287, 289, 290, 294, 296, 302, 313, 316, 317, 341, 342, 369
Cud 100, 139, 232, 300, 301, 303, 304, 306
Cytoplasm 8

D

Darwin 75, 313, 314, 315, 325, 326, 352, 353
Days of creation 135, 278, 279, 281, 290
Death 96, 297, 354
Dedifferentiation 80, 352
Definition of life 1, 108, 294
Descendance .. 120, 152, 260, 268, 270, 326, 340
Developmental biology .. 5, 34, 75, 127, 135
Dinosaurs 55, 56, 59, 64, 128, 130, 137, 138, 229, 230, 251, 339, 356
DNA 5, 6, 12, 22, 25, 75, 76, 80, 81, 102, 109, 127, 147, 148, 149, 150, 151, 152, 153, 154, 159, 161, 261, 262, 263, 272, 314, 316, 325, 327, 338, 340, 352

E

Earth 2, 9, 12, 15, 17, 22, 31, 32, 37, 43, 46, 50, 56, 89, 92, 93, 94, 95, 96, 99, 105, 113, 116, 117, 118, 122, 126, 127, 128, 129, 139, 140, 141, 142, 143, 145, 151, 153, 159, 166, 167, 203, 206, 218, 228, 230, 233, 250, 251, 253, 260, 271, 276, 278, 280, 281, 282, 283, 284, 285, 286, 287, 289, 291, 292, 293, 294, 297, 302, 306, 313, 314, 315, 316, 317, 323, 325, 331, 332, 336, 339, 340, 358, 369
Endoplasmic reticulum 8
Energy 5, 7, 8, 12, 30, 31, 32, 34, 36, 49, 100, 122, 123, 148, 160, 195, 196, 261, 315, 317, 323, 337
Enzymes 8, 11, 75, 149, 150, 153
Eukaryotes 8, 19, 21, 102, 149, 151, 152, 158, 159, 261, 262, 314
Eve 282, 284, 285, 286, 296
Evolution .. 2, 3, 4, 6, 7, 1, 6, 22, 30, 37, 51, 66, 114, 138, 144, 152, 227, 260, 271, 276, 281, 283, 290, 292, 312, 313, 314, 315, 316, 317, 320, 324, 325, 326, 327, 328, 331, 339, 340, 351, 357
Extraterrestrial life 94

F

Faith 6, 99, 103, 275, 276, 281, 284, 313, 341
Feathers ... 65, 118, 164, 221, 222, 223, 283
Fish 51, 53, 64, 65, 73, 91, 101, 102, 111, 128, 129, 131, 138, 139, 142, 154, 218, 219, 220, 221, 227, 232, 251, 271, 279, 280, 281, 302, 303, 306, 307, 308, 322, 324, 335, 356
Flowers 81, 104, 168, 174, 175, 203, 206, 207, 209, 210, 215, 295
Fluid layers ... 34, 35, 37, 41, 42, 43, 44, 78, 107, 127, 131, 132, 135, 144, 180, 181, 193, 276, 283, 332, 333, 335, 338
Fluorine 10, 11, 124, 140, 141
Fruits 134, 135, 211, 214
Fungi .. 6, 4, 5, 8, 16, 18, 19, 51, 62, 84, 85, 101, 106, 176, 257, 258, 262, 292, 332, 352

G

Garden of Eden 284, 286, 296
Genes 210, 352
Genesis 16, 37, 276, 278, 279, 280, 281, 282, 283, 285, 296, 300, 358
Germ layers 78
Germination 81, 169
Giraffes ... 55, 56, 65, 92, 129, 137, 138, 251, 339, 354, 358
God. 2, 1, 6, 16, 23, 26, 29, 37, 94, 105, 112, 113, 135, 272, 275, 276, 278, 279, 280, 281, 282, 283, 284, 285, 286, 287, 289, 290, 291, 292, 293, 294, 295, 296, 297, 302, 303, 305, 307, 313, 316, 317, 325, 331, 341, 342, 344, 370

INDEX

Golgi complex 8

H

Heterotrophs 12, 18
Hinduism 37
Hoof 102, 232, 300, 301, 303, 305, 306
Human beings.. 6, 9, 10, 11, 12, 15, 20, 50, 51, 64, 65, 76, 85, 93, 94, 96, 97, 98, 99, 101, 102, 104, 106, 107, 110, 111, 112, 113, 117, 118, 119, 120, 121, 122, 123, 124, 127, 128, 131, 132, 141, 143, 144, 145, 154, 155, 161, 179, 230, 232, 233, 250, 251, 252, 253, 271, 276, 280, 281, 282, 283, 284, 285, 286, 287, 290, 291, 292, 293, 295, 296, 297, 298, 300, 304, 314, 320, 321, 322, 323, 324, 334, 336, 337, 338, 341, 369
Hydrogen 9, 12, 123, 124, 131, 140, 147, 149, 151, 282, 339

I

Inflorescence 41, 50, 168, 175, 176, 183, 190, 203, 206, 207, 208, 209, 210, 333
Insects 227, 353
Intermittence 35, 83
Iron 10, 11, 124, 140, 141
Islam 37, 313
Israel 354, 355

J

Judeo-Christian 37, 313
Jupiter 31, 45, 89, 127, 159, 324

L

Lanthanoid 9
Law of calibration 123
Law of mission of life 116
Law of species belonging 321
Leaves .5, 41, 42, 43, 47, 48, 49, 50, 62, 64, 67, 69, 70, 71, 72, 81, 85, 97, 118, 131, 132, 138, 163, 168, 169, 170, 171, 174, 175, 176, 179, 180, 181, 182, 183, 184, 188, 190, 191, 192, 193, 194, 195, 196, 197, 198, 199, 201, 202, 209, 272, 292, 296, 307, 332, 333, 334, 335
Leviticus. 128, 280, 296, 300, 302, 303, 305, 309
Lysosomes 8

M

Macromolecules 6, 7, 76, 108, 147, 150, 153, 154, 155, 157, 158, 159, 181, 262, 332, 340
Magnesium ... 10, 11, 12, 124, 140, 141
Mammals 231, 232, 282
Mars 31, 127, 159, 324
Mary 294, 358
Mercury 31, 37, 127, 135, 159, 324
Meristems ... 19, 66, 76, 106, 162, 173, 174, 182
Messianic 134, 293, 309
Methuselah 285
Microvilli 8
Milky Way 317, 369
Mitochondria 8, 12, 157
Monkeys 65, 131, 233, 283, 308
Moon 32, 37, 43, 46, 89, 122, 276, 278, 281, 283, 297, 317, 369
Morphogenesis 77
Moses 37, 276, 284, 300
Muslims 313

N

Nathanael-Israel Israel... 2, 4, 3, 56, 57, 58, 59, 63, 74, 189, 192, 194, 195, 196, 197, 198, 208, 222, 227, 263, 343, 344, 345, 346, 347, 348, 349, 354, 355, 369
National Academy of Sciences 312
Natural Selection 314, 352
Neptune 31, 45, 324
Nitrogen 9, 21, 111, 123, 149, 150, 151, 260, 339
Noah 285
Noble gas 9

Nonliving 97, 145
Nucleic acids 148, 149
Nucleotides 147, 149, 150, 151, 316
Nucleus 8

O

On the Origin of Species. 313, 314, 326, 352, 353
Organelles 6, 7, 8, 19, 21, 114, 123, 143, 144, 148, 153, 156, 157, 158, 159, 160, 161, 165, 181, 260, 261, 262, 324, 332, 340
Organisms 5, 4, 5, 6, 7, 8, 9, 12, 15, 16, 17, 18, 19, 20, 21, 22, 27, 29, 34, 36, 37, 42, 49, 50, 51, 53, 54, 55, 56, 61, 62, 64, 65, 67, 75, 76, 77, 79, 80, 81, 84, 85, 86, 87, 89, 96, 97, 99, 101, 102, 104, 105, 106, 107, 108, 109, 110, 111, 112, 113, 114, 116, 117, 118, 119, 120, 121, 122, 123, 124, 125, 126, 127, 128, 129, 130, 131, 132, 133, 134, 135, 137, 138, 140, 141, 142, 143, 144, 146, 147, 148, 149, 150, 151, 152, 153, 154, 156, 157, 158, 159, 160, 161, 162, 163, 165, 166, 167, 168, 169, 176, 180, 181, 215, 218, 219, 220, 221, 228, 230, 232, 233, 250, 251, 252, 257, 258, 260, 261, 262, 263, 268, 270, 271, 272, 276, 280, 281, 282, 283, 288, 289, 290, 291, 292, 293, 296, 300, 302, 303, 305, 306, 307, 313, 314, 315, 316, 317, 319, 320, 321, 322, 323, 324, 325, 326, 327, 328, 331, 332, 334, 335, 336, 337, 338, 339, 340, 341, 355, 357
Organs 6, 5, 7, 12, 53, 62, 63, 76, 78, 79, 81, 100, 101, 108, 113, 114, 119, 120, 122, 123, 124, 125, 132, 135, 143, 144, 148, 153, 156, 159, 160, 161, 162, 163, 164, 165, 167, 168, 173, 175, 176, 181, 182, 193, 199, 201, 207, 208, 210, 211, 215, 219, 230, 233, 271, 281, 291, 296, 303, 322, 332, 337, 338, 340
Origin of Chemical Particles 354
Origin of Life 355
Ostrich 56, 111, 112, 131, 222
Oxygen . 9, 16, 111, 122, 123, 124, 140, 147, 149, 151, 164, 219, 260, 282, 323, 339

P

Pedicel 208
Peduncle 208
Peroxisomes 8
Petiole 73, 175, 176, 182, 191, 194, 195, 196, 197, 198, 199, 201, 335
Philosophy 22, 23, 26, 28, 94, 252, 279, 345
Phosphorus 9, 123, 339
Photosynthesis 12, 16, 20, 100, 174, 183, 190
Phylogenetic trees 289, 316
Phylogeny 316
Planets ... 31, 32, 35, 36, 42, 43, 45, 47, 48, 50, 81, 83, 91, 124, 127, 129, 135, 144, 145, 151, 152, 159, 160, 179, 269, 316, 324, 327, 328, 333, 334, 338, 369
Plants 6, 4, 5, 7, 8, 9, 12, 16, 18, 19, 20, 21, 37, 41, 42, 43, 47, 49, 50, 51, 57, 61, 62, 64, 65, 66, 67, 69, 70, 73, 76, 78, 80, 84, 85, 97, 100, 101, 105, 106, 108, 111, 112, 117, 118, 119, 120, 122, 129, 130, 131, 134, 135, 136, 138, 139, 141, 142, 143, 162, 165, 166, 167, 168, 169, 170, 171, 173, 174, 175, 176, 178, 179, 180, 181, 182, 183, 188, 190, 191, 192, 194, 196, 197, 198, 199, 201, 202, 206, 207, 209, 210, 211, 214, 215, 218, 219, 223, 257, 258, 259, 260, 270, 272, 280, 282, 285, 286, 288, 289, 292, 293, 295, 321, 323, 324, 332, 334, 335, 337, 338, 340
Plastids 8

Nathanael-Israel Israel: Has had the honor to be acknowledged the First Human Being that Scientifically Reconciled Science and Biblical Creation

INDEX

Pluto 31, 159, 324
Plutonium 89
Potassium 10, 11, 124, 140, 141
Precursors...5, 7, 30, 31, 34, 35, 36, 37, 41, 42, 43, 44, 45, 46, 48, 49, 50, 54, 56, 62, 63, 64, 65, 66, 69, 70, 73, 76, 78, 79, 80, 81, 83, 84, 105, 107, 108, 113, 117, 119, 124, 126, 127, 129, 130, 131, 132, 133, 135, 136, 137, 138, 139, 140, 141, 142, 143, 144, 145, 148, 149, 150, 151, 152, 153, 157, 158, 159, 160, 161, 162, 163, 164, 165, 168, 169, 170, 174, 175, 176, 179, 180, 181, 182, 183, 189, 190, 191, 192, 193, 194, 195, 196, 197, 198, 201, 206, 207, 208, 209, 210, 211, 215, 218, 219, 221, 223, 226, 227, 228, 230, 233, 252, 257, 258, 259, 260, 261, 262, 263, 264, 268, 269, 270, 281, 282, 283, 289, 291, 302, 303, 304, 305, 315, 317, 324, 326, 327, 328, 332, 333, 334, 335, 336, 337, 338, 339, 340
Pregnancy 144
Prokaryotes 8, 16, 20, 21, 102, 148, 149, 151, 152, 158, 159, 261, 262, 314
Proteins ... 8, 11, 12, 75, 109, 147, 150, 152, 153, 154, 157, 210, 272, 314, 315, 316, 340
Protists ...6, 4, 5, 8, 18, 20, 84, 85, 106, 257, 258, 259, 260, 262, 332

R

Rachis 183, 209
Relativity 29, 131
Reproduction..... 21, 51, 117, 134, 157, 161, 166, 174, 214, 230, 293, 327
Reptiles 229
Resurrection 95, 294, 295, 296, 341
Ribosomes 8
RNA 5, 22, 109, 147, 148, 149, 150, 151, 152, 153, 154, 263, 264, 272, 314, 315, 316, 325, 340

Root 42, 47, 48, 50, 62, 66, 76, 81, 135, 136, 161, 168, 169, 170, 171, 176, 179, 181, 202, 209, 296, 333, 334, 338
Ruminants 21, 100, 303, 304, 306

S

Satellites .30, 31, 35, 36, 37, 42, 43, 44, 45, 47, 48, 69, 83, 84, 91, 129, 138, 145, 159, 160, 179, 180, 190, 200, 201, 228, 269, 316, 324, 328, 333, 334, 335, 369
Saturn 31, 45, 127, 324
Science.................................. 2
Science180 Academy.......... 2, 347, 369
Science180 Cosmology 3, 2
Seeds 16, 81, 85, 86, 157, 168, 169, 194, 201, 208, 211, 214, 215, 278, 282, 285
Silicon 10, 11, 124, 140, 141
Sin 285, 286, 287, 291, 296
Snakes 85, 91, 100, 110, 111, 137, 138, 220, 227, 229, 230, 233, 285, 306, 308
Sodium................ 10, 12, 124, 140, 141
Solar System 30, 31, 35, 36, 37, 42, 43, 45, 46, 48, 50, 83, 91, 124, 128, 129, 135, 151, 159, 160, 174, 190, 268, 269, 282, 283, 317, 324, 333, 338
Spirit 93, 95, 97, 106, 140, 287, 292, 293, 294, 295, 322, 341
Spirits........................ 293, 294
Spiritual 5, 6, 22, 28, 34, 93, 94, 95, 96, 97, 99, 105, 106, 108, 109, 116, 117, 118, 119, 120, 123, 124, 127, 132, 140, 141, 142, 143, 145, 165, 251, 283, 284, 285, 287, 292, 294, 295, 296, 315, 323, 332, 336, 341
Spiritual World 355
Split-gathering ..6, 7, 34, 35, 36, 37, 41, 43, 44, 54, 64, 69, 70, 77, 80, 81, 82, 83, 84, 107, 108, 109, 110, 112, 113, 114, 126, 127, 128, 129, 130, 132, 133, 137, 139, 142, 143, 145, 152,

157, 158, 159, 160, 163, 165, 168, 176, 180, 181, 182, 183, 190, 191, 193, 195, 196, 206, 207, 208, 209, 210, 218, 220, 228, 261, 262, 264, 269, 282, 292, 317, 324, 332, 333, 334, 335, 336, 338, 339, 340
Stars . 35, 36, 81, 83, 84, 128, 129, 152, 159, 219, 316, 327, 328, 369
Struggles 25, 292, 309, 315
Subatomic particles.. 31, 34, 36, 84, 86, 89
Sugars 147, 148, 149, 151
Sulfur 9, 12, 111, 339
Sun ... 30, 31, 32, 37, 42, 43, 46, 47, 48, 89, 91, 116, 118, 122, 127, 129, 135, 152, 159, 180, 219, 269, 276, 278, 280, 281, 289, 297, 317, 333, 338, 369
Supernatural .. 3, 95, 96, 106, 108, 133, 145, 251, 275, 313, 341
Survival. 6, 62, 119, 138, 292, 309, 315, 320, 321
Synthetic element 9

T

Taxonomy 22, 354
Tissues 6, 7, 12, 76, 78, 79, 81, 114, 123, 139, 143, 144, 153, 156, 161, 162, 163, 165, 173, 178, 179, 181, 332, 340
Trunk 37, 41, 42, 43, 47, 48, 50, 51, 53, 54, 64, 81, 135, 168, 174, 175, 176, 177, 178, 179, 180, 181, 182, 190, 196, 201, 202, 209, 219, 223, 228, 232, 258, 272, 332, 333, 334
Turbulence ... 5, 2, 7, 27, 28, 29, 30, 34, 35, 36, 41, 42, 45, 49, 50, 51, 52, 54, 55, 56, 61, 62, 63, 64, 65, 66, 73, 75, 77, 80, 84, 99, 108, 109, 112, 114, 116, 119, 120, 121, 122, 124, 125, 126, 127, 129, 130, 131, 132, 133, 137, 138, 139, 141, 142, 143, 148, 149, 150, 151, 152, 153, 154, 157, 159, 160, 161, 165, 166, 168, 169,
175, 176, 178, 179, 180, 181, 182, 183, 190, 191, 192, 193, 194, 195, 196, 197, 198, 199, 201, 202, 206, 207, 208, 209, 210, 211, 215, 218, 219, 221, 222, 223, 226, 230, 232, 233, 250, 257, 259, 261, 262, 263, 264, 268, 272, 279, 281, 283, 291, 302, 303, 315, 317, 322, 323, 324, 327, 328, 331, 332, 333, 334, 335, 336, 337, 338, 339, 340, 341, 369
Turbulence zones 130
Turbulent arms 64
Turbulent belly 62
Turbulent branches 43
Turbulent head 61
Turbulent intermittence 84
Turbulent law of calibration........... 123
Turbulent law of mission 116
Turbulent leaves 45
Turbulent legs 64
Turbulent neck 5, 52, 53, 54, 61, 73, 134, 165, 198, 201, 207, 208, 219, 226, 230, 332, 334, 338
Turbulent Origin Formula 3, 369
Turbulent program of life..... 6, 99, 106, 107, 108, 109, 114, 116, 119, 120, 122, 124, 130, 131, 134, 142, 143, 149, 150, 152, 154, 157, 158, 159, 160, 161, 163, 168, 170, 182, 190, 215, 218, 219, 233, 250, 261, 262, 280, 281, 335, 336, 337, 338, 339, 340, 341
Turbulent programming of life 126
Turbulent skeleton 63
Turbulent spine 63
Turbulent tail 65
Turbulent tree 41, 42, 43, 45, 46, 48, 51, 174
Turbulent vertebrate 63

U

Unclean animals 302
Universal holistic being 105
Universal pool of qualities 99

Nathanael-Israel Israel: Has had the honor to be acknowledged the First Human Being that Scientifically Reconciled Science and Biblical Creation

INDEX

Universe 2, 3, 4, 2, 7, 25, 26, 28, 29, 30, 35, 36, 37, 41, 42, 43, 50, 51, 76, 83, 91, 105, 116, 120, 129, 135, 152, 162, 182, 193, 201, 270, 276, 324, 325, 333, 339, 341, 344, 369, 370
Uranium 10, 89
Uranus 31, 324
USA 354, 355

V

Venation 69

Venus 31, 127, 159, 324
Viruses .. 6, 4, 5, 7, 8, 22, 121, 257, 263, 264, 315, 328, 332, 357

W

Water 280
Wisdom 2, 111

Z

Zinc 10, 11, 124, 141

TURBULENT ORIGIN OF LIFE

ABOUT THE AUTHOR

Dr. Nathanael-Israel Israel is the founder of Science180 (www.Science180.com), the American company which mission is to improve the current and future state of human beings by accurately decoding and teaching them the real origin and formation of the universe, of life, and of chemicals, and meaningfully engaging business, nonprofit, political, academic, civil society leaders and followers to properly shape local and global agendas that authentically value the truth. As the creator of the Universe Turbulent Origin Formula™, Dr. Nathanael-Israel Israel has revolutionized the way billions of people around the world think about the origin of the universe, of life, and of chemicals. He was born just to decode the origin of the universe and to impart that knowledge onto human beings. Hence, verily, nobody understands and teaches the formation of everything in the universe (e.g. the Milky Way Galaxy, the Sun, the Earth, the Moon, and all other galaxies, stars, planets, satellites, and asteroids) better than Nathanael-Israel Israel. Individuals and organizations across the globe have been calling him so he helps them scientifically unlock the code of the universe-formation, helping veterans and rookies to have the real keys to decrypt the universe and turbulence (one of the top biggest unsolved mysteries in science) from the historic, unique, accurate, simple, easy-to-understand, nonconformist, trailblazing perspective that anybody can quickly learn at Science180 Academy (Science180Academy.com). Science180 Academy delivers outstanding value, insight, and lessons to assist people to accurately understand the true origin of the universe, chemicals, and life, so they can tap into that knowledge to improve lives perpetually. Nathanael-Israel's goal is to give you practicable and undeniable proofs of the formation of the universe so you can be fired up to become the best version of you, and to cause positive changes to your initiatives that will profit you today and forever. For Nathanael-Israel, accurately decoding and teaching the origin of the universe and everything in it is not a job, but his life mission, and helping others to fully understand that brings him closer to his assignment.

Dr. Israel earned his PhD in Plant, Insect, and Microbial Sciences in the USA, where he graduated first of his class of hundreds of PhD candidates. This Beninese-American is a member of the American Chemical Society, American Association for the Advancement of Science, American Society of Agricultural and Biological Engineers, American Society for Microbiology, American Society of Biochemistry

ABOUT THE AUTHOR

and Molecular Biology, Ecological Society of America, American Society of Agronomy, Crop Science Society of America, and Soil Science Society of America. A scientist, a mathematician, a consultant, and the owner of Global Diaspora News (www.GlobalDiasporaNews.com), a news company in the USA, Dr. Israel is the author of the popular books:

- Turbulent Origin of Chemical Particles
- Turbulent Origin of Life
- From Science to Bible's Conclusions
- How Baby Universe Was Born
- How God Created Baby Universe
- Science180 Accurate Scientific Proof of God
- Turbulent Origin of the Universe
- Reconciling Science and Creation Accurately
- Origin of the Spiritual World

If you want to accurately understand the origin of anything, then be sure to get a copy of these amazing books. You cannot afford to ignore the greater, better, faster, simpler, cheaper, easier, and accurate formulas unlocked in these important books that successfully cracked the origin of the universe, of life, and of chemicals in a language that scientists, laypeople, adults, children, believers, skeptics, and anybody else can properly understand and enjoy.

Visit Israel120.com today to connect with this historic discoverer of the all-in-one proven and uncomplicated formula that accurately decoded the origin of the universe, of life, and of chemicals.